Cadmium and Trace Elements Toxicity

Cadmium and Trace Elements Toxicity

Guest Editors

Roberto Madeddu
Soisungwan Satarug
Peter Massányi

MDPI

Basel • Beijing • Wuhan • Barcelona • Belgrade • Novi Sad • Cluj • Manchester

Guest Editors
Roberto Madeddu
Department of
Biomedical Sciences
University of Sassari
Sassari
Italy

Soisungwan Satarug
Centre for Kidney
Disease Research
Translational
Research Institute
Brisbane
Australia

Peter Massányi
Institute of Applied Biology
Slovak University of
Agriculture in Nitra
Nitra
Slovakia

Editorial Office
MDPI AG
Grosspeteranlage 5
4052 Basel, Switzerland

This is a reprint of the Special Issue, published open access by the journal *Toxics* (ISSN 2305-6304), freely accessible at: https://www.mdpi.com/journal/toxics/special_issues/4O114TB5O7.

For citation purposes, cite each article independently as indicated on the article page online and as indicated below:

Lastname, A.A.; Lastname, B.B. Article Title. *Journal Name* **Year**, *Volume Number*, Page Range.

ISBN 978-3-7258-6580-2 (Hbk)
ISBN 978-3-7258-6581-9 (PDF)
https://doi.org/10.3390/books978-3-7258-6581-9

Contents

About the Editors

Roberto Madeddu

Roberto Madeddu received his Medical Degree (M.D.) from the University of Sassari, Italy, summa cum laude. He has his medical license as a doctor and surgeon. He received his Ph.D. from the University of Granada, Spain, with European Doctorate distinction. He is a Professor at the Faculty of Medicine and Surgery, University of Sassari, and is President of the International Society for Research on Cadmium and Trace Elements Toxicity. He is also Director of the Heavy Metals Research and Study Center, University of Sassari.

Soisungwan Satarug

Soisungwan Satarug received a B.S. degree in Medical Technology from Chiang Mai University in Thailand; an M.S. degree in Biochemistry from Mahidol University in Thailand; an M.C.H degree in Nutrition from the University of Queensland in Australia; and a Ph.D. degree in Biochemistry from the University of Arizona in the USA. She received postdoctoral fellowship training in metabolic activation of carcinogens and carcinogenesis in the USA, Japan, Germany, and France. She was a research scientist at the National Research Centre for Environmental Toxicology in Brisbane, Australia, where she investigated environmental exposure to toxic metals as part of the Australia National Cadmium Minimization Program. Currently, she is a research advisor at the Kidney Disease Research Collaborative, Translational Research Institute, University of Queensland, in Brisbane, Australia. Her research interests revolve around the interplay of nutrition, genetics, and the environment in human disease.

Peter Massányi

Peter Massányi is the head of the Institute of Applied Biology and a professor at the Institute of Applied Biology, Faculty of Biotechnology and Food Science, Slovak University of Agriculture in Nitra. His scientific research activities are focused on reproductive biology, toxicology, pathophysiology, cytology, and cytotoxicology in connection with the action of environmental factors in the biological system. His research has culminated in significant scientific outputs and rich national and transnational cooperation with many institutions. He is a member of various Scientific Boards, a member of the commissions for the defense of the PhD. and a member of the academic senate of the faculty. He is also a member of the Editorial Board of CC journals, including *Journal of Environmental Science and Health, Part A; Toxic/Hazardous Substances and Environmental Engineering; Acta Veterinaria Hungarica*; and *Biological Trace Elements Research.* His publishing activities include 207 documents (SCOPUS) with 4479 citations and an h-index of 37.

toxics

MDPI

Editorial

Editorial for the Special Issue "Cadmium and Trace Elements Toxicity"

Roberto Madeddu

Department of Biomedical Sciences, University of Sassari, 07100 Sassari, Italy; rmadeddu@uniss.it

1. Introduction

Cadmium (Cd) and other trace elements represent a significant global environmental and public health concern due to their persistence, bioaccumulation potential, and widespread distribution in ecosystems [1–5]. These toxic metals enter the environment through both natural processes and anthropogenic activities, including industrial emissions, mining operations, agricultural practices, and the extensive use of phosphate fertilizers [6–10]. The non-biodegradable nature of cadmium, coupled with its high soil-to-plant transfer rates, makes it particularly hazardous to human health and environmental integrity [11–13].

The toxicological implications of cadmium and trace element exposure are multifaceted, affecting virtually every organ system in humans and other organisms. Chronic exposure has been linked to kidney dysfunction, cardiovascular diseases, bone disorders, neurodegenerative conditions, and various forms of cancer. Essential trace elements, while necessary for normal physiological functions, can also become toxic at elevated concentrations, disrupting cellular metabolism and inducing oxidative stress. The complex interactions between toxic and essential elements further complicate our understanding of their health effects and environmental fate [14,15].

This Special Issue, entitled "Cadmium and Trace Elements Toxicity," brings together ten high-quality research articles that advance our understanding of the sources, mechanisms, and health impacts of cadmium and trace element exposure. The contributions span multiple disciplines, from molecular toxicology and environmental monitoring to plant science and risk assessment, providing a comprehensive overview of current research in this critical field. The articles address key topics including novel toxicity mechanisms, biomonitoring approaches, protective strategies, and environmental quality criteria.

2. Contributions to This Special Issue

The first contribution by Schäfer et al. (Contribution 1) provides a comprehensive review of the role of trace elements in cardiovascular diseases. This article examines both essential and non-essential trace elements, emphasizing how chronic exposure to toxic metals such as cadmium, lead, and mercury can contribute to cardiovascular pathology through multiple mechanisms, including oxidative stress, inflammatory responses, and disruption of essential metal homeostasis. The review highlights the growing body of evidence linking environmental metal exposure to increased cardiovascular morbidity and mortality, underscoring the need for preventive measures and public health interventions.

Satchanska (Contribution 2) presents an extensive review of the mechanisms of cadmium toxicity in living organisms, tracing the historical recognition of cadmium as a

 https://doi.org/10.3390/toxics13121056

toxicant from the mid-19th century to contemporary understanding. The article comprehensively describes cadmium's toxicokinetics and toxicodynamics, its propensity for long-term biological retention, and its preferential accumulation in soft tissues. The review emphasizes the shift in primary cadmium pollution sources from industrial activities to phosphate fertilizers, which now represent the main contamination pathway affecting soil, water, and ultimately human health through the food chain.

Referring to gender studies, Satarug et al. (Contribution 3) demonstrate gender differences in the severity of cadmium nephropathy, examining the relationship between cadmium body burden and renal dysfunction in 448 residents from both polluted and non-polluted regions of Thailand. The study reveals that while cadmium body burden was similar between men and women, significant gender differences emerged in the severity of renal effects. The research demonstrates that cadmium-induced tubulopathy and reduced glomerular filtration rate occurred at similar exposure levels in both sexes, but the prevalence odds ratios differed, suggesting potential gender-specific susceptibility factors that warrant further investigation in risk assessment frameworks.

Building on the theme of organ-specific toxicity, Tokumoto et al. (Contribution 4) elucidate a novel mechanism based on cadmium-induced hepatic iron deficiency through a long-term animal study. Using mice chronically exposed to dietary cadmium for up to 21 months, the researchers demonstrate that cadmium causes marked decreases in liver iron concentration by suppressing the expression of iron transport-related genes in the proximal duodenum. This study reveals a previously underappreciated mechanism of cadmium toxicity, showing how chronic exposure can disrupt essential metal homeostasis and lead to secondary nutritional deficiencies, with potential implications for understanding cadmium-induced anemia and related metabolic disorders.

Forte et al. (Contribution 5) present novel findings on toxic metal and essential element concentrations in pancreatic ductal adenocarcinoma (PDAC) patients. By analyzing blood and tissue samples from PDAC patients and healthy controls using inductively coupled plasma mass spectrometry, the study reveals significantly altered levels of chromium, copper, and copper-to-zinc ratios in patient blood, as well as increased concentrations of copper, selenium, iron, and zinc in cancer tissue. These findings suggest that metal dysregulation may play a role in PDAC pathogenesis and that metal profiling could potentially serve as a diagnostic or monitoring tool for this highly aggressive malignancy.

In contrast to the human health focus of preceding articles, Vannini et al. (Contribution 6) offer an environmental perspective by comparing cadmium accumulation and release dynamics in the lichen *Evernia prunastri* and wood-derived biochar, exploring the potential use of biochar as an environmental biomonitor. The controlled laboratory experiments demonstrate that the lichen exhibits superior cadmium adsorption capacity (46.5% higher than biochar samples) but releases only about 6% of the accumulated metal. The study identifies surface area and cation exchange capacity as key determinants of cadmium sequestration ability, suggesting that biochar could serve as an alternative or complementary tool for monitoring atmospheric cadmium deposition and potentially water body contamination.

Yang et al. (Contribution 7) develop hardness-dependent freshwater quality criteria for cadmium protection of aquatic organisms in China. Based on a comprehensive toxicity database comprising 249 acute and 62 chronic toxicity data points across 52 species, the study establishes region-specific water quality criteria that account for the significant influence of water hardness on cadmium toxicity. The research identifies the most sensitive species for both short-term and long-term exposure scenarios and provides critical tools for environmental regulators to protect aquatic ecosystems from cadmium pollution while considering local environmental conditions.

Using a seed priming approach, Mudassar et al. (Contribution 8) investigate the protective effects of triacontanol against cadmium toxicity in mung bean (*Vigna radiata* L.) and demonstrate that triacontanol treatment significantly improves plant growth, photosynthetic performance, nutrient uptake, and stress tolerance under cadmium stress conditions. Notably, the optimal concentration of triacontanol (20 μM) reduced cadmium accumulation in plants by 3-fold while increasing the metal tolerance index by 6.6-fold. These findings offer promising agricultural strategies for cultivating crops in cadmium-contaminated soils while minimizing metal uptake into the food chain.

Urbano et al. (Contribution 9) examine exposure to cadmium and other trace elements among individuals with mild cognitive impairment (MCI). This cross-sectional study of 128 MCI patients from Northern Italy measured trace element concentrations in both serum and cerebrospinal fluid, providing unique insights into the potential role of environmental chemicals in cognitive decline. The study reveals complex relationships between trace elements and biomarkers of neurodegeneration, particularly involving tau proteins and β-amyloid, suggesting that both toxic and essential trace elements at dysregulated concentrations may contribute to the pathogenesis of cognitive impairment.

Finally, Wang et al. (Contribution 10) elucidate the molecular mechanisms of cadmium-induced kidney apoptosis through the IRE1α-XBP1 signaling pathway and demonstrate the protective effects of quercetin. Using a rat model, the study shows that cadmium exposure activates endoplasmic reticulum stress through the IRE1α-XBP1 pathway, leading to renal cell apoptosis and tissue damage. Importantly, quercetin supplementation significantly attenuates these effects by inhibiting the IRE1α-XBP1 pathway, suggesting that natural flavonoid compounds may offer therapeutic potential for preventing or mitigating cadmium-induced nephrotoxicity.

3. Conclusions and Future Perspectives

The collection of articles in this Special Issue provides valuable insights into the diverse aspects of cadmium and trace element toxicity, from molecular mechanisms to ecosystem-level impacts.

Collectively, the studies in this Special Issue advance the field by bridging molecular mechanisms with population-level health assessments and by introducing innovative approaches for environmental monitoring and agricultural remediation.This collection highlights a paradigm shift from merely documenting exposure to understanding complex inter-element interactions and developing practical intervention strategies.

Several overarching themes emerge from these contributions:

First, the articles collectively demonstrate that cadmium toxicity operates through multiple, interconnected pathways affecting virtually all biological systems. Understanding these mechanisms at the molecular level, such as the IRE1α-XBP1 signaling pathway in kidney cells or the disruption of iron homeostasis in the liver, is crucial for developing targeted interventions and identifying vulnerable populations.

Second, the research highlights significant gaps in our understanding of inter-individual variability in susceptibility to cadmium and trace element toxicity. Gender differences, genetic factors, nutritional status, and co-exposure to other elements all appear to modulate toxic responses, suggesting the need for more personalized approaches to risk assessment and prevention.

Third, the studies emphasize the importance of developing practical tools for environmental monitoring and remediation. From establishing region-specific water quality criteria to exploring novel biomonitoring approaches using biochar, these efforts are essential for protecting both human health and ecosystem integrity.

Fourth, the identification of protective strategies, whether through natural compounds like quercetin and triacontanol or through nutritional interventions, offers hope for mitigating the health impacts of cadmium exposure, particularly in populations where contamination is difficult to avoid completely.

Looking forward, several research priorities emerge from this collection. There is a critical need for long-term epidemiological studies that can better characterize the health effects of low-level, chronic cadmium exposure across diverse populations. Mechanistic studies should continue to explore the interactions between cadmium and essential trace elements, as these relationships appear central to understanding toxic outcomes. Additionally, research into cost-effective remediation strategies and protective interventions is urgently needed, particularly for agricultural systems and populations in heavily contaminated regions.

Climate change and evolving agricultural practices may alter cadmium bioavailability and exposure patterns, necessitating ongoing vigilance and adaptive management strategies. The development of early biomarkers of cadmium-induced organ damage, as well as more sophisticated modeling approaches to predict tissue-specific accumulation and toxicity, will be crucial for improving risk assessment frameworks.

These studies should serve as catalysts for future research initiatives and proposals. For instance, the findings of Satarug et al. and Urbano et al. highlight the critical need for prospective cohort studies incorporating biomonitoring data alongside genetic and nutritional covariates to elucidate individual susceptibility to chronic low-level cadmium exposure. Similarly, the promising protective effects of quercetin (Wang et al.) and triacontanol (Mudassar et al.) warrant systematic investigation into additional natural compounds and their potential synergistic mechanisms in ameliorating cadmium-induced toxicity across diverse species and tissue types.

Furthermore, certain limitations identified in some studies may serve as valuable starting points for future research endeavors. For instance, while the biochar study by Vannini et al. presents a promising novel tool for cadmium remediation, its applicability under diverse field conditions remains to be fully validated through large-scale trials and long-term monitoring programs.

In conclusion, this Special Issue advances our understanding of cadmium and trace element toxicity while highlighting the continued challenges posed by these persistent environmental pollutants. The multidisciplinary approaches represented here, combining environmental chemistry, molecular toxicology, epidemiology, and plant science, exemplify the collaborative effort required to address this complex public health issue. We hope that these contributions will stimulate further research and inform evidence-based policies to minimize human and environmental exposure to these hazardous substances.

Funding: This research received no external funding.

Data Availability Statement: No new data were created or analyzed in this study. Data sharing is not applicable to this article.

Conflicts of Interest: The authors declare no conflicts of interest.

List of Contributions:

1. Schäfer, K.; Kyriakopoulos, A.M. The Role of Trace Elements in Cardiovascular Diseases. *Toxics* **2023**, *11*, 956.
2. Satchanska, G. The Mechanisms of Cadmium Toxicity in Living Organisms. *Toxics* **2024**, *12*, 875.
3. Satarug, S.; Gobe, G.C.; Vesey, D.A.; Phelps, K.R. Gender Differences in the Severity of Cadmium Nephropathy. *Toxics* **2023**, *11*, 616.

4. Tokumoto, M.; Lee, J.-Y.; Fujiwara, Y.; Satoh, M. Long-Term Exposure to Cadmium Causes Hepatic Iron Deficiency through the Suppression of Iron-Transport-Related Gene Expression in the Proximal Duodenum. *Toxics* **2023**, *11*, 641.
5. Forte, G.; Pisano, A.; Bocca, B.; Fenu, G.; Farace, C.; Etzi, F.; Perra, T.; Sabalic, A.; Porcu, A.; Madeddu, R. Toxic Metal and Essential Element Concentrations in the Blood and Tissues of Pancreatic Ductal Adenocarcinoma Patients. *Toxics* **2024**, *12*, 32.
6. Vannini, A.; Pagano, L.; Bartoli, M.; Fedeli, R.; Malcevschi, A.; Sidoli, M.; Magnani, G.; Pontiroli, D.; Riccò, M.; Marmiroli, M.; et al. Accumulation and Release of Cadmium Ions in the Lichen *Evernia prunastri* (L.) Ach. and Wood-Derived Biochar: Implication for the Use of Biochar for Environmental Biomonitoring. *Toxics* **2024**, *12*, 66.
7. Yang, D.; Liu, S.; Wang, J.; Wang, Y.; Li, Y.; Huang, Y.; Pan, B. Hardness-Dependent Freshwater Quality Criteria for the Protection of Aquatic Organisms for Cadmium in China. *Toxics* **2024**, *12*, 892.
8. Mudassar, S.; Ahmed, S.; Sardar, R.; Yasin, N.A.; Jabbar, M.; Lackner, M. Exogenously Applied Triacontanol Mitigates Cadmium Toxicity in *Vigna radiata* L. by Optimizing Growth, Nutritional Orchestration, and Metal Accumulation. *Toxics* **2024**, *12*, 911.
9. Urbano, T.; Vinceti, M.; Carbone, C.; Wise, L.A.; Malavolti, M.; Tondelli, M.; Bedin, R.; Vinceti, G.; Marti, A.; Chiari, A.; et al. Exposure to Cadmium and Other Trace Elements Among Individuals with Mild Cognitive Impairment. *Toxics* **2024**, *12*, 933.
10. Wang, L.; Cao, W.; Wu, T. Cadmium-Induced Kidney Apoptosis Based on the IRE1α-XBP1 Signaling Pathway and the Protective Effect of Quercetin. *Toxics* **2025**, *13*, 129.

References

1. WHO (World Health Organization). *Guidelines for Drinking-Water Quality—Cadmium in Drinking Water*; WHO: Geneva, Switzerland, 2022; Available online: https://cdn.who.int/media/docs/default-source/wash-documents/wash-chemicals/cadmium.pdf (accessed on 21 March 2022).
2. IARC (International Agency for Research on Cancer). *Beryllium, Cadmium, Mercury, and Exposures in the Glass Manufacturing Industry*; IARC: Lyon, France, 1993; Volume 58.
3. IARC (International Agency for Research on Cancer). *Cadmium and Cadmium Compounds (Group 1 Carcinogen)*; IARC: Lyon, France, 2012; Volume 100C. Available online: https://publications.iarc.who.int/ (accessed on 21 October 2025).
4. EFSA (European Food Safety Authority). Scientific Opinion on Cadmium in Food (2009)—EFSA Panel on Contaminants in the Food Chain (CONTAM). *EFSA J.* **2009**, *7*, 980. [CrossRef]
5. EFSA (European Food Safety Authority). Cadmium Dietary Exposure in the European Population. *EFSA J.* **2012**, *10*, 2551. [CrossRef]
6. FAO; WHO; JECFA (Joint Expert Committee on Food Additives). *Toxicological Evaluation of Certain Food Additives and Contaminants—Cadmium*; WHO: Geneva, Switzerland.
7. FAO; WHO; JECFA (Joint Expert Committee on Food Additives). *Food Additives Series No. 24 (1989) and Updates*; WHO: Geneva, Switzerland, 1989.
8. FAO; WHO; JECFA (Joint Expert Committee on Food Additives). *Monographs—Heavy Metals Assessment*; JECFA: Geneva, Switzerland; Available online: https://inchem.org/documents/jecfa/jecmono/v024je09.htm (accessed on 21 October 2025).
9. ATSDR (Agency for Toxic Substances and Disease Registry). *Toxicological Profile for Cadmium*; ATSDR (Agency for Toxic Substances and Disease Registry): Atlanta, GA, USA; Department of Health and Human Services: Washington, DC, USA, 2012. Available online: https://www.atsdr.cdc.gov/toxprofiles/ (accessed on 21 October 2025).
10. US EPA (Environmental Protection Agency). Integrated Risk Information System (IRIS)–Cadmium and Cadmium Compounds. Reference Dose (RfD) for Cadmium in Drinking Water and Dietary Exposure. Available online: https://www.epa.gov/iris (accessed on 21 October 2025).
11. NTP (National Toxicology Program). *Cadmium and Cadmium Compounds*, 15th ed.; Report on Carcinogens; U.S. Department of Health and Human Services: Washington, DC, USA. Available online: https://ntp.niehs.nih.gov/ (accessed on 21 October 2025).
12. UNEP (United Nations Environment Programme). *Global Assessment of Cadmium—Lead and Cadmium Mandates*; UNEP: Nairobi, Kenya, 2010.
13. UNEP (United Nations Environment Programme). Heavy Metals Programme—Mercury, Cadmium and Lead Assessment. Available online: https://www.unep.org/topics/chemicals-and-pollution-action/pollution-and-health/heavy-metals (accessed on 21 October 2025).

14. UNECE (United Nations Economic Commission for Europe). Protocol on Heavy Metals (1998, Amended 2012)—Convention on Long-Range Transboundary Air Pollution. Covers Cadmium, Lead, and Mercury Emissions. Available online: https://unece.org/ (accessed on 21 October 2025).
15. *CODEX STAN 193-1995*; General Standard for Contaminants and Toxins in Food and Feed. Food and Agriculture Organization of the United Nations: Rome, Italy; World Health Organization: Geneva, Switzerland, 1995; Revised 1997, 2006, 2008, 2009. Amended 2010, 2012, 2013, 2014, 2015, 2016, 2017, 2018, 2019, 2021, 2022, 2023, 2024. Available online: https://www.fao.org/fao-who-codexalimentarius/sh-proxy/fr/?lnk=1&url=https://workspace.fao.org/sites/codex/Standards/CXS+193-1995/CXS_193e.pdf (accessed on 21 October 2025).

toxics

MDPI

Review

The Role of Trace Elements in Cardiovascular Diseases

Christian Wechselberger [1,*], Barbara Messner [2] and David Bernhard [1,3,*]

1 Division of Pathophysiology, Institute of Physiology and Pathophysiology, Medical Faculty, Johannes Kepler University Linz, 4020 Linz, Austria
2 Cardiac Surgery Research Laboratory, Department of Cardiac Surgery, Medical University of Vienna, 1090 Vienna, Austria; barbara.messner@meduniwien.ac.at
3 Clinical Research Institute for Cardiovascular and Metabolic Diseases, Medical Faculty, Johannes Kepler University Linz, 4020 Linz, Austria
* Correspondence: christian.wechselberger@jku.at (C.W.); david.bernhard@jku.at (D.B.)

Abstract: Essential trace elements play an important role in human physiology and are associated with various functions regulating cellular metabolism. Non-essential trace elements, on the other hand, often have well-documented toxicities that are dangerous for the initiation and development of diseases due to their widespread occurrence in the environment and their accumulation in living organisms. Non-essential trace elements are therefore regarded as serious environmental hazards that are harmful to health even in low concentrations. Many representatives of these elements are present as pollutants in our environment, and many people may be exposed to significant amounts of these substances over the course of their lives. Among the most common non-essential trace elements are heavy metals, which are also associated with acute poisoning in humans. When these elements accumulate in the body over years of chronic exposure, they often cause severe health damage in a variety of tissues and organs. In this review article, the role of selected essential and non-essential trace elements and their role in the development of exemplary pathophysiological processes in the cardiovascular system will be examined in more detail.

Keywords: cadmium; trace elements; cardiovascular disease

1. Introduction

Heavy metal trace elements are found as natural components of the Earth's crust. However, especially since the age of the industrial revolution starting in the mid-18th century, anthropogenic activities have created a major problem with regard to the aspects of environmental pollution and the poisoning of food chains [1,2]. In particular, the accumulation of heavy metals that enter the organism as part of people's daily activities, e.g., through food, cosmetics or the abovementioned increasing pollution of the environment, represents an ever-increasing problem [3–5].

Inorganic elements that are present in low concentrations in body tissues or fluids are colloquially called trace elements. For several of these trace elements, at least one important physiological function in human metabolism is described [6]. Deficiency in essential trace elements often leads to serious pathophysiological conditions that may lead to growth impairment, reduced physical and mental power, and an increased risk of developing chronic diseases, to name only a few [7–10]. However, it has to be noted that the true metabolic demands remain unknown for most of the essential trace elements, and it was only for safety reasons that upper intake levels have been defined for most of the essential trace elements [11]. Upon metabolization, such elements are able to enter various biochemical pathways, and their interaction or conjugation with essential components like DNA or proteins can ultimately lead to structural and functional glitches [12]. Very common byproducts of these interactions represent free radicals like reactive oxygen species (ROS), which can subsequently damage a variety of cellular and molecular components in a wide range of pathways and are ultimately also responsible for the development of cancerous lesions, neurological disorders, or endocrine abnormalities, to name only a few [13–16]. However, mechanistically the generation of such radicals varies for each type of metal and, therefore, the radical species that are generated also depend on the respective metal involved [17–19]. Potential toxic trace elements can be classified according to their importance for physiological and biochemical processes as either essential (e.g., selenium (Se), zinc (Zn), copper (Cu), manganese (Mn), chromium (Cr), iron (Fe), molybdenum (Mo), and nickel (Ni)) or non-essential/toxic (e.g., aluminum (Al), cadmium (Cd), and lead (Pb)). The toxicity of trace elements are dose-dependent, and while non-essential elements may cause harmful effects at minute concentrations, essential elements exhibit toxicity only at higher concentrations not usually reached under physiological conditions.

The first indications that trace elements could play a role in the development of cardiovascular diseases already existed at the end of the 1960s [20]. Statistical analyses in selected countries like Japan, the UK, Sweden, Ireland, and others revealed correlations between the hardness of drinking water and the incidence of certain cardiovascular diseases [21–23]. This observation was later confirmed when striking differences were found in the mineral content of infarct tissue of the heart [24]. Today, the pathologies of the cardiovascular system are considered to be diseases of global relevance. In the addition, several epidemiological studies have also been published recently that deal with the topic of heavy metal trace elements in connection with the development of various diseases or, more specifically, cardiovascular disorders [25–27].

The pathophysiology of different heavy metal toxidromes is relatively similar. In most cases, heavy metals bind to oxygen, nitrogen, and sulfhydryl groups in various proteins, ultimately leading to alterations in enzymatic activity. Some metals also generate free radicals that can damage and degrade important cellular proteins, membranes, and organelles. The result of these various mechanisms is damage to various organ systems, such as the kidneys, nervous, respiratory, and cardiovascular systems [28–31]. In addition to directly interfering with the metabolic pathways of cells, the direct introduction of epigenetic changes upon trace element exposure by provoking both hyper- and hypo-methylation events has also been described [32–34]. In the case of Ni, the introduction of changes has been attributed to the direct inhibition of DNA methyltransferase enzymes [35]. Furthermore, new findings have been presented recently that show how exposure to metals can also cause epitranscriptomic dysregulation. These represent a possibly previously unrecognized new mechanism that could additionally be responsible for metal toxicity and carcinogenesis [36].

The prognosis and treatment of such diseases are often very lengthy and treatment-intensive for patients. Despite the availability of medical therapies and surgical measures, these diseases are very often fatal and therefore represent a major burden for the healthcare systems of many countries. This is just one reason why medical research in this field is of such great importance. This review therefore provides an overview of the latest developments, mainly in the last decade, in relation to the physiological functions and pathophysiological aspects of selected trace elements in connection with the development of cardiovascular diseases.

2. Essential Trace Elements

Trace elements are of great importance for a multitude of cellular functions at the biological, chemical, and molecular levels. They make essential biochemical reactions possible by acting as cofactors for enzymes and by serving to stabilize the three-dimensional structures of enzymes and proteins. Trace elements, however, have a dual importance. In normal concentrations, they are indispensable for proper physiological functions. However, disturbed concentrations can stimulate alternative metabolic pathways and even cause disease [37,38].

2.1. Selenium (Se)

Se is an essential trace element in the body. The Se-containing protein glutathione peroxidase exhibits central roles in regulating the physiological antioxidant status and plays additional roles in the thyroid metabolism and regulation of the immune response [39].

Reduced glutathione peroxidase activity could be related to the generation of toxic lipid peroxides, leading to endothelial dysfunction and arterial stiffness [40–42].

2.2. Zinc (Zn)

Zn is an essential trace element and its main sources in food are, e.g., meat, milk, eggs, and fish. Zn is required for a multitude of physiological processes, which is reflected by the fact that there are more than 300 enzymes that require Zn for their catalytic action. It plays a central role in nucleic acid and protein synthesis as well as regulating the transcription of genes as a co-factor for Zn-sensitive transcription factors [43]. It therefore represents an essential trace element that exerts essential roles in physiological and biochemical processes [44]. The increased intracellular accumulation of Ca^{2+} ions leads to stiffened arteries, and the respective deficiency could reduce vasodilatation at the receptor level [45–47].

2.3. Copper (Cu)

Cu is, after Fe und Zn, the third most abundant essential trace element in the human body. It is also widely distributed in nature and usually found in the form of minerals, rarely in a more native state. Cu exists in two oxidation states as Cu(I) and Cu(II). Its ability to gain or lose an electron characterizes its role in energy transfer processes in, e.g., cellular respiratory chain reactions. In addition, Cu functions as a cofactor for many enzymes and is involved in neurotransmitter synthesis and energy metabolism, as well as the cross-linking of extracellular matrix proteins like collagen and elastin [48]. Due to its role as an essential component of Cu-metalloenzymes, multiple functions in the hematologic, vascular, antioxidant and neurologic pathologies have been described. These include, e.g., increasing the cholesterol content of blood serum and disturbing the crosslinking of elastin fibers in blood vessels [49,50].

2.4. Manganese (Mn)

Mn functions as a cofactor of many enzymes, e.g., hydrolases, ligases, and lyases, and is central in metabolic processes such as protein glycosylation and lipid synthesis. Mn is a cofactor of the enzyme superoxide dismutase and therefore plays a central role in the detoxification of free radicals [51]. There are several reports that Mn exerts its vasculoprotective activities via a reduction in oxidative stress [33,34].

2.5. Chromium (Cr)

Cr occurs in multiple oxidation states [52]. On the one hand, Cr(VI) is mostly associated with different pathologies; on the other hand, Cr(III) is necessary during the metabolism of lipids and proteins and also acts as a cofactor for the action of insulin [53,54]. Cr can cause a multitude of pathologies through bioaccumulation in tissues and organs [55]. The association between Cr(VI) toxicity and lung cancer in steel industry workers is well established [56].

2.6. Iron (Fe)

Fe is the most abundant trace element in the human organism and almost 70% of the body's total Fe is contained in heme proteins such as hemo- and myoglobin. Besides playing a central role in the transport of oxygen, Fe is also involved in numerous physiological processes such as mitochondrial respiration and oxidative phosphorylation and acts as an essential cofactor of many enzymes and functional proteins, e.g., cytochrome C oxidase, cytochrome P-450, and catalase [57]. Inappropriate Fe overload or deficiency correlates to a wide range of cardiovascular diseases (CVDs). Fe deficiency can impair cardiomyocyte mitochondrial function and energy supplement, leading to cardiac dysfunction [58,59].

2.7. Molybdenum (Mo)

The element Mo is complexed by a special cofactor to gain catalytic activity and is therefore an essential part of several enzymes (e.g., sulphite oxidase, xanthine dehydrogenase, aldehyde oxidase) [60]. High uric acid levels correlating with a high serum Mo activity are described as a marker of inflammation and endothelial dysfunction [61,62].

2.8. Nickel (Ni)

Traces of Ni are proven to be essential, at least for animal physiology, exerting a wide range of effects, including growth, senescence and Fe-uptake. However, a deficiency state in humans has not yet been clearly defined [63,64]. Ni as an immunotoxic and carcinogenic agent can cause a variety of health effects, such as contact dermatitis, cardiovascular disease, asthma, lung fibrosis and respiratory tract cancer [65].

3. Non-Essential/Toxic Trace Elements

3.1. Cadmium (Cd)

Cd is a toxic non-essential transition metal that poses severe health risks for both humans and animals and exerts no physiological functions [66]. It is present naturally in the environment and its concentration is increased in areas with industrial activities. The main use of Cd today is in batteries, but it is also an essential component in paints and a stabilizing agent in various plastic parts. One of the major source for increased Cd uptake occurs through the smoking of tobacco products [67,68]. For nonsmokers, Cd-rich food is the main source of uptake [69]. After uptake, Cd is distributed throughout the body and accumulates in different organs to varying degrees. Egger et al., showed the average concentration of Cd in the human body in a variety of organs and revealed new Cd pools and identified adipose and muscle tissue, although accumulating lower concentrations, as important pools due to their mass. This study was conducted on four fresh, non-embalmed human bodies (two males and two females) with a maximum post-mortem time of 48 h. Measured Cd concentrations in samples (μg kg^{-1}) of the four donors in tissues of the cardiovascular system were as follows: heart muscle, 35; artery (radial), 63; aorta (ascendens), 94; and aorta (abdominal), 200 [70]. Cd is considered a highly toxic metal, and Cd and its compounds are classified by the International Agency for Research on Cancer (IARC) as a Group 1 carcinogen [71]. In recent decades, elevated Cd blood levels have been linked to various diseases in the human body, including pathological changes in the cardiovascular system, for example, in the development of atherosclerosis [72,73] and heart fibrosis [74]. In 2001, Abu-Hayyeh et al., showed that the aortic wall of smokers was, until that point, an underestimated target for Cd accumulation in the human body. In this study, total cadmium content was associated with smoking, assessed as pack-years, but was similar in aneurysmal and undilated aortas. However, a strong correlation between medial Cd content and the pack-years of smoking was detected. In vitro assays using aortic smooth muscle cells cultured on fibrillar collagen demonstrated that Cd was able to inhibit DNA synthesis and collagen synthesis and diminish cell numbers (IC(50) 2 μmol/L, 6 μmol/L, and 6 μmol/L, respectively), but higher concentrations of cadmium were required for the upregulation of metallothionein (EC(50) 23 μmol/L) [75]. The majority of negative effects on the cardiovascular system are based on the cytotoxic effects of Cd. Bernhard et al., showed in 2006 that the exposure of human primary arterial endothelial cells leads to a change in the cellular gene expression influencing cell shape and immune defense. In this study, 56 individuals were grouped into four subgroups based on their smoking habits. The data from this study suggest that increased levels of Cd are the result of the direct delivery of Cd to the human body by cigarette smoke and the Cd accumulates in the vasculature and changes arterial endothelial cell transcription [68]. Since that time, further papers have been published which substantiate that Cd is a major contributor to the development of atherosclerosis. Atherosclerosis, in turn, is involved in the development of various vascular diseases. Through a human study, an animal study and in vitro experiments with endothelial cells, Messner et al., were able to identify Cd as a novel and independent risk factor for the development of atherosclerosis. In the 195 young, healthy women of the Atherosclerosis Risk Factors in Female Youngsters (ARFY) study, Cd level was independently associated with early atherosclerotic vessel wall thickening. Similarly, Cd-fed ApoE knockout mice yielded a significantly increased aortic plaque surface compared to controls (9.5 versus 26.0 mm (^{2}), $p < 0.004$). In vitro results indicated that physiological doses of Cd increased vascular endothelial permeability up to six-fold via (1) the inhibition of endothelial cell proliferation and (2) the induction of a caspase-independent but Bcl-xL-inhibitable form of cell death more than 72 h after Cd addition. Both phenomena have been shown to be preceded by Cd-induced DNA strand breaks and a cellular DNA damage response. The pro-atherosclerotic effect of Cd was therefore shown to be due to the inhibition of cell proliferation as well as the induction of DNA damage-induced cell death [72]. A study published in 2011 showed that Cd also promotes vascular inflammation, thus contributing to the progression of atherosclerosis. In this study, an in-depth histological analysis using light and scanning electron microscopy was performed on 18 sections taken from six cadmium-fed ApoE$^{-/-}$ mice and 12 sections from five littermates not exposed to Cd. The Cd-fed mice showed a marked increase in lesion load (plaque area) and severity, and several inflammatory markers studied (CD68, CD3, CD25, vascular cell adhesion molecule 1 (VCAM-1), and heat shock protein 60 (Hsp60)) yielded a higher expression in Cd-fed mice. Statistical differences were identified for VCAM-1 and Hsp60 ($p = 0.03$ and $p = 0.02$). [73]. A further study by Messner et al., indicated that the mode of Cd-induced cytotoxicity is much more complex than previously anticipated. In summary, the incubation of endothelial cells triggers complex signaling pathways involving autophagy and apoptosis signaling that ultimately culminates in cell necrosis [76,77]. As mentioned above, the development of vascular atherosclerosis is a prerequisite for different cardiovascular events. The influence of Cd and other trace elements on various cardiovascular diseases is discussed in more detail in Section 4. Some of these diseases, but not all of them, are associated with the development of atherosclerosis and the abovementioned processes.

3.2. *Aluminum (Al)*

Al is the third most abundant element in the Earth's crust. It has been used for centuries in the form of clay, glass and alum, but its industrial use began only at the end of the late 19th century. Despite its wide occurrence, Al is not essential for any physiological processes characteristic for life [78]. Al is known to inhibit more than 300 biological relevant reactions involving kinases or phosphatases, mostly due to its ability to bind to phosphates [79,80]. It mediates the extra mitochondrial release of free oxygen radicals, resulting in Fe-induced lipid peroxidation and protein denaturation [81,82].

3.3. *Lead (Pb)*

Pb is ubiquitous in our environment. However, no physiologic role in biological systems has yet been described [83]. Pb directly interferes with selected enzymatic activities, leads to competitive inhibition of absorption of essential trace elements, deactivates sulfhydryl antioxidant pools, and leads to an increase in arterial stiffness. Pb represents an exemplary environmental pollutant that exerts high toxic effects on many tissues and organs of exposed organisms. Pb toxicity is exerted via molecular mimicry with cellular cations and the generation of ROS. Due to its ability to replace Zn and Ca in proteins, it has the ability to interfere with essential physiological processes [84]. The main effects of Pb exposure include neurological, respiratory, and cardiovascular disorders [85,86]. These are usually based on inflicting disturbances during immune modulation as well as oxidative and inflammatory mechanisms and are associated with a multitude of diseases [87–89]. Regarding the influence of Pb on cardiovascular disease, the study by Zeller et al., has to be mentioned, in which the authors were able to show that Pb is a novel and importantly independent risk factor for vascular intimal hyperplasia, a prerequisite for growing atherosclerosis [90]. In a study that investigated the relationship between lead exposure and cardiovascular disease, mortality was evaluated in a cohort of 15,036 adults. It was reported that the estimated risk of dying from cardiovascular disease in association with blood lead levels was 3.76, 8.11, and 14.77 per 1000 person-years for patients in low, moderate, and high blood level cohorts, respectively [91].

3.4. Arsenic (As)

Inorganic As is present in groundwater used for drinking in several countries all over the world, whereas organic As compounds are contaminants primarily found in fish, representing the most relevant source of human exposure [92]. In addition, energy production from fossil fuel as well as the smelting of non-ferrous metals are major industrial processes that ultimately lead to the contamination of the environment by As [93]. While the concentrations in rural areas range from <1 to 4 ng/m^3, concentrations in cities can reach values of 200 ng/m^3. Even higher values of >1000 ng/m^3 have been reported in special industrial or mining areas [92].

3.5. Mercury (Hg)

Hg is a common heavy metal pollutant found in the natural environment in different forms [94]. Inorganic Hg can be converted to organic compounds such as methylmercury, the latter being a very stable compound that can easily accumulate in the food chain. Therefore, the primary form of exposure for humans is via food, with fish and shellfish being the main source of methylmercury exposure [95]. The mechanism by which mercury exhibits toxic effects on the cardiovascular system is not yet fully understood, but the mechanism is believed to ultimately lead to an increase in oxidative stress levels within the cardiac tissues [96].

4. The Role of Trace Elements in Selected Cardiovascular Diseases

4.1. Abdominal Aortic Aneurysm

Abdominal aortic aneurysm (AAA) represents a serious health problem, is diagnosed in approximately 1.4% of people aged between 50 and 84, and is found to be almost six times more common in men than in women [97]. The dominating cause for developing an AAA is atherosclerosis and—to a lesser extent—also chronic inflammation or trauma [98–101]. Starting with the formation of atherosclerotic plaques, a weakening of the inner membrane of the aortic vessel due to a decrease in the density of elastin and collagen fibers can be detected. This results in a vessel wall that is susceptible to stretch-forces exerted by blood pressure, finally leading to the development of an aneurysm. Factors involved throughout this process include the activity of matrix metalloproteinases and the appearance of fibroblast apoptosis and inflammatory processes in the aortic wall [102–104].

In a recent study, the content of trace elements in the wall of AAA in respect to a coexisting iliac artery aneurysm (IAA) was evaluated. In this work, an association with a lower content of Ni as well as a significantly higher content of Cd was demonstrated. The levels of remaining trace elements like Cu, Zn, Mn, Mg, and Ca were detected at similar concentrations [105]. The authors of this study linked the higher Cd concentrations to an increased exposure to tobacco smoke [68,106]. However, the variation of the Ni contents in this study were found not to be statistically relevant. This was most likely due to the small size of the study group, as claimed by the authors, since the content of trace elements was assessed in samples of AAA walls harvested intraoperatively in only 19 consecutive patients [105].

One study investigated the influence of dietary habits and smoking on the level of Se and Pb in blood, the aortic wall and parietal thrombus of patients with aortic abdominal aneurysms [107]. Forty-nine patients with AAA prior to surgical procedures aged 42–81 years and a control group of twenty-two healthy volunteers aged 31–72 years as well as seventeen aortic wall samples from deceased individuals were included in this study. The authors reported significantly ($p < 0.008$) reduced mean Se levels in serum of patients with AAA (60.37 $\pm$ 21.2 cm/L) compared to healthy volunteers (75.87 $\pm$ 22.4 cm/L). There was also a significant correlation between the serum content of Se and the parietal thrombus of the examined patients ($r = 0.69$, $p < 0.0001$). It was also shown that the Se concentration in the aortic walls was inversely correlated to the concentration of Pb.

Significantly lower concentrations ($p < 0.05$) of Se (39.14 ± 37.1 cm/g) and significantly higher ($p < 0.05$) concentrations of Pb (202.69 ± 180.6 cm/g) were detected in samples from the aortic wall of smoking patients compared to non-smoking patients (77.56 ± 70.0 cm/g and 73.09 ± 49.8 cm/g, respectively). The conclusion from this study was that the measured Se serum levels are lower in patients with AAA than in healthy control individuals. In aortic wall samples, Se concentration was demonstrated to be inversely correlated with the Pb concentration. Based on previous data, it was concluded that dietary habits as well as smoking exhibit a considerable influence on the Se and the Pb status in patients with AAA [108,109].

Another study demonstrated that intraluminal thrombus thickness is not related to the concentrations of trace elements in the wall of infrarenal abdominal aortic aneurysms [110]. It was shown that concentrations of Mg, Zn, Mn, and Pb in the wall of AAA were significantly increased in respect to the intraluminal thrombus (ILT) samples. Only Cu concentration was lower in the AAA wall compared with the thrombus. Moreover, the concentration of Ca, Zn, Pb, Cu, and Mg have been shown increase with ILT thickness. The authors of the study concluded that the ILT participates in the progression of AAA via mechanisms independent of trace element supply to the wall of the aneurysm sack.

Several model systems have initially supported the notion that Cu might also play a role in the development of aortic aneurysms. One example is the Blotchy mouse, characterized by a defect of Cu metabolism that includes a reduced activity of the Cu containing enzyme lysyl oxidase that is essential for cross-linking elastin [111]. In addition, it has been demonstrated that Cu deficiency in pigs can lead to spontaneous arterial rupture [112]. Furthermore, the Menkes' syndrome is characterized by a reduction in elastic fibers in arterial walls as well as an abnormal Cu metabolism [113].

A study that addressed the question whether such a Cu deficiency was indeed associated with aortic aneurysms analyzed the respective Cu levels in liver and aortic wall samples from patients with aortic aneurysms. However, in this study, the concentrations of Cu were not different between the patients and the control group and an association with human aortic aneurysms could therefore not be supported [114].

4.2. Thoracic Aortic Dissection

Aortic dissection (AD) represents a life-threatening condition that is caused when the intimal layer of the aorta is ruptured. This results in the separation or dissection of the individual layers of the aortic wall and blood in the vessel spills into the media [115]. After the formation of a dissection, aortic inflammation enhances the dilation and subsequent rupture of the aorta due to the infiltration of inflammatory cells into the adventitia of the aortic wall and the concomitant degradation of ECM structures [116]. Due to the dramatic nature of this event, disease progress is rapid and the fatality rate is very high, reaching almost 50% within 24 h after the initial dissection and increasing with time [117,118].

There are several studies available indicating an association between Zn serum concentrations and the occurrence and respective outcome of cardiovascular diseases [119–121]. For example, it has been demonstrated that the Zn content of aortic wall tissue in patients with thoracic aortic aneurysm (TAA) was lower than in healthy individuals. In this study, a total of 108 patients (47 with abdominal aortic aneurysm (AAA), 61 patients with thoracic aortic aneurysm (TAA), and a control group of 20 abdominal aortic (AA) and 20 thoracic aortic (TA) wall samples from deceased patients) were studied. It was demonstrated that the mean concentration of Zn in the aortic wall of patients with TAA and AAA (12.9 ± 4.05 µg/g and 18.54 ± 12.3 µg/g, respectively) was significantly lower ($p < 0.05$) than in the control group samples [122]. This was confirmed by other studies demonstrating Zn deficiencies in serum as well as aortic tissues from patients with the corresponding aneurysms [123,124].

Zn exhibits a multitude of physiological effects, e.g., the induction of lymphopenia and compromised immune responses, caused by the presence of only low levels of this trace element [125,126]. A recent study aimed to the characterize the influence of Zn on the progression of thoracic aortic dissection through the process of the inhibition of inflammation [127]. The motivation for this study were observations that an inflammatory response is involved in the formation of the TAD and patients with serious symptoms exhibited higher activities of inflammatory cells in the aortic tissue compared to healthy individuals [128]. The hypothesis was that the accumulation of activated inflammatory cells within the vascular wall usually induces aortic weakening through the degeneration of extracellular matrix material. If Zn deficiency reduced tissue inflammation, an amelioration of AD should be observable. To achieve this goal, a β-aminopropionitrile monofumarate (BAPN)-induced TAD model was used to determine the effects of low Zn levels on the dissection process. And indeed, low Zn treatment was shown to attenuate the progression of TAD by improving AD formation and rupture, thereby reducing mortality. This was due to a down-regulation of aortic inflammation by attenuating the infiltration of macrophages, suppressing the switch of VSMCs from contractile to synthetic phenotypes, and eventually inhibiting TAD development.

A recent study deals with the finding that low levels of circulating and aortic tissue Fe are associated with vascular smooth muscle cells (VSMC) dysfunction and aortic instability [129]. Fe deficiency was reported to increase the incidence and severity of aortic dissection (AD) as well as dysregulation of VSMCs, most likely mediated through the integrin pathways. Moreover, it was also presented that congenital Fe deficiency was causative of vascular developmental disorders. Regarding Cd, no increased concentration was found in the tissues of aortic dissection patients [124].

4.3. *Aortic Valve Sclerosis/Stenosis*

Aortic valve sclerosis (AVS) is described as the calcification and concomitant thickening of aortic valve cusps without the obstruction of the ventricular outflow [130,131]. A recent study investigated the relationship between the development of AVS and the levels of selected trace elements like Fe, Zn, Se, and Cu. The authors were able to demonstrate a significant difference in the prevalence of diabetes and blood pressure levels as well as the body mass index between patients and healthy controls [132]. In this study, it was also shown that serum Zn concentrations in AVS patients are significantly reduced compared to those in the healthy control group, a finding in line with other studies that demonstrated lower Zn concentrations in patients with coronary artery disease, sclerotic heart valves, rheumatic heart disease, and heart failure [133–135].

Aortic valve stenosis (AS), on the other hand, represents the most common valvulopathy among adults [136]. It is characterized by inflammation, the remodeling of the extracellular matrix, and subsequent calcification that leads to a narrowing of the valve and the consequential obstruction of the cardiac outflow [137]. A recent study aiming to characterize valvular Fe in relation to pathological changes associated with AS and the effects on valvular interstitial cells (VIC) in terms of Fe uptake and Fe-induced responses [138]. Based on the findings of this study, the authors suggested that VIC and smooth muscle cells share the characteristics of an inducible Fe storage under pathological conditions, also linking Fe accumulation with increased VIC proliferation as well as promoting Wnt/β-catenin signaling in other cell types. This has been confirmed by a study demonstrating that Fe loading had an effect on Wnt signaling using the mutant adenomatous polyposis coli (APC) gene cell lines Caco-2 and SW480. In contrast, wild-type APC and beta-catenin-containing lines, HEK 293, and human primary fibroblasts were not responsive to iron loading. The verification of this finding was possible with SW480 cells that no longer induced iron-mediated Wnt signaling when transfected with wild-type APC. In addition, using the cell line LS174T, wild-type APC but mutant beta-catenin also proved responsive, suggesting that the role of Fe is to regulate beta-catenin. The authors therefore speculated that excess Fe could exacerbate tumorigenesis against the background of APC loss, a situation commonly observed in tumors [139–141].

In another study, the authors investigated the effects of Fe overload on the aorta of rats since excessive Fe has already been recognized as a risk factor for tissue damage. In one study, twenty 8-week-old male Sprague Dawley rats were randomly divided into two groups: (1) the Fe-overload group (n = 10), which received an intraperitoneal injection of Fe-Dextran (250 mg/kg of body weight) 5 days a week for 4 weeks, and (2) a control group (n = 10) that received an intraperitoneal injection of the same dose of 0.9% NaCl solution. Upon analysis, it was demonstrated that iron levels in serum (88.165 ± 15.830 µmol/L) and aortic tissue (10.494 ± 3.636 µmol/g protein) were higher in the iron-overload group than in the control group (17.338 ± 2.289, 6.507 ± 1.259) ($p < 0.05$) [142,143]. The authors demonstrated that these excessive Fe levels also led to subsequent renal and hepatic damage. In line with other published data, Fe metabolism-related factors were significantly changed during Fe overload, also leading to Ca deposition in the aorta, implying a key role in the pathophysiological process of vascular calcification by inducing osteoblast differentiation factors and downregulating the inhibitory factor for calcification [144].

The goal of another recent study focusing on trace elements in calcified valves in patients with acquired severe aortic valve stenosis was to determine the concentration of no fewer than 21 metals and trace elements [145]. Authors have been able to identify significantly higher concentrations of Ba, Ca, Cr, Mg, P, Pb, Se, Sn, Sr, and Zn as well as lower concentrations of Cd, Cu, Mo, S, and V in calcified aortic valves than in healthy controls. In addition, positive correlations of pairs of trace elements (Ca-P, Cu-S, and Se-S) as well as negative correlations (Mg-Se, P-S, and Ca-S) were demonstrated in affected valves structures. The authors concluded that some exposures might very well increase their accumulation and environmental burden over time, and the actual aortic valve calcification process should not be ruled out.

Interestingly, calcified aortic valves—as a prerequisite for aortic valve stenosis—contained lower amounts of Cd compared to healthy aortic valves (age-unadjusted model). Yet, analyzing sex differences revealed that female patients with aortic valve calcification had significantly higher Cd levels. However, it has to be kept in mind that the female patients in this study were significantly older than the male group [145].

Another mechanism through which Hg exerts toxic effects on the cardiovascular system is through the inactivation of paraoxonase, an extracellular antioxidative enzyme related to high-density lipoprotein (HDL). [146,147] This enzyme also plays an important role as an antioxidant of LDL, which directly links this process to the development of atherosclerosis and the increased risk of acute myocardial infarction, cardiovascular disease, coronary heart disease, and carotid artery stenosis. [148]

4.4. Heart Failure

Damage to myocardial tissue occurs mainly due to a lack of oxygen supply in the course of a myocardial infarction. However, heart failure is not only triggered by oxygen deprivation as a vast number of other harmful substances can also trigger the death of heart muscle cells. Cardiotoxicity characterized by the metal-induced death of cardiomyocytes has gained more attention in recent years [149]. This is also due to the fact that the cardiovascular system, and thus the heart, has long not been considered a significant target of metal ion toxicity. This view changed after the study by Egger et al., which showed that muscle tissue, including the heart muscle, was an (at the time) underestimated target for Cd deposition [70]. Recently, Ćirović et al., showed that patients with secondary cardiomyopathy had significantly higher concentrations of Pb, Ni, Mn, and Cu in their heart tissue than the group without cardiomyopathy. However, as the authors stated in their publication, cardiomyopathy is a complex diseases and this status cannot solely be explained by the presence and toxicity of the mentioned elements [150].

Pb, whose increased concentration in the left ventricle was noted by Ćirović et al. [150], was identified as severely cardiotoxic in an animal study [151]. In detail, Klinova et al., observed higher levels of angiotensin-converting enzyme (ACE) and an amplified T-wave amplitude during an electrocardiograph examination. In addition, the cardiomyocyte thickness of the lead-exposed group (5.38 ± 0.12 µm) was found to be higher than that of the control group (4.74 ± 0.08 µm); although this difference was relatively small, it was still statistically significant ($p < 0.01$) [151]. Moreover, they recorded a reduction in the maximal velocity of thin filament slide over myosin, leading to a reduced contraction of the cardiomyocytes. In vitro, the exposure of cardiomyocyte to Pb-induced toxicity through the induction of apoptosis [152]. In 2013, Turdi et al., also observed abnormalities in the contractile function of cardiomyocytes after Cd treatment [153].

Cu, known for its actually beneficial effect at lower concentrations, exerts toxic effects at increased concentrations and cardiac-specific accumulation. Pan et al., were able to show that the long-term exposure of mice with Cu resulted in the mitochondria-mediated apoptosis of cardiomyocytes and, therefore, heart damage. In detail, the authors were able to characterize a negative effect of Cu on the extracellular matrix within the heart as well as damage to the mitochondrial membrane, leading to the induction of cardiomyocyte death [154] Toxic effects have also been reported with increased exposure to Mn and Ni, with the latter suspected being responsible for congenital heart defects in offspring [155,156].

The vast majority of data linking metals and cardiotoxicity are available for Cd, with the number of publications having increased significantly, especially in the last 10 years. Below we quote some of the most important of these studies, although we cannot claim that this list is exhaustive. In 2015, Borné et al., published a population based prospective cohort study consisting of 4378 participants without a history of heart failure or atrial fibrillation (aged 46–67 years, 60% women) showing that elevated blood Cd levels are associated with an increased incidence of heart failure [157]. In 2015, our own group was able to show that high cholesterol levels together with elevated Cd exposure is a risk factor for significant heart fibrosis in ApoE knockout mice, mainly due to cardiomyocyte cell death following scar tissue formation. For in vitro and in vivo experiments, the HL-1 cardiomyocyte cell line as well as female C57BL/6J mice and female $ApoE^{-/-}$ mice were used. [74]. Similar effects have been demonstrated by Ghosh et al., in albino Wistar rats. The intragastric administration of Cd-induced inflammatory responses, apoptosis, and distorted myofibril arrangement as well as the vacuolization and congestion of vessels within cardiac tissue. Amongst others, these processes include the induction of oxidative DNA damage, damage to proteins and lipids, and decreased overall antioxidant activity [158]. In addition, in vitro studies using cardiomyocyte culture (with either primary cells or cell lines) reached similar conclusions. Chen et al., demonstrated that the incubation of cardiomyocytes with Cd resulted in ER stress and negatively influenced the energy homoeostasis of the cells [74,159].

Oluranti et al., also demonstrated an adverse effect of Cd on cardiomyocyte metabolism in a rat study. In detail, Cd leads to a reduction in fasting serum insulin levels and the reduced activity of pyruvate and hexokinase, while the activity of pyruvate dehydrogenase significantly increased after treatment [160].

As mentioned previously, the administration of Cd to rats, in addition to its effect on cardiomyocyte physiology, also leads to significant and pathological changes in the extracellular matrix of cardiac tissue. Das et al., in 2021, examined a Cd-induced imbalance in the MMP-TIMP system, potentially induced by inflammatory signals. Consequently, the authors hypothesized that this processes potentially contributes to various cardiovascular pathologies [161]. Chou et al., describing the detailed analyses of cardiac tissue that revealed inter alia focal necrosis, perivascular and interstitial fibrosis, and irregular sarcomere structures, recently found similar effects. As reported by many other studies, they observed the apoptosis of cardiomyocytes and increased expression of MMPs (in this case MMP-14). The authors of this study then extended their analysis to the aortic tissue and included in vitro cell experiments. Through this, the authors demonstrated damage to the intima and media of the aorta, potentially induced by the reduced viability of smooth muscle cells, as shown by in vitro experiments [162].

Interestingly, other metals can also neutralize the effect of Cd. For example, Feng et al., demonstrated that Se is able to protect against Cd-induced cardiac injury by disturbing the cell death signal pathway induced by Cd. In this study, 40 rabbits were randomly divided into four groups: the control group, the Se (0.5 $mg\,kg^{-1}$·body weight (BW)) group, the Cd (1 $mg\,kg^{-1}$·BW) group, and the Se + Cd group. After 30 days of feeding, morphological changes, the levels of oxidative stress and myocardial enzymes, the content of cardiac troponin T, programmed cell death (pyroptosis, autophagy, and apoptosis), and PI3K/AKT/PTEN transduction capacity were observed. These study results showed that Cd impaired the physiological balance of trace elements and caused myocardial damage, increased cardiac oxidative damage, and led to programmed cell death. The co-administration of Se prominently ameliorated histological lesions and improved the cardiac function of hearts in Cd-induced rabbits [163]. Finally, the study by Fitch et al., is worth mentioning, as they exposed female and male C57Bl/6J mice to Cd (5 mg/L ad libitum) and analyzed cardiac function after 8 weeks of treatment. The results of this study were surprising, as they showed that male mice featured a reduced left ventricular ejection fraction and a fractional shortening after treatment, together with an increased ventricular volume at end-systole and a decreased inter-ventricular septal thickness at end-systole. Mechanistically, the authors were able to show an influence of Cd on the ER, i.e., the decreased protein expression levels of sarco/endoplasmatic reticulum Ca^{2+}-ATPase 2a (SERCA2a). The surprising results, however, were not the Cd-induced changes in the heart but that these occurred only in the male mice and not in the female mice [164].

As can be seen from this section on Cd and its negative effects on cardiac muscle tissue, numerous studies have already been published that demonstrate both the effects and the mechanisms behind it. Nevertheless, further studies in this area are absolutely necessary, as not all effects and their consequences are yet fully understood, as shown by the recent study by Fitch et al. [164].

5. Conclusions

Cardiovascular disease is the leading cause of death worldwide, and an increasing number of studies point to the central role that trace elements play in the pathophysiology of symptom development and progression. However, much more work is needed to better understand the pathophysiological processes down to the molecular details. An additional important topic that will become even more relevant in the near future in connection with legalization efforts concerns the chapter on the consumption of marijuana or cannabis. The use of these substances is on the rise worldwide, which is significant in that the marijuana plant is known to accumulate Cd and Pb from the soil in very high quantities. A recent article addressed this issue and reported on general correlations between internal metal levels and exclusive marijuana use [165]. This makes the issue of exposure to such metals all the more relevant in the context of cardiovascular disease and is likely to become increasingly important to human health in the coming decades. It also remains critical to further elucidate and describe the specific mechanisms by which trace elements interfere with normal physiological processes, especially in light of recent findings that trace elements are able to induce changes at the epigenomic and the epitranscriptomic levels. This underlines the importance of the omnipresent environmental pollution caused by nanoparticles and microplastics in combination with trace elements, which has become even more significant in recent years. Such particles dispersed in the environment have long been recognized as an environmental problem. However, the various effects of the interaction between plastics and the environment remain difficult to reconstruct and further research in this area is urgently needed. This could ultimately help to correctly classify the specific hazards of the role of trace elements in environmental pollution. In addition, the mechanisms derived from these findings could also help to identify further therapeutically relevant targets for future medical applications with regard to the prevention and treatment of cardiovascular diseases and other pathophysiological processes.

Author Contributions: Conceptualization, writing, review, and editing, C.W., B.M. and D.B. All authors have read and agreed to the published version of the manuscript.

Funding: Published with the support of the Johannes Kepler University's Publication Fund.

Conflicts of Interest: The authors declare no conflict of interest.

References

1. Bai, X.; Tian, H.; Zhu, C.; Luo, L.; Hao, Y.; Liu, S.; Guo, Z.; Lv, Y.; Chen, D.; Chu, B.; et al. Present Knowledge and Future Perspectives of Atmospheric Emission Inventories of Toxic Trace Elements: A Critical Review. *Environ. Sci. Technol.* **2023**, *57*, 1551–1567. [CrossRef] [PubMed]
2. Thanigaivel, S.; Vickram, S.; Dey, N.; Jeyanthi, P.; Subbaiya, R.; Kim, W.; Govarthanan, M.; Karmegam, N. Ecological disturbances and abundance of anthropogenic pollutants in the aquatic ecosystem: Critical review of impact assessment on the aquatic animals. *Chemosphere* **2023**, *313*, 137475. [CrossRef] [PubMed]
3. Pasinszki, T.; Prasad, S.S.; Krebsz, M. Quantitative determination of heavy metal contaminants in edible soft tissue of clams, mussels, and oysters. *Environ. Monit. Assess.* **2023**, *195*, 1066. [CrossRef] [PubMed]
4. Sarath, K.V.; Shaji, E.; Nandakumar, V. Characterization of trace and heavy metal concentration in groundwater: A case study from a tropical river basin of southern India. *Chemosphere* **2023**, *338*, 139498. [CrossRef] [PubMed]
5. Lawi, D.J.; Abdulwhaab, W.S.; Abojassim, A.A. Potential Health Risks of Zn, Fe, and Pb in Medical Skin Creams and Cosmetic Products Derived from Plants in Iraq. *Biol. Trace Elem. Res.* **2023**, *201*, 4167–4176. [CrossRef]
6. Pais, I.; Jones, J.B. *The Handbook of Trace Elements*; St. Lucie Press: Boca Raton, FL, USA, 1997; ISBN 1884015344.
7. Chan, S.; Gerson, B.; Subramaniam, S. The role of copper, molybdenum, selenium, and zinc in nutrition and health. *Clin. Lab. Med.* **1998**, *18*, 673–685. [CrossRef] [PubMed]
8. Zimmermann, M.B. Iodine deficiency. *Endocr. Rev.* **2009**, *30*, 376–408. [CrossRef]
9. Wong, T. Parenteral trace elements in children: Clinical aspects and dosage recommendations. *Curr. Opin. Clin. Nutr. Metab. Care* **2012**, *15*, 649–656. [CrossRef]
10. Abbaspour, N.; Hurrell, R.; Kelishadi, R. Review on iron and its importance for human health. *J. Res. Med. Sci.* **2014**, *19*, 164–174.
11. Institute of Medicine (U.S.). *Dietary Reference Intakes: Guiding Principles for Nutrition Labeling and Fortification*; National Academies Press: Washington, DC, USA, 2010; ISBN 030952962X.
12. Azeh Engwa, G.; Udoka Ferdinand, P.; Nweke Nwalo, F.; Unachukwu, M.N. Mechanism and Health Effects of Heavy Metal Toxicity in Humans. In *Poisoning in the Modern World—New Tricks for an Old Dog?* Arslan, B., Karcioglu, O., Eds.; IntechOpen: London, UK, 2019; ISBN 978-1-83880-785-6.
13. Tchounwou, P.B.; Yedjou, C.G.; Patlolla, A.K.; Sutton, D.J. Heavy metal toxicity and the environment. *Exp. Suppl.* **2012**, *101*, 133–164. [CrossRef]
14. Dettwiler, M.; Flynn, A.C.; Rigutto-Farebrother, J. Effects of Non-Essential "Toxic" Trace Elements on Pregnancy Outcomes: A Narrative Overview of Recent Literature Syntheses. *Int. J. Environ. Res. Public Health* **2023**, *20*, 5536. [CrossRef] [PubMed]
15. Järup, L. Hazards of heavy metal contamination. *Br. Med. Bull.* **2003**, *68*, 167–182. [CrossRef] [PubMed]
16. Zhu, Y.; Costa, M. Metals and molecular carcinogenesis. *Carcinogenesis* **2020**, *41*, 1161–1172. [CrossRef] [PubMed]
17. Valko, M.; Morris, H.; Cronin, M.T.D. Metals, toxicity and oxidative stress. *Curr. Med. Chem.* **2005**, *12*, 1161–1208. [CrossRef] [PubMed]
18. Jomova, K.; Valko, M. Advances in metal-induced oxidative stress and human disease. *Toxicology* **2011**, *283*, 65–87. [CrossRef] [PubMed]
19. Kozlowski, H.; Kolkowska, P.; Watly, J.; Krzywoszynska, K.; Potocki, S. General aspects of metal toxicity. *Curr. Med. Chem.* **2014**, *21*, 3721–3740. [CrossRef]
20. Masironi, R. Trace elements and cardiovascular diseases. *Bull. World Health Organ.* **1969**, *40*, 305–312.
21. Schroeder, H.A. Relations between hardness of water and death rates from certain chronic and degenerative diseases in the United States. *J. Chronic Dis.* **1960**, *12*, 586–591. [CrossRef]
22. Morris, J.N.; Crawford, M.D.; Heady, J.A. Hardness of local water-supplies and mortality from cardiovascular disease in the County Boroughs of England and Wales. *Lancet* **1961**, *1*, 860–862. [CrossRef]
23. Crawford, M.D. Hardness of drinking-water and cardiovascular disease. *Proc. Nutr. Soc.* **1972**, *31*, 347–353. [CrossRef]
24. Schroeder, H.A. Municipal drinking water and cardiovascular death rates. *JAMA* **1966**, *195*, 81–85. [CrossRef] [PubMed]
25. Lamas, G.A.; Ujueta, F.; Navas-Acien, A. Lead and Cadmium as Cardiovascular Risk Factors: The Burden of Proof Has Been Met. *J. Am. Heart Assoc.* **2021**, *10*, e018692. [CrossRef]
26. Solenkova, N.V.; Newman, J.D.; Berger, J.S.; Thurston, G.; Hochman, J.S.; Lamas, G.A. Metal pollutants and cardiovascular disease: Mechanisms and consequences of exposure. *Am. Heart J.* **2014**, *168*, 812–822. [CrossRef] [PubMed]
27. Lamas, G.A.; Bhatnagar, A.; Jones, M.R.; Mann, K.K.; Nasir, K.; Tellez-Plaza, M.; Ujueta, F.; Navas-Acien, A. Contaminant Metals as Cardiovascular Risk Factors: A Scientific Statement From the American Heart Association. *J. Am. Heart Assoc.* **2023**, *12*, e029852. [CrossRef]
28. Hollenberg, P.F. Introduction: Mechanisms of metal toxicity special issue. *Chem. Res. Toxicol.* **2010**, *23*, 292–293. [CrossRef] [PubMed]
29. Balali-Mood, M.; Naseri, K.; Tahergorabi, Z.; Khazdair, M.R.; Sadeghi, M. Toxic Mechanisms of Five Heavy Metals: Mercury, Lead, Chromium, Cadmium, and Arsenic. *Front. Pharmacol.* **2021**, *12*, 643972. [CrossRef] [PubMed]
30. Jaishankar, M.; Tseten, T.; Anbalagan, N.; Mathew, B.B.; Beeregowda, K.N. Toxicity, mechanism and health effects of some heavy metals. *Interdiscip. Toxicol.* **2014**, *7*, 60–72. [CrossRef]
31. Lamas, G.A.; Navas-Acien, A.; Mark, D.B.; Lee, K.L. Heavy Metals, Cardiovascular Disease, and the Unexpected Benefits of Chelation Therapy. *J. Am. Coll. Cardiol.* **2016**, *67*, 2411–2418. [CrossRef]

32. Brocato, J.; Costa, M. Basic mechanics of DNA methylation and the unique landscape of the DNA methylome in metal-induced carcinogenesis. *Crit. Rev. Toxicol.* **2013**, *43*, 493–514. [CrossRef]
33. Arita, A.; Costa, M. Epigenetics in metal carcinogenesis: Nickel, arsenic, chromium and cadmium. *Metallomics* **2009**, *1*, 222–228. [CrossRef]
34. Marsit, C.J. Influence of environmental exposure on human epigenetic regulation. *J. Exp. Biol.* **2015**, *218*, 71–79. [CrossRef] [PubMed]
35. Lee, Y.W.; Broday, L.; Costa, M. Effects of nickel on DNA methyltransferase activity and genomic DNA methylation levels. *Mutat. Res.* **1998**, *415*, 213–218. [CrossRef] [PubMed]
36. Yang, C.; Wang, Z. The Epitranscriptomic Mechanism of Metal Toxicity and Carcinogenesis. *Int. J. Mol. Sci.* **2022**, *23*, 11830. [CrossRef] [PubMed]
37. Islam, M.R.; Akash, S.; Jony, M.H.; Alam, M.N.; Nowrin, F.T.; Rahman, M.M.; Rauf, A.; Thiruvengadam, M. Exploring the potential function of trace elements in human health: A therapeutic perspective. *Mol. Cell Biochem.* **2023**, *478*, 2141–2171. [CrossRef] [PubMed]
38. Mohammadifard, N.; Humphries, K.H.; Gotay, C.; Mena-Sánchez, G.; Salas-Salvadó, J.; Esmaillzadeh, A.; Ignaszewski, A.; Sarrafzadegan, N. Trace minerals intake: Risks and benefits for cardiovascular health. *Crit. Rev. Food Sci. Nutr.* **2019**, *59*, 1334–1346. [CrossRef] [PubMed]
39. Zhang, F.; Li, X.; Wei, Y. Selenium and Selenoproteins in Health. *Biomolecules* **2023**, *13*, 799. [CrossRef] [PubMed]
40. Mistry, H.D.; Wilson, V.; Ramsay, M.M.; Symonds, M.E.; Broughton Pipkin, F. Reduced selenium concentrations and glutathione peroxidase activity in preeclamptic pregnancies. *Hypertension* **2008**, *52*, 881–888. [CrossRef]
41. Detopoulou, P.; Letsiou, S.; Nomikos, T.; Karagiannis, A.; Pergantis, S.A.; Pitsavos, C.; Panagiotakos, D.B.; Antonopoulou, S. Selenium, Selenoproteins and 10-year Cardiovascular Risk: Results from the ATTICA Study. *Curr. Vasc. Pharmacol.* **2023**, *21*, 346–355. [CrossRef]
42. Dabravolski, S.A.; Sukhorukov, V.N.; Melnichenko, A.A.; Khotina, V.A.; Orekhov, A.N. The Role of Selenium in Atherosclerosis Development, Progression, Prevention and Treatment. *Biomedicines* **2023**, *11*, 2010. [CrossRef]
43. Tuerk, M.J.; Fazel, N. Zinc deficiency. *Curr. Opin. Gastroenterol.* **2009**, *25*, 136–143. [CrossRef]
44. Taheri, S.; Asadi, S.; Nilashi, M.; Ali Abumalloh, R.; Ghabban, N.M.A.; Mohd Yusuf, S.Y.; Supriyanto, E.; Samad, S. A literature review on beneficial role of vitamins and trace elements: Evidence from published clinical studies. *J. Trace Elem. Med. Biol.* **2021**, *67*, 126789. [CrossRef] [PubMed]
45. Ripa, S.; Ripa, R. Zinco e pressione arteriosa. *Minerva Med.* **1994**, *85*, 455–459. [PubMed]
46. Kitala, K.; Tanski, D.; Godlewski, J.; Krajewska-Włodarczyk, M.; Gromadziński, L.; Majewski, M. Copper and Zinc Particles as Regulators of Cardiovascular System Function-A Review. *Nutrients* **2023**, *15*, 3040. [CrossRef] [PubMed]
47. Nazari, M.; Ashtary-Larky, D.; Nikbaf-Shandiz, M.; Goudarzi, K.; Bagheri, R.; Dolatshahi, S.; Omran, H.S.; Amirani, N.; Ghanavati, M.; Asbaghi, O. Zinc supplementation and cardiovascular disease risk factors: A GRADE-assessed systematic review and dose-response meta-analysis. *J. Trace Elem. Med. Biol.* **2023**, *79*, 127244. [CrossRef] [PubMed]
48. Wachnik, A. The physiological role of copper and the problems of copper nutritional deficiency. *Nahrung* **1988**, *32*, 755–765. [CrossRef] [PubMed]
49. Ghalibaf, A.M.; Soflaei, S.S.; Ferns, G.A.; Saberi-Karimian, M.; Ghayour-Mobarhan, M. Association between dietary copper and cardiovascular disease: A narrative review. *J. Trace Elem. Med. Biol.* **2023**, *80*, 127255. [CrossRef] [PubMed]
50. Yang, L.; Yang, P.; Lip, G.Y.H.; Ren, J. Copper homeostasis and cuproptosis in cardiovascular disease therapeutics. *Trends Pharmacol. Sci.* **2023**, *44*, 573–585. [CrossRef]
51. Chen, P.; Chakraborty, S.; Mukhopadhyay, S.; Lee, E.; Paoliello, M.M.B.; Bowman, A.B.; Aschner, M. Manganese homeostasis in the nervous system. *J. Neurochem.* **2015**, *134*, 601–610. [CrossRef]
52. Boros-Lajszner, E.; Wyszkowska, J.; Kucharski, J. Evaluation and Assessment of Trivalent and Hexavalent Chromium on Avena sativa and Soil Enzymes. *Molecules* **2023**, *28*, 4693. [CrossRef]
53. Cefalu, W.T.; Hu, F.B. Role of chromium in human health and in diabetes. *Diabetes Care* **2004**, *27*, 2741–2751. [CrossRef]
54. Tsave, O.; Yavropoulou, M.P.; Kafantari, M.; Gabriel, C.; Yovos, J.G.; Salifoglou, A. The adipogenic potential of Cr(III). A molecular approach exemplifying metal-induced enhancement of insulin mimesis in diabetes mellitus II. *J. Inorg. Biochem.* **2016**, *163*, 323–331. [CrossRef] [PubMed]
55. Fang, Z.; Zhao, M.; Zhen, H.; Chen, L.; Shi, P.; Huang, Z. Genotoxicity of tri- and hexavalent chromium compounds in vivo and their modes of action on DNA damage in vitro. *PLoS ONE* **2014**, *9*, e103194. [CrossRef] [PubMed]
56. Gad, S.C. Acute and chronic systemic chromium toxicity. *Sci. Total Environ.* **1989**, *86*, 149–157. [CrossRef] [PubMed]
57. Zhang, D.-L.; Ghosh, M.C.; Rouault, T.A. The physiological functions of iron regulatory proteins in iron homeostasis—An update. *Front. Pharmacol.* **2014**, *5*, 124. [CrossRef] [PubMed]
58. Von Haehling, S.; Ebner, N.; Evertz, R.; Ponikowski, P.; Anker, S.D. Iron Deficiency in Heart Failure: An Overview. *JACC Heart Fail.* **2019**, *7*, 36–46. [CrossRef]
59. Li, S.; Zhang, X. Iron in Cardiovascular Disease: Challenges and Potentials. *Front. Cardiovasc. Med.* **2021**, *8*, 707138. [CrossRef] [PubMed]
60. Mendel, R.R. The molybdenum cofactor. *J. Biol. Chem.* **2013**, *288*, 13165–13172. [CrossRef]

61. Gutierrez, C.T.; Loizides, C.; Hafez, I.; Biskos, G.; Loeschner, K.; Brostrøm, A.; Roursgaard, M.; Saber, A.T.; Møller, P.; Sørli, J.B.; et al. Comparison of acute phase response in mice after inhalation and intratracheal instillation of molybdenum disulphide and tungsten particles. *Basic. Clin. Pharmacol. Toxicol.* **2023**, *133*, 265–278. [CrossRef]
62. Vyskočil, A.; Viau, C. Assessment of molybdenum toxicity in humans. *J. Appl. Toxicol.* **1999**, *19*, 185–192. [CrossRef]
63. Sager, M. Nickel—A Trace Element Hardly Considered. *IJHAF* **2019**, *3*, 75–90. [CrossRef]
64. Barceloux, D.G. Nickel. *J. Toxicol. Clin. Toxicol.* **1999**, *37*, 239–258. [CrossRef] [PubMed]
65. Genchi, G.; Carocci, A.; Lauria, G.; Sinicropi, M.S.; Catalano, A. Nickel: Human Health and Environmental Toxicology. *Int. J. Environ. Res. Public Health* **2020**, *17*, 679. [CrossRef]
66. Genchi, G.; Sinicropi, M.S.; Lauria, G.; Carocci, A.; Catalano, A. The Effects of Cadmium Toxicity. *Int. J. Environ. Res. Public Health* **2020**, *17*, 3782. [CrossRef] [PubMed]
67. Kim, J.; Song, H.; Lee, J.; Kim, Y.J.; Chung, H.S.; Yu, J.M.; Jang, G.; Park, R.; Chung, W.; Oh, C.-M.; et al. Smoking and passive smoking increases mortality through mediation effect of cadmium exposure in the United States. *Sci. Rep.* **2023**, *13*, 3878. [CrossRef] [PubMed]
68. Bernhard, D.; Rossmann, A.; Henderson, B.; Kind, M.; Seubert, A.; Wick, G. Increased serum cadmium and strontium levels in young smokers: Effects on arterial endothelial cell gene transcription. *Arterioscler. Thromb. Vasc. Biol.* **2006**, *26*, 833–838. [CrossRef] [PubMed]
69. Fatima, G.; Raza, A.M.; Hadi, N.; Nigam, N.; Mahdi, A.A. Cadmium in Human Diseases: It's More than Just a Mere Metal. *Indian. J. Clin. Biochem.* **2019**, *34*, 371–378. [CrossRef] [PubMed]
70. Egger, A.E.; Grabmann, G.; Gollmann-Tepeköylü, C.; Pechriggl, E.J.; Artner, C.; Türkcan, A.; Hartinger, C.G.; Fritsch, H.; Keppler, B.K.; Brenner, E.; et al. Chemical imaging and assessment of cadmium distribution in the human body. *Metallomics* **2019**, *11*, 2010–2019. [CrossRef]
71. International Agency for Research on Cancer; IARC Working Group on the Evaluation of Carcinogenic Risks to Humans. *Beryllium, Cadmium, Mercury, and Exposures in the Glass Manufacturing Industry*; IARC World Health Organization: Geneva, Switzerland, 1993; ISBN 9789283212584.
72. Messner, B.; Knoflach, M.; Seubert, A.; Ritsch, A.; Pfaller, K.; Henderson, B.; Shen, Y.H.; Zeller, I.; Willeit, J.; Laufer, G.; et al. Cadmium is a novel and independent risk factor for early atherosclerosis mechanisms and in vivo relevance. *Arterioscler. Thromb. Vasc. Biol.* **2009**, *29*, 1392–1398. [CrossRef]
73. Knoflach, M.; Messner, B.; Shen, Y.H.; Frotschnig, S.; Liu, G.; Pfaller, K.; Wang, X.; Matosevic, B.; Willeit, J.; Kiechl, S.; et al. Non-toxic cadmium concentrations induce vascular inflammation and promote atherosclerosis. *Circ. J.* **2011**, *75*, 2491–2495. [CrossRef]
74. Türkcan, A.; Scharinger, B.; Grabmann, G.; Keppler, B.K.; Laufer, G.; Bernhard, D.; Messner, B. Combination of cadmium and high cholesterol levels as a risk factor for heart fibrosis. *Toxicol. Sci.* **2015**, *145*, 360–371. [CrossRef]
75. Abu-Hayyeh, S.; Sian, M.; Jones, K.G.; Manuel, A.; Powell, J.T. Cadmium accumulation in aortas of smokers. *Arterioscler. Thromb. Vasc. Biol.* **2001**, *21*, 863–867. [CrossRef] [PubMed]
76. Messner, B.; Ploner, C.; Laufer, G.; Bernhard, D. Cadmium activates a programmed, lysosomal membrane permeabilization-dependent necrosis pathway. *Toxicol. Lett.* **2012**, *212*, 268–275. [CrossRef] [PubMed]
77. Messner, B.; Türkcan, A.; Ploner, C.; Laufer, G.; Bernhard, D. Cadmium overkill: Autophagy, apoptosis and necrosis signalling in endothelial cells exposed to cadmium. *Cell. Mol. Life Sci.* **2016**, *73*, 1699–1713. [CrossRef] [PubMed]
78. Nayak, P. Aluminum: Impacts and disease. *Environ. Res.* **2002**, *89*, 101–115. [CrossRef] [PubMed]
79. Furumo, N.C.; Viola, R.E. Aluminum-adenine nucleotides as alternate substrates for creatine kinase. *Arch. Biochem. Biophys.* **1989**, *275*, 33–39. [CrossRef] [PubMed]
80. Kawahara, M.; Tanaka, K.I.; Kato-Negishi, M. Neurotoxicity of aluminum and its link to neurodegenerative diseases. *Met. Res.* **2021**, *1*, rev-47–rev-65. [CrossRef]
81. Mehrpour, O.; Jafarzadeh, M.; Abdollahi, M. A systematic review of aluminium phosphide poisoning. *Arh. Hig. Rada Toksikol.* **2012**, *63*, 61–73. [CrossRef]
82. Carbonara, C.E.M.; Roza, N.A.V.; Quadros, K.R.S.; França, R.A.; Esteves, A.B.A.; Pavan, C.R.; Barreto, J.; Dos Reis, L.M.; Jorgetti, V.; Sposito, A.C.; et al. Effect of aluminum accumulation on bone and cardiovascular risk in the current era. *PLoS ONE* **2023**, *18*, e0284123. [CrossRef]
83. Wani, A.L.; Ara, A.; Usmani, J.A. Lead toxicity: A review. *Interdiscip. Toxicol.* **2015**, *8*, 55–64. [CrossRef]
84. Garza, A.; Vega, R.; Soto, E. Cellular mechanisms of lead neurotoxicity. *Med. Sci. Monit.* **2006**, *12*, RA57-65.
85. Rouzi, L.; Elhamri, H.; Kalouch, S.; Salam, S.; El Moutawakil, B.; Chaoui, H.; Badrane, N.; Fekhaoui, M.; Jouhadi, Z. Lead poisoning with encephalic and neuropathic involvement in a child: Case report. *Pan Afr. Med. J.* **2022**, *42*, 276. [CrossRef] [PubMed]
86. Ferreira, G.; Santander, A.; Chavarría, L.; Cardozo, R.; Savio, F.; Sobrevia, L.; Nicolson, G.L. Functional consequences of lead and mercury exposomes in the heart. *Mol. Asp. Med.* **2022**, *87*, 101048. [CrossRef] [PubMed]
87. Kataba, A.; Yohannes, Y.B.; Nakata, H.; Yabe, J.; Toyomaki, H.; Muzandu, K.; Zyambo, G.; Ikenaka, Y.; Choongo, K.; Ishizuka, M.; et al. Association between Chronic Environmental Lead (Pb) Exposure and Cytokines in Males and Females of Reproductive Age from Kabwe, Zambia. *Int. J. Environ. Res. Public Health* **2023**, *20*, 5596. [CrossRef] [PubMed]
88. Halmo, L.; Nappe, T.M. *StatPearls: Lead Toxicity*; StatPearls Publishing: Treasure Island, FL, USA, 2023.

89. Dhar, S.; Garg, D. Lead Poisoning. *N. Engl. J. Med.* **2023**, *388*, e63. [CrossRef] [PubMed]
90. Zeller, I.; Knoflach, M.; Seubert, A.; Kreutmayer, S.B.; Stelzmüller, M.E.; Wallnoefer, E.; Blunder, S.; Frotschnig, S.; Messner, B.; Willeit, J.; et al. Lead contributes to arterial intimal hyperplasia through nuclear factor erythroid 2-related factor-mediated endothelial interleukin 8 synthesis and subsequent invasion of smooth muscle cells. *Arterioscler. Thromb. Vasc. Biol.* **2010**, *30*, 1733–1740. [CrossRef] [PubMed]
91. Cook, M.K.; Zhang, J.; Wei, Y. Blood Lead Levels and Risk of Deaths from Cardiovascular Disease. *Am. J. Cardiol.* **2022**, *173*, 132–138. [CrossRef] [PubMed]
92. World Health Organization. *Arsenic and Arsenic Compounds*, 2nd ed.; World Health Organization: Geneva, Switzerland, 2001; ISBN 9789240687622.
93. Schlesinger, W.H.; Klein, E.M.; Vengosh, A. The Global Biogeochemical Cycle of Arsenic. *Glob. Biogeochem. Cycles* **2022**, *36*, e2022GB007515. [CrossRef]
94. Li, R.; Wu, H.; Ding, J.; Fu, W.; Gan, L.; Li, Y. Mercury pollution in vegetables, grains and soils from areas surrounding coal-fired power plants. *Sci. Rep.* **2017**, *7*, 46545. [CrossRef]
95. Hong, Y.-S.; Kim, Y.-M.; Lee, K.-E. Methylmercury exposure and health effects. *J. Prev. Med. Public. Health* **2012**, *45*, 353–363. [CrossRef]
96. Genchi, G.; Sinicropi, M.S.; Carocci, A.; Lauria, G.; Catalano, A. Mercury Exposure and Heart Diseases. *Int. J. Environ. Res. Public Health* **2017**, *14*, 74. [CrossRef]
97. Kent, K.C.; Zwolak, R.M.; Egorova, N.N.; Riles, T.S.; Manganaro, A.; Moskowitz, A.J.; Gelijns, A.C.; Greco, G. Analysis of risk factors for abdominal aortic aneurysm in a cohort of more than 3 million individuals. *J. Vasc. Surg.* **2010**, *52*, 539–548. [CrossRef] [PubMed]
98. Nordon, I.M.; Hinchliffe, R.J.; Loftus, I.M.; Thompson, M.M. Pathophysiology and epidemiology of abdominal aortic aneurysms. *Nat. Rev. Cardiol.* **2011**, *8*, 92–102. [CrossRef]
99. Hirsch, A.T.; Haskal, Z.J.; Hertzer, N.R.; Bakal, C.W.; Creager, M.A.; Halperin, J.L.; Hiratzka, L.F.; Murphy, W.R.C.; Olin, J.W.; Puschett, J.B.; et al. ACC/AHA 2005 Practice Guidelines for the management of patients with peripheral arterial disease (lower extremity, renal, mesenteric, and abdominal aortic): A collaborative report from the American Association for Vascular Surgery/Society for Vascular Surgery, Society for Cardiovascular Angiography and Interventions, Society for Vascular Medicine and Biology, Society of Interventional Radiology, and the ACC/AHA Task Force on Practice Guidelines (Writing Committee to Develop Guidelines for the Management of Patients With Peripheral Arterial Disease): Endorsed by the American Association of Cardiovascular and Pulmonary Rehabilitation; National Heart, Lung, and Blood Institute; Society for Vascular Nursing; TransAtlantic Inter-Society Consensus; and Vascular Disease Foundation. *Circulation* **2006**, *113*, e463–e654. [CrossRef] [PubMed]
100. Doppler, C.; Messner, B.; Mimler, T.; Schachner, B.; Rezk, M.; Ganhör, C.; Wechselberger, C.; Müller, M.; Puh, S.; Pröll, J.; et al. Noncanonical atherosclerosis as the driving force in tricuspid aortic valve associated aneurysms—A trace collection. *J. Lipid Res.* **2023**, *64*, 100338. [CrossRef] [PubMed]
101. Stern, C.; Scharinger, B.; Tuerkcan, A.; Nebert, C.; Mimler, T.; Baranyi, U.; Doppler, C.; Aschacher, T.; Andreas, M.; Stelzmueller, M.-E.; et al. Strong Signs for a Weak Wall in Tricuspid Aortic Valve Associated Aneurysms and a Role for Osteopontin in Bicuspid Aortic Valve Associated Aneurysms. *Int. J. Mol. Sci.* **2019**, *20*, 4782. [CrossRef] [PubMed]
102. McCormick, M.L.; Gavrila, D.; Weintraub, N.L. Role of oxidative stress in the pathogenesis of abdominal aortic aneurysms. *Arterioscler. Thromb. Vasc. Biol.* **2007**, *27*, 461–469. [CrossRef]
103. Michel, J.-B.; Martin-Ventura, J.-L.; Egido, J.; Sakalihasan, N.; Treska, V.; Lindholt, J.; Allaire, E.; Thorsteinsdottir, U.; Cockerill, G.; Swedenborg, J. Novel aspects of the pathogenesis of aneurysms of the abdominal aorta in humans. *Cardiovasc. Res.* **2011**, *90*, 18–27. [CrossRef]
104. Rizas, K.D.; Ippagunta, N.; Tilson, M.D. Immune cells and molecular mediators in the pathogenesis of the abdominal aortic aneurysm. *Cardiol. Rev.* **2009**, *17*, 201–210. [CrossRef]
105. Ziaja, D.; Chudek, J.; Sznapka, M.; Kita, A.; Biolik, G.; Sieroń-Stołtny, K.; Pawlicki, K.; Domalik, J.; Ziaja, K. Trace elements in the wall of abdominal aortic aneurysms with and without coexisting iliac artery aneurysms. *Biol. Trace Elem. Res.* **2015**, *165*, 119–122. [CrossRef]
106. Nandi, M.; Slone, D.; Jick, H.; Shapiro, S.; Lewis, G.P. Cadmium content of cigarettes. *Lancet* **1969**, *2*, 1329–1330. [CrossRef]
107. Socha, K.; Borawska, M.H.; Gacko, M.; Guzowski, A. Diet and the content of selenium and lead in patients with abdominal aortic aneurysm. *Vasa* **2011**, *40*, 381–389. [CrossRef] [PubMed]
108. Unkiewicz-Winiarczyk, A.; Gromysz-Kałkowska, K.; Szubartowska, E. Aluminium, cadmium and lead concentration in the hair of tobacco smokers. *Biol. Trace Elem. Res.* **2009**, *132*, 41–50. [CrossRef] [PubMed]
109. Ellingsen, D.G.; Thomassen, Y.; Rustad, P.; Molander, P.; Aaseth, J. The time-trend and the relation between smoking and circulating selenium concentrations in Norway. *J. Trace Elem. Med. Biol.* **2009**, *23*, 107–115. [CrossRef] [PubMed]
110. Ziaja, D.; Kita, A.; Janowska, J.; Pawlicki, K.; Mikuła, B.; Sznapka, M.; Chudek, J.; Ziaja, K. Intraluminal thrombus thickness is not related to lower concentrations of trace elements in the wall of infrarenal abdominal aortic aneurysms. *J. Trace Elem. Med. Biol.* **2014**, *28*, 28–31. [CrossRef] [PubMed]
111. Rowe, D.W.; McGoodwin, E.B.; Martin, G.R.; Grahn, D. Decreased lysyl oxidase activity in the aneurysm-prone, mottled mouse. *J. Biol. Chem.* **1977**, *252*, 939–942. [CrossRef] [PubMed]

112. Shields, G.S.; Coulson, W.F.; Kimball, D.A.; Carnes, W.H.; Cartwright, G.E.; Wintrobe, M.M. Studies on copper metabolism. 32. Cardiovascular lesions in copper-deficient swine. *Am. J. Pathol.* **1962**, *41*, 603–621. [PubMed]
113. Levinson, B.; Vulpe, C.; Elder, B.; Martin, C.; Verley, F.; Packman, S.; Gitschier, J. The mottled gene is the mouse homologue of the Menkes disease gene. *Nat. Genet.* **1994**, *6*, 369–373. [CrossRef]
114. Senapati, A.; Carlsson, L.K.; Fletcher, C.D.; Browse, N.L.; Thompson, R.P. Is tissue copper deficiency associated with aortic aneurysms? *Br. J. Surg.* **1985**, *72*, 352–353. [CrossRef]
115. Nienaber, C.A.; Clough, R.E.; Sakalihasan, N.; Suzuki, T.; Gibbs, R.; Mussa, F.; Jenkins, M.P.; Thompson, M.M.; Evangelista, A.; Yeh, J.S.M.; et al. Aortic dissection. *Nat. Rev. Dis. Primers* **2016**, *2*, 16071. [CrossRef]
116. Erbel, R.; Aboyans, V.; Boileau, C.; Bossone, E.; Di Bartolomeo, R.; Eggebrecht, H.; Evangelista, A.; Falk, V.; Frank, H.; Gaemperli, O.; et al. 2014 ESC Guidelines on the diagnosis and treatment of aortic diseases: Document covering acute and chronic aortic diseases of the thoracic and abdominal aorta of the adult. *Eur. Heart J.* **2014**, *35*, 2873–2926. [CrossRef]
117. Nienaber, C.A.; Clough, R.E. Management of acute aortic dissection. *Lancet* **2015**, *385*, 800–811. [CrossRef] [PubMed]
118. Clough, R.E.; Nienaber, C.A. Management of acute aortic syndrome. *Nat. Rev. Cardiol.* **2015**, *12*, 103–114. [CrossRef] [PubMed]
119. Pilz, S.; Dobnig, H.; Winklhofer-Roob, B.M.; Renner, W.; Seelhorst, U.; Wellnitz, B.; Boehm, B.O.; März, W. Low serum zinc concentrations predict mortality in patients referred to coronary angiography. *Br. J. Nutr.* **2009**, *101*, 1534–1540. [CrossRef] [PubMed]
120. Giacconi, R.; Caruso, C.; Malavolta, M.; Lio, D.; Balistreri, C.R.; Scola, L.; Candore, G.; Muti, E.; Mocchegiani, E. Pro-inflammatory genetic background and zinc status in old atherosclerotic subjects. *Ageing Res. Rev.* **2008**, *7*, 306–318. [CrossRef] [PubMed]
121. Singh, R.B.; Gupta, U.C.; Mittal, N.; Niaz, M.A.; Ghosh, S.; Rastogi, V. Epidemiologic study of trace elements and magnesium on risk of coronary artery disease in rural and urban Indian populations. *J. Am. Coll. Nutr.* **1997**, *16*, 62–67. [CrossRef] [PubMed]
122. Socha, K.; Karwowska, A.; Kurianiuk, A.; Markiewicz-Żukowska, R.; Guzowski, A.; Gacko, M.; Hirnle, T.; Borawska, M.H. Estimation of Selected Minerals in Aortic Aneurysms-Impaired Ratio of Zinc to Lead May Predispose? *Biol. Trace Elem. Res.* **2021**, *199*, 2811–2818. [CrossRef] [PubMed]
123. Kurianiuk, A.; Socha, K.; Gacko, M.; Błachnio-Zabielska, A.; Karwowska, A. The Relationship between the Concentration of Cathepsin A, D, and E and the Concentration of Copper and Zinc, and the Size of the Aneurysmal Enlargement in the Wall of the Abdominal Aortic Aneurysm. *Ann. Vasc. Surg.* **2019**, *55*, 182–188. [CrossRef]
124. Edvinsson, M.; Ilbäck, N.-G.; Frisk, P.; Thelin, S.; Nyström-Rosander, C. Trace Element Changes in Thoracic Aortic Dissection. *Biol. Trace Elem. Res.* **2016**, *169*, 159–163. [CrossRef]
125. Rink, L.; Haase, H. Zinc homeostasis and immunity. *Trends Immunol.* **2007**, *28*, 1–4. [CrossRef]
126. Murakami, M.; Hirano, T. Intracellular zinc homeostasis and zinc signaling. *Cancer Sci.* **2008**, *99*, 1515–1522. [CrossRef]
127. Zhu, L.; An, P.; Zhao, W.; Xia, Y.; Qi, J.; Luo, J.; Luo, Y. Low Zinc Alleviates the Progression of Thoracic Aortic Dissection by Inhibiting Inflammation. *Nutrients* **2023**, *15*, 1640. [CrossRef] [PubMed]
128. Kuehl, H.; Eggebrecht, H.; Boes, T.; Antoch, G.; Rosenbaum, S.; Ladd, S.; Bockisch, A.; Barkhausen, J.; Erbel, R. Detection of inflammation in patients with acute aortic syndrome: Comparison of FDG-PET/CT imaging and serological markers of inflammation. *Heart* **2008**, *94*, 1472–1477. [CrossRef] [PubMed]
129. Li, B.; Wang, Z.; Hong, J.; Che, Y.; Chen, R.; Hu, Z.; Hu, X.; Wu, Q.; Hu, J.; Zhang, M. Iron deficiency promotes aortic medial degeneration via destructing cytoskeleton of vascular smooth muscle cells. *Clin. Transl. Med.* **2021**, *11*, e276. [CrossRef] [PubMed]
130. Agmon, Y.; Khandheria, B.K.; Meissner, I.; Sicks, J.R.; O'Fallon, W.M.; Wiebers, D.O.; Whisnant, J.P.; Seward, J.B.; Tajik, A.J. Aortic valve sclerosis and aortic atherosclerosis: Different manifestations of the same disease? Insights from a population-based study. *J. Am. Coll. Cardiol.* **2001**, *38*, 827–834. [CrossRef] [PubMed]
131. Milin, A.C.; Vorobiof, G.; Aksoy, O.; Ardehali, R. Insights into aortic sclerosis and its relationship with coronary artery disease. *J. Am. Heart Assoc.* **2014**, *3*, e001111. [CrossRef] [PubMed]
132. Al-Taesh, H.; Çelekli, A.; Sucu, M.; Taysi, S. Trace elements in patients with aortic valve sclerosis. *Ther. Adv. Cardiovasc. Dis.* **2021**, *15*, 1753944720985985. [CrossRef] [PubMed]
133. Kosar, F.; Sahin, I.; Acikgöz, N.; Aksoy, Y.; Kucukbay, Z.; Cehreli, S. Significance of serum trace element status in patients with rheumatic heart disease: A prospective study. *Biol. Trace Elem. Res.* **2005**, *107*, 1–10. [CrossRef]
134. Koşar, F.; Sahin, I.; Taşkapan, C.; Küçükbay, Z.; Güllü, H.; Taşkapan, H.; Cehreli, S. Trace element status (Se, Zn, Cu) in heart failure. *Anadolu Kardiyol. Derg.* **2006**, *6*, 216–220.
135. Nyström-Rosander, C.; Lindh, U.; Friman, G.; Lindqvist, O.; Thelin, S.; Ilbäck, N.-G. Trace element changes in sclerotic heart valves from patients are expressed in their blood. *Biometals* **2004**, *17*, 121–128. [CrossRef]
136. Nkomo, V.T.; Gardin, J.M.; Skelton, T.N.; Gottdiener, J.S.; Scott, C.G.; Enriquez-Sarano, M. Burden of valvular heart diseases: A population-based study. *Lancet* **2006**, *368*, 1005–1011. [CrossRef]
137. Yutzey, K.E.; Demer, L.L.; Body, S.C.; Huggins, G.S.; Towler, D.A.; Giachelli, C.M.; Hofmann-Bowman, M.A.; Mortlock, D.P.; Rogers, M.B.; Sadeghi, M.M.; et al. Calcific aortic valve disease: A consensus summary from the Alliance of Investigators on Calcific Aortic Valve Disease. *Arterioscler. Thromb. Vasc. Biol.* **2014**, *34*, 2387–2393. [CrossRef]
138. Laguna-Fernandez, A.; Carracedo, M.; Jeanson, G.; Nagy, E.; Eriksson, P.; Caligiuri, G.; Franco-Cereceda, A.; Bäck, M. Iron alters valvular interstitial cell function and is associated with calcification in aortic stenosis. *Eur. Heart J.* **2016**, *37*, 3532–3535. [CrossRef] [PubMed]

139. Armstrong, A.; Mandala, A.; Malhotra, M.; Gnana-Prakasam, J.P. Canonical Wnt Signaling in the Pathology of Iron Overload-Induced Oxidative Stress and Age-Related Diseases. *Oxid. Med. Cell. Longev.* **2022**, *2022*, 7163326. [CrossRef] [PubMed]
140. Xu, S.; Gotlieb, A.I. Wnt3a/β-catenin increases proliferation in heart valve interstitial cells. *Cardiovasc. Pathol. Off. J. Soc. Cardiovasc. Pathol.* **2013**, *22*, 156–166. [CrossRef] [PubMed]
141. Brookes, M.J.; Boult, J.; Roberts, K.; Cooper, B.T.; Hotchin, N.A.; Matthews, G.; Iqbal, T.; Tselepis, C. A role for iron in Wnt signalling. *Oncogene* **2008**, *27*, 966–975. [CrossRef] [PubMed]
142. Song, Y.; Yang, N.; Si, H.; Liu, T.; Wang, H.; Geng, H.; Qin, Q.; Guo, Z. Iron overload impairs renal function and is associated with vascular calcification in rat aorta. *Biometals* **2022**, *35*, 1325–1339. [CrossRef] [PubMed]
143. Talvio, K.; Kanninen, K.M.; White, A.R.; Koistinaho, J.; Castrén, M.L. Increased iron content in the heart of the Fmr1 knockout mouse. *Biometals* **2021**, *34*, 947–954. [CrossRef]
144. Morvan, M.; Arangalage, D.; Franck, G.; Perez, F.; Cattan-Levy, L.; Codogno, I.; Jacob-Lenet, M.-P.; Deschildre, C.; Choqueux, C.; Even, G.; et al. Relationship of Iron Deposition to Calcium Deposition in Human Aortic Valve Leaflets. *J. Am. Coll. Cardiol.* **2019**, *73*, 1043–1054. [CrossRef]
145. Tomášek, A.; Maňoušek, J.; Kuta, J.; Hlásenský, J.; Křen, L.; Šindler, M.; Zelený, M.; Kala, P.; Němec, P. Metals and Trace Elements in Calcified Valves in Patients with Acquired Severe Aortic Valve Stenosis: Is There a Connection with the Degeneration Process? *J. Pers. Med.* **2023**, *13*, 320. [CrossRef]
146. Gonzalvo, M.C.; Gil, F.; Hernández, A.F.; Villanueva, E.; Pla, A. Inhibition of paraoxonase activity in human liver microsomes by exposure to EDTA, metals and mercurials. *Chem. Biol. Interact.* **1997**, *105*, 169–179. [CrossRef]
147. Salonen, J.T.; Malin, R.; Tuomainen, T.P.; Nyyssönen, K.; Lakka, T.A.; Lehtimäki, T. Polymorphism in high density lipoprotein paraoxonase gene and risk of acute myocardial infarction in men: Prospective nested case-control study. *BMJ* **1999**, *319*, 487–489; discussion 490. [CrossRef] [PubMed]
148. Kulka, M. A review of paraoxonase 1 properties and diagnostic applications. *Pol. J. Vet. Sci.* **2016**, *19*, 225–232. [CrossRef] [PubMed]
149. Sevim, Ç.; Doğan, E.; Comakli, S. Cardiovascular disease and toxic metals. *Curr. Opin. Toxicol.* **2020**, *19*, 88–92. [CrossRef]
150. Ćirović, A.; Buha Đorđević, A.; Ćirović, A.; Jevtić, J.; Tasić, D.; Janković, S.; Antonijević, B.; Petrović, Z.; Orisakwe, O.E.; Tasić, N. Trace Element Concentrations in Autopsied Heart Tissues from Patients with Secondary Cardiomyopathy. *Biol. Trace Elem. Res.* 2023; *online ahead of print*. [CrossRef]
151. Klinova, S.V.; Minigalieva, I.A.; Protsenko, Y.L.; Sutunkova, M.P.; Gurvich, V.B.; Ryabova, J.V.; Valamina, I.E.; Gerzen, O.P.; Nabiev, S.R.; Balakin, A.A.; et al. Changes in the Cardiotoxic Effects of Lead Intoxication in Rats Induced by Muscular Exercise. *Int. J. Mol. Sci.* **2022**, *23*, 4417. [CrossRef] [PubMed]
152. Li, C.; Shi, L.; Peng, C.; Yu, G.; Zhang, Y.; Du, Z. Lead-induced cardiomyocytes apoptosis by inhibiting gap junction intercellular communication via autophagy activation. *Chem. Biol. Interact.* **2021**, *337*, 109331. [CrossRef] [PubMed]
153. Turdi, S.; Sun, W.; Tan, Y.; Yang, X.; Cai, L.; Ren, J. Inhibition of DNA methylation attenuates low-dose cadmium-induced cardiac contractile and intracellular Ca(2+) anomalies. *Clin. Exp. Pharmacol. Physiol.* **2013**, *40*, 706–712. [CrossRef] [PubMed]
154. Pan, M.; Cheng, Z.-W.; Huang, C.-G.; Ye, Z.-Q.; Sun, L.-J.; Chen, H.; Fu, B.-B.; Zhou, K.; Fang, Z.-R.; Wang, Z.-J.; et al. Long-term exposure to copper induces mitochondria-mediated apoptosis in mouse hearts. *Ecotoxicol. Environ. Saf.* **2022**, *234*, 113329. [CrossRef]
155. Jiang, Y.; Zheng, W. Cardiovascular toxicities upon manganese exposure. *Cardiovasc. Toxicol.* **2005**, *5*, 345–354. [CrossRef]
156. Zhang, N.; Chen, M.; Li, J.; Deng, Y.; Li, S.; Guo, Y.; Li, N.; Lin, Y.; Yu, P.; Liu, Z.; et al. Metal nickel exposure increase the risk of congenital heart defects occurrence in offspring: A case-control study in China. *Med. Baltim.* **2019**, *98*, e15352. [CrossRef]
157. Borné, Y.; Barregard, L.; Persson, M.; Hedblad, B.; Fagerberg, B.; Engström, G. Cadmium exposure and incidence of heart failure and atrial fibrillation: A population-based prospective cohort study. *BMJ Open* **2015**, *5*, e007366. [CrossRef]
158. Ghosh, K.; Indra, N. Cadmium treatment induces echinocytosis, DNA damage, inflammation, and apoptosis in cardiac tissue of albino Wistar rats. *Environ. Toxicol. Pharmacol.* **2018**, *59*, 43–52. [CrossRef]
159. Chen, C.-Y.; Zhang, S.-L.; Liu, Z.-Y.; Tian, Y.; Sun, Q. Cadmium toxicity induces ER stress and apoptosis via impairing energy homoeostasis in cardiomyocytes. *Biosci. Rep.* **2015**, *35*, e00214. [CrossRef] [PubMed]
160. Oluranti, O.I.; Agboola, E.A.; Fubara, N.E.; Ajayi, M.O.; Michael, O.S. Cadmium exposure induces cardiac glucometabolic dysregulation and lipid accumulation independent of pyruvate dehydrogenase activity. *Ann. Med.* **2021**, *53*, 1108–1117. [CrossRef] [PubMed]
161. Das, S.C.; Varadharajan, K.; Shanmugakonar, M.; Al-Naemi, H.A. Chronic Cadmium Exposure Alters Cardiac Matrix Metalloproteinases in the Heart of Sprague-Dawley Rat. *Front. Pharmacol.* **2021**, *12*, 663048. [CrossRef] [PubMed]
162. Chou, S.-H.; Lin, H.-C.; Chen, S.-W.; Tai, Y.-T.; Jung, S.-M.; Ko, F.-H.; Pang, J.-H.S.; Chu, P.-H. Cadmium exposure induces histological damage and cytotoxicity in the cardiovascular system of mice. *Food Chem. Toxicol.* **2023**, *175*, 113740. [CrossRef] [PubMed]
163. Feng, J.; Yang, F.; Wu, H.; Xing, C.; Xue, H.; Zhang, L.; Zhang, C.; Hu, G.; Cao, H. Selenium protects against cadmium-induced cardiac injury by attenuating programmed cell death via PI3K/AKT/PTEN signaling. *Environ. Toxicol.* **2022**, *37*, 1185–1197. [CrossRef] [PubMed]

164. Fitch, M.L.; Kabir, R.; Ebenebe, O.V.; Taube, N.; Garbus, H.; Sinha, P.; Wang, N.; Mishra, S.; Lin, B.L.; Muller, G.K.; et al. Cadmium exposure induces a sex-dependent decline in left ventricular cardiac function. *Life Sci.* **2023**, *324*, 121712. [CrossRef] [PubMed]
165. McGraw, K.E.; Nigra, A.E.; Klett, J.; Sobel, M.; Oelsner, E.C.; Navas-Acien, A.; Hu, X.; Sanchez, T.R. Blood and Urinary Metal Levels among Exclusive Marijuana Users in NHANES (2005–2018). *Environ. Health Perspect.* **2023**, *131*, 87019. [CrossRef]

toxics

MDPI

Review

The Mechanisms of Cadmium Toxicity in Living Organisms

Slavena Davidova [1,2], Viktor Milushev [1,2] and Galina Satchanska [1,2,*]

[1] UPIZ Educational and Research Laboratory of Biology-MF-NBU, New Bulgarian University, 1618 Sofia, Bulgaria; stdavidova@nbu.bg (S.D.); f113742@students.nbu.bg (V.M.)
[2] Department of Natural Sciences, New Bulgarian University, Montevideo Blvd., 1618 Sofia, Bulgaria
* Correspondence: gsatchanska@nbu.bg

Abstract: Cadmium (Cd) is a toxic metal primarily found as a by-product of zinc production. Cd was a proven carcinogen, and exposure to this metal has been linked to various adverse health effects, which were first reported in the mid-19th century and thoroughly investigated by the 20th century. The toxicokinetics and dynamics of Cd reveal its propensity for long biological retention and predominant storage in soft tissues. Until the 1950s, Cd pollution was caused by industrial activities, whereas nowadays, the main source is phosphate fertilizers, which strongly contaminate soil and water and affect human health and ecosystems. Cd enters the human body mainly through ingestion and inhalation, with food and tobacco smoke being the primary sources. It accumulates in various organs, particularly the kidney and liver, and is known to cause severe health problems, including renal dysfunction, bone diseases, cardiovascular problems, and many others. On a cellular level, Cd disrupts numerous biological processes, inducing oxidative stress generation and DNA damage. This comprehensive review explores Cd pollution, accumulation, distribution, and biological impacts on bacteria, fungi, edible mushrooms, plants, animals, and humans on a molecular level. Molecular aspects of carcinogenesis, apoptosis, autophagy, specific gene expression, stress protein synthesis, and ROS formation caused by Cd were discussed as well. This paper also summarizes how Cd is removed from contaminated environments and the human body.

Keywords: Cd pollution; Cd toxicity; bacteria; plants; animals; human; molecular mechanisms

1. Introduction

Cadmium (Cd, atomic number: 48, atomic mass: 112.4, period number: 5, Group 12, electron configuration: $4d^{10}5s^2$, isotopes: 7) was discovered in 1817 by the German chemist Friedrich Stromeyer. This chemical element was named after one of the first heroes in Greek mythology, Cadmus, the legendary founder of Thebes, known for his immense strength before Heracles. The Latin word cadmia and the Greek word καδμεία, which was once used to refer to the standard zinc (Zn) ore calamine, are the sources of the name [1]. Cd was not known to be a separate element until 1817, as in nature, Cd is always found in the presence of zinc. Cd and zinc are chemically and physically similar. Cd is a by-product of zinc and lead production. The Cd/Zn ratio is 0.5% in nonferrous ores, with the actual recovery of cadmium metal estimated at 50–65% of that present in raw ore [2]. Cd is unique among metals due to its diverse toxic effects, long biological half-life, low rate of excretion, and predominant storage in soft tissues rather than bone [3]. In Table 1 are described some of Cd's physical and chemical properties.

Cd was initially found to have negative health consequences in 1858. Those who used polishing agents containing Cd suffered from respiratory and gastrointestinal problems. The earliest toxicological experiments were conducted in 1919. In humans, emphysema and proteinuria were first documented in the 1940s in workers exposed to Cd dust [3]. Following World War II, a bone ailment known as itai-itai disease—a type of Cd-induced renal osteomalacia—was reported in Japan, causing fractures and excruciating agony [3,4]. The toxicokinetics and toxicodynamics of Cd were then discussed, along with how it binds to the metallothionein protein [5]. In the 1970s, health advisories regarding the dangers of Cd pollution were distributed globally [6]. The World Health Organization's International Program on Chemical Safety recognized renal failure as a critical consequence of Cd and offered a rough quantitative evaluation. In the 1990s and 2000s, various epidemiological studies found adverse health impacts in demographic groups in Japan, China, Europe, and the United States, sometimes with minimal ambient Cd exposure. The early discovery of metallothionein's important role in Cd toxicology laid the groundwork for recent studies that used biomarkers of susceptibility to the development of Cd-related renal dysfunction, such as metallothionein gene expression in peripheral lymphocytes and metallothionein autoantibodies in blood plasma [7].

Table 1. Physical and chemical properties of Cd [4].

Atomic number	48
Atomic weight	112.41 u
Atomic radius	155 pm
Electronic configuration	$[Kr]4d^{10}5s^2$
Melting point	321.07 °C
Boiling point	767.3 °C
Density at 20 °C	8.65 g/cm^3
Reduction potential $Cd^{2+} + 2e^- \rightarrow Cd(s)$	−0.40 E°
Heat of fusion	6.21 kJ/mol
Heat of vaporization	99.6 kJ/mol
Electronegativity (Pauling scale)	1.69
First ionization energy	867.8 kJ/mol
Second ionization energy	1631.4 kJ/mol

2. The Origin of Cd Pollution

Cd is an environmental contaminant classified eighth on the Top 20 Hazardous Substances Priority List because of its high toxicity and sluggish metabolism [8]. The primary source of Cd is stack dust, produced during zinc purification by distillation and deposited in all fractions [9]. Plants can absorb Cd directly from the soil [10]. Phosphate fertilizers and atmospheric deposition have been the primary sources of Cd intake into soils. Phosphorite and apatite rocks mainly used in the production of phosphate fertilizers contain Cd and several other heavy metals [11]. The amount of Cd accumulated in soil due to environmental contamination depends on the magnitude of emissions, transit, and retention. The fate of heavy metal contaminants in soil is mainly determined by the balance of sorption, leaching, and plant uptake. Soil variables like pH, redox state, organic matter, clay, hydrous oxides, and free carbonates significantly impact these processes. Metal destiny varies significantly amongst soil types, including forest and heavily developed agricultural land [12]. Global production and use of Cd are significant; for instance, Cd pigments consumption surpasses 2500 tons annually [13]. For thousands of years, it has also been used as a pigment due to its ability to produce almost all the rainbow colors: brilliant yellow, orange, and red, interacting with other chemical elements. Thus, a bright orange color is made when the bright yellow pigment cadmium sulfide (CdS) is mixed with cadmium selenite (CdSe). Cadmium sulphoselenide (Cd_2SSe) creates a pigment called cadmium red. The pigments mentioned above are still used nowadays as plastic colorants. In principle, Cd is part of various chemical compounds as a divalent cation [14]. Human activities associated with Cd emissions include industrial production, crop farming, animal breeding, aquaculture, and wastewater treatment [15]. Low amounts of Cd are naturally found in the lithosphere (0.15 mg/kg in the Earth's crust and 1.1×10^{-4} mg/L in seawater) [16]. Still, various industrial processes, including mining and smelting, have increased the element's availability in the environment and increased human exposure to it. Thousands of tons of garbage polluted with Cd are dumped into the environment globally each year [16]. In Figure 1 is presented the distribution of heavy metals in the environment.

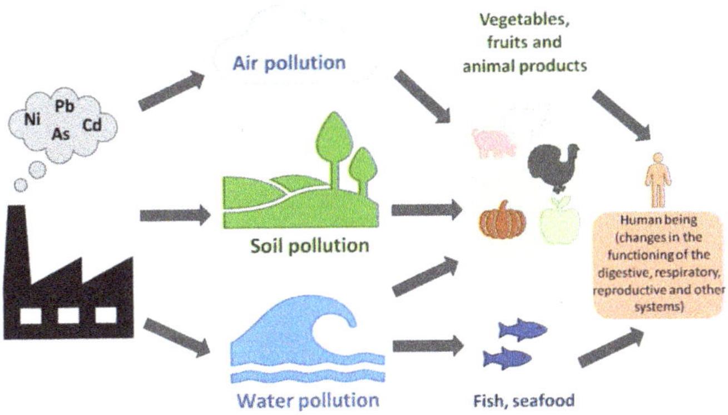

Figure 1. Distribution of heavy metals in the environment [17].

The Jinding lead-zinc mine is currently the largest lead-zinc deposit proven in China, one of only a few with over 10-million-ton reserves worldwide. Its mining area is 6.8 km^2, with more than 80% open-pit mining. The large-scale mining of the Jinding Pb-Zn deposit began in the 1980s, causing significant environmental problems [18]. Descriptive statistics revealed that the 1.1% As, 7.3% Cd, 0.3% Pb, and 0.2% Hg contents exceeded China's Soil Environmental Quality Management Standard (GB 15618-2018, in Chinese) [19]. Furthermore, 32.8% of As, 74.4% of Cd, 89.2% of Pb, 45.0% of Cr, and 13.7% of Hg concentrations in soil samples exceeded the background soil concentrations of heavy metals in this location, with Cd and Pb having the highest levels, which were 11.64- and 21.47-fold the background values. On the other hand, compared to regular farming areas, industrial and mining enterprises, sewage irrigation, and urban samples from those areas had considerably greater concentrations of As, Cd, Pb, Cr, or Hg in their soil [20]. Surface waters, especially the lakes are a significant global source of freshwater, and Cd pollution of water bodies is becoming more of an issue as industry, agriculture, and other human activities advance [21]. Cd pollution in lakes poses a significant risk to water quality, drinking water supplies, the food chain, and freshwater ecosystems. Sediments have been observed to act as both a sink and a source of Cd in water bodies. Yearlong monitoring in Meiliang Bay (northern part of Taihu Lake, National Wetland Park, eastern China) revealed that the mobility of Cd in sediments varied widely and significantly impacted the Cd pollution level in the overlying water [8]. Tourism has been shown to contribute to Cd pollution (primarily through hotel wastewater and increased traffic) and vice versa. Cd pollution of beaches, coastal waterways, food, urban parks, and other areas poses risks to tourists and increases human exposure to this poisonous metal [22].

2.1. *Natural Sources of Cd*

2.1.1. Cd in Soil Water and Groundwater

Generally, Cd concentration in the earth is around 0.1–0.5 ppm, and this metal mainly accumulates in sedimentary rocks. Natural activities such as erosion, weathering of rocks, volcano eruptions, and wildfires release large amounts of Cd into soils and rivers and, respectively, seas and oceans. Phosphorites and marine phosphates contain high amounts of Cd, as much as 500 ppm. Maximum permissible Cd concentrations are 5 μg/g in soil and water and 1 μg/L in groundwater [21]. Large quantities of Cd (about 15,000 mt (metric tons) are transported into rivers by erosion and weathering of rock materials. Volcanic eruptions release about 820 metric tons of Cd, while forest fires release up to 70 metric tons [23]. Detailed information on Cd distribution in European top soils is available on the European Commission's official webpage: https://esdac.jrc.ec.europa.eu/search/node/cadmium [24] (accessed on 20 November 2024). As seen from Figure 2, out of the total, 72.6% of the samples have Cd values < 0.07 mg kg^{-1}, 21.6% in the range 0.07–1 mg kg^{-1}, and the remaining 5.5% higher than the threshold of 1 mg kg^{-1}, which is generally considered the limit for risk assessment [25].

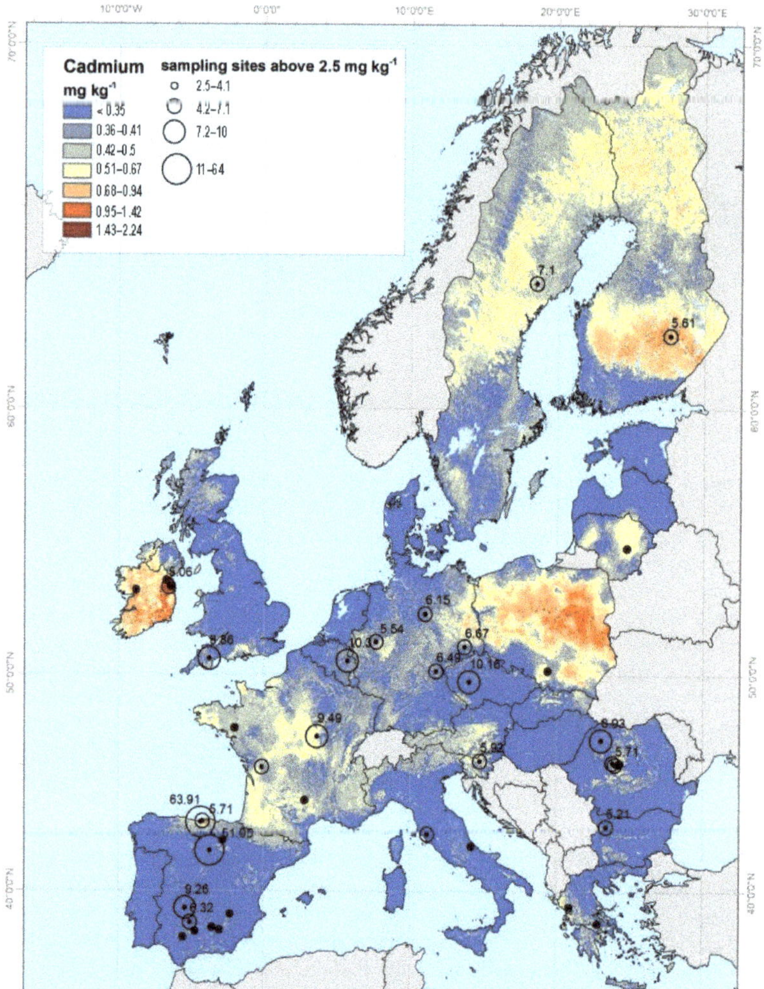

Figure 2. Cd concentration for 2024 in Europe with the outliers overlaid as black circles. The dimension of the circle indicates the concentration for the given outlier; concentrations above 5 mg/kg^{-1} are also indicated by their numerical value [25]. As seen from the figure, the highest Cd concentration was registered in northern Spain.

Data about all heavy metal concentrations are permanently deposited at the European Soil Data Centre (ESDAC, https://esdac.jrc.ec.europa.eu/, accessed on 28 October 2024) EU, European Commission, Joint Research Centre. Other data on the topic can be found in the Agency for Toxic Substances and Disease Registry (ATSDR), EFSA CONTAM Panel reports, and the Organisation for Economic Co-operation and Development (OECD).

The quality of surface water in Europe is regulated by Directive 2000/60/EC [26]. There are 100,000 surface water bodies in Europe, but only 40% are in good condition. As Kubier et al. [21] highlighted, WHO Guidelines for Drinking Water Quality recommend a value of up to 3 µg/L.

Groundwater in Pakistan has typical Cd contents of 10 µg/L from Jurassic sulfide-bearing sedimentary rocks. In contrast, in Germany, groundwater concentrations of Cd range from 0.11 µg/L in loess aquifers beneath arable land to 2.7 µg/L in sandy aquifers beneath forested regions [21].

A comparison of crucial aquifer systems reveals a link between rock type, groundwater environment, and Cd contents. Cd's 90th percentile background levels in groundwater varied from less than 0.1 µg/L in Paleozoic, Triassic, and Jurassic aquifers to more than 1 µg/L in Rotliegend, Cretaceous, and Cenozoic aquifers. Aside from calcium carbonate Cretaceous aquifers, limestone-dominated aquifer basins exhibited low Cd levels in groundwater. Most of them are in alkaline aquifer systems [21]. Cd concentrations above 1 µg/L were found in groundwater in sandstone aquifers and unconsolidated sand and gravel aquifer systems in the western United States. However, in most samples collected from 3124 wells in the US, Cd was below 1 µg/L.

Groundwater, but from a glacial aquifer system in the United States, had Cd values ranging from 0.018 µg/L to 1.0 µg/L. However, 84% of groundwater samples (N = 847) were below the detection limit. A survey of groundwater near garbage sites in the United States revealed Cd values of up to 6000 µg/L. Municipal solid waste dumps in the European Union can produce leachates with Cd values up to 2700 µg/L. As a result, Cd concentrations that exceed the natural background can be caused by both natural and anthropogenic mechanisms [21].

2.1.2. Air

Natural Cd emissions come mainly from rock weathering, airborne soil particles from deserts, sea spray, forest fires, biogenic material, volcanoes, and hydrothermal vents. In Europe, the leading standard regulating the Cd concentration in ambient air is the Directive 2008/50/EC (https://ec.europa.eu/environment/air/quality/directive.htm, accessed on 28 October 2024) [27]. According to this directive, the maximum permission level is 5 ng/m^3, which has been valid since 2013 [27]. In the USA, this restriction is even more substantial, and the regulatory bodies such as NAAQS (National Ambient Air Quality Standards), USEPA (United States Environmental Protection Energy Agency), and NIOSH (National Institute of Occupational Safety and Health) limited Cd in the air to 200 pg/m^3. It is known that $PM_{2.5}$ can transport heavy metals such as Cd, Hg, Pb, Cr, and Mn. In research conducted in Beijing, China, the immediate impacts of $PM_{2.5}$ exposure on blood Cd levels were studied. The findings revealed that the average blood Cd concentration was 0.64 µg/L. A significant correlation was observed between $PM_{2.5}$ exposure and blood Cd level ($p < 0.05$), as authors reported [28].

Soil particles are the most common source of natural emissions into the atmosphere, followed by forest and bushfires, sea salt, volcanic emissions, and meteoric dust [29]. Wildfires increase Cd concentrations in soils and ashes. Long-term behavior analysis revealed decreasing Cd concentrations in the solid phase, as rainfall and pH decrease with time following fire, resulting in desorption and mobility of Cd and other heavy metals [21]. Wildfires in California, for example, raised the average Cd concentration in the runoff by more than two orders of magnitude. Cd concentrations in biomass ash can reach 30 mg/kg, providing an additional method for increasing Cd concentrations in soil because such ash is commonly used as fertilizer. In the short term, the bioavailable pool of Cd remains low due to an ash-induced pH increase, but as pH rises due to rainfall, Cd bioavailability increases [21].

2.2. Anthropogenic Cd Sources

Anthropogenic Cd inputs into soil, groundwater, and atmosphere come from mining—lead, zinc smelters [30], uranium mill tailings [31–33] as reported by Satchanska et al., nonferrous metal manufacturing, fossil fuel combustion, phosphate fertilizer and pesticide manufacturing, iron, steel, cement production, road dust, plastics production, wildfires, and municipal and sewage sludge incineration. Environmental pollution with Cd stems from its widespread use in the production of alloys and batteries, as a pigment in plastics, paints, and ceramics, and corrosion-resistant coatings of metal products. This heavy metal is primarily found and abundant in lead, copper, and zinc ores. Over the past 50 years, anthropogenic Cd emissions have fallen by more than 90% [34]. Like with uranium (U), using phosphate fertilizers with Cd as an impurity is a common cause of high Cd concentrations in soil and groundwater. This Cd addition pathway to groundwater was explored in the United States, Canada, Britain, Norway, Sweden, Finland, Denmark, Germany, Australia, and New Zealand. The findings indicate that P fertilizer application alters soil chemistry. Furthermore, Cd can enter the food chain and be hazardous to living organisms. Cd sources can be both local and diffuse [30].

Local sources such as mines, industrial sites, and abandoned mining deposits cause increased Cd concentrations, albeit generally on a limited spatial scale [35]. Atmospheric emissions, wastewater reuse, and agricultural operations can all act as diffuse sources, resulting in the widespread distribution of Cd in the environment [36].

The screens of mobile phones from different generations showed a significant decrease in the quantities of Cd (from 1.0 μg/g to undetectable levels) and of Pb (from 35.0 μg/g to 2.0 μg/g) from feature phones to smartphones [37]. Cd is also commonly present in children's toys. As reported by Igweze et al. [38], cheap toys purchased from Port Harcourt, Nigeria, stores were determined to contain three toxic metals (Pb, Cd, and As). The present heavy metals in all the toys were below the limits set by the EU [38].

Heavy metals, including arsenic, cadmium, chromium, and lead, are present in goods made from leathers, synthetic leathers used in shoe production, and textiles. Bielak et al. [39] found that children's footwear made from sheepskins contained As, Ba, Cd, Cr, Hg, Pb, Sb, and Se. In 2006, leathers used for insoles, shoe uppers, and clothing underwent testing in Turkey for Co, Cr, Cu, Pb, Ni, and Zn due to their close contact with the body. In Bielak et al.'s paper, it is described that in 2015, various types of fibers used in the textile industry in Turkey, such as cotton, acrylic, polyester, nylon, viscose, and polypropylene, were examined for the presence of Al, Cd, Co, Cr, Cu, Fe, Mn, Ni, Pb, Tl, and Zn. Additionally, in Italy in 2009, Cr was identified in a wool top [39].

The utilization of cadmium for creating affordable jewelry has recently become of concern. Kern et al. conducted a study in 2020 and found that seven out of the nine undamaged jewelry items tested released cadmium in amounts that surpassed the recommended maximum of 18 μg. They concluded that the undamaged items had a maximum extractable cadmium amount of 6230 μg, nearly three hundred times greater than the 18 μg limit [40].

Nickel-cadmium batteries are the most common source of Cd in dump sites with municipal solid waste worldwide [41]. In European municipal solid wastes, Cd levels range from 0.3 to 12 mg/kg. Pigments, coatings and platings, PVC stabilizers, and alloys are also Cd-containing items [21].

3. Cd Toxicity on Living Organisms

Heavy metals, respectively Cd, exist in soils in two forms—immobilized (organically bound) and mobile (in the soil solution). The immobilized ones are considered nonbioavailable, while the mobile is considered bioavailable. The access of heavy metals to organisms is also affected by different soil characteristics such as pH, oxido-reductive balance, clay, iron oxide, and organic matter content. It is well known that the presence of Cd and other heavy metals in the environment influences the growth and survival of microorganisms and also shifts the microbial community in these habitats, resulting in poorer soil fertility [29,41]. As a rule, Gram-positive bacteria are more sensitive than Gram-negative bacteria [42].

3.1. Cd Accumulation and Toxicity on Bacteria

Several studies on the interactions of Cd with microorganisms have been reported [43]. In some bacteria, such as *Ralstonia metallidurans* CH34, isolated near a zinc production plant in Belgium, a gene called the *czc*-gene, responsible for the induction of resistance to Cd, was found [44]. Heavy-metal resistance genes were found to be localized outside this microorganism's bacterial chromosome in two small (3–5 Kb) self-replicating circular DNA molecules called plasmids pMOL28 and pMOL30. Overview results on bioremediation of Cd by the Gram(−) bacterium *Pseudomonas aeruginosa* are published by Chellaiah [43].

3.1.1. Resistance Mechanisms of Bacteria

Bacteria inhabiting high-metal environments evolved several mechanisms of resistance, fighting for survival. Usually, Cd(II) forms stable toxic complexes in the cell. The concentration of Cd ions in the cell is regulated by cellular homeostasis through these specific resistance mechanisms [45].

Resistance mechanisms of bacteria function as energy-dependent effluxes of metal ions that export them out of the bacterial cell. In most cases, the resistance is carried out by plasmids that contain genes encoding resistance to Cd [37]. Besides Cd^{2+}, the plasmid-mediated resistance system removes other metal ions such as Ag^{+}, AsO_2^{-}, AsO_4^{3-}, Co^{2+}, CrO_4^{2-}, Cu^{2+}, Hg^{2+}, Ni^{2+}, Pb^{2+}, TeO_3^{2-}, and Zn^{2+}. Three mechanisms of resistance can be distinguished.

Proteins Conducting the Export of Heavy Metals

The first group of bacterial transport proteins is proteins associated with resistance, budding, and cell division (resistance-nodulation-cell division), shortly named "RND proteins". These proteins were first found in the widely studied multi-resistant bacterium *Ralstonia metallidurans*, which are polyresistant to several heavy metals. This bacterium was isolated near a zinc smelter in Belgium and described by Mergeay et al. [44]. Besides the heavy metal efflux, RND proteins play a role in the budding of *Mesorhisobium lotti* and the cell division of *E. coli*.

RND proteins are divided into seven notable families found in organisms of all kingdoms. Most important are the families of inner-membrane, periplasmic, and outer-membrane efflux proteins. They form a single efflux—a protein complex that can evacuate Cd from the cytoplasm via the cytoplasmic membrane, cell wall, and outer membrane (typical for the Gram-negatives) into the extracellular space. Unlike the ATP-dependent one, this type of transport system is energy-independent, known as the CBA-efflux transport system.

The Cd-Co-Zn- CBA-efflux system in polyresistant *R. metallidurans* is widely discussed. This beta-proteobacterium contains two large plasmids that determine its resistance to heavy metals, including Cd: pMOL30 and pMOL28. The first plasmid, pMOL30, encodes resistance to Co^{2+}, Zn^{2+}, and Cd^{2+} (named *czc* genes). When genes are expressed, they increase the MIC of Zn^{2+}, Co^{2+}, and Cd^{2+} from 7 to 50 times [37]. The structural genes responsible for resistance to Cd and other heavy metals in the pMOL30 plasmid were cloned and sequenced at least 20 years ago. Four open reading frames (ORFs) were detected in the czc-DNA sequence. They are arranged on the czc operon as follows: first is the *czcC* gene (encodes a CzcC protein of 346 amino acids (AA) followed by the *czcB* gene (CzcB protein composed of 521 AA), the *czcA* gene (CzcA protein composed of 1064 AA), and the *czcD* gene (CzcD protein consisting of 200 AA) [45]. The second plasmid, pMOL28, encodes resistance to Ni^{2+} and Co^{2+}, increasing the MIC of Co^{2+} by 16 times. This type of resistance is called the *cnr* gene, which encodes the CnrA protein. CzcA and CnrA proteins were the first proteins identified in the RND family. The resistance to Cd, along with cobalt and zinc, in the bacterium *R. metallidurans* (former name *Alcaligenes eutrophus*) is shown in Figure 3.

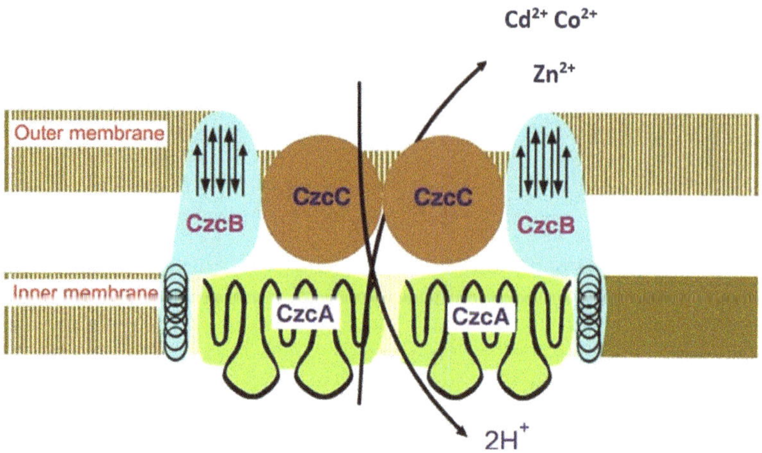

Figure 3. Czc model for Cd^{2+}, Zn^{2+}, and Co^{2+} efflux system functioning as proton/cation antiporter consisting of inner membrane (CzcA), outer membrane (CzcC), and membrane fusion (CzcB) proteins functioning as a dimer [45].

Proteins Facilitating the Transport of Heavy Metals

These proteins (cation diffusion facilitators, or CDF) have been found in many prokaryotes and eukaryotes. Their substrate is predominantly cadmium but also zinc, cobalt, nickel, and iron cations [45]. Ion transport is carried out via both concentration and chemiosmotic gradients. Bacteria usually possess one or more CDF genes. Such have been found in the bacterium *R. metallidurans* and yeast *Saccharomyces cerevisiae.*

P-Type ATP-Ase

P-type ATP-ases play a protective role on the bacterial cell against heavy metals. These enzymes belong to the transport proteins that pump out from the cell the metal cations using energy from ATP hydrolysis, i.e., their action is energy dependent. Among the heavy metals, their substrates are the Cd and silver cations. P-type ATPase can import its substrate from the external environment into the cytoplasm and export it from the cytoplasm outside the cell [45]. According to this principle, P-type ATP-ases are divided into importing and exporting categories. It should be noted that P-type ATPase is precious for two reasons: it imports micronutrients such as Mg^{2+}, but at the same time, it can also import heavy metals, and secondly, it exports heavy metal cations and thus neutralizes them, driving the bacterial cell to life.

3.1.2. Bacteria in Cd-Contaminated Soil

In a study by Salam et al., Illumina shotgun sequencing of DNA extracted from two Cd-contaminated agricultural soils showed the predominance of the phyla, classes, genera, and species of *Proteobacteria* (37.38%), *Actinobacteria* (35.02%), *Prevotella* (6.93%), and *Conexibacter woesei* (8.93%) in the first sample, and *Proteobacteria* (50.50%), *Alphaproteobacteria* (22.28%), *Methylobacterium* (9.14%), and *Methylobacterium radiotolerans* (12.80%) in the second one. The study concluded that contamination with Cd has a significant impact on the structure and function of the soil microbial community, changes the resistome of heavy metals, modifies the physicochemical properties of the soil, and leads to the substantial decline of certain native community members that are not accustomed to Cd stress [46].

The presence of Cd has a crucial impact on bacterial diversity, and there are notable differences in microbial communities between areas with high Cd pollution and those with low Cd pollution. Yu et al. reported that higher concentrations of Cd significantly increased the abundance of *Proteobacteria* and *Gemmatimonas* and decreased the abundance of *Nitrospirae*. Members of the genera *Burkholderia* and *Bacillus* were reported to develop a resistance to Cd and may play an essential role in the bioremediation of Cd-contaminated soils [47]. In a study by Luo et al., high Cd-level sites displayed lower diversity indices than low Cd-level sites. The dominant phyla observed by the authors in paddy soil samples are *Proteobacteria, Chloroflexi, Acidobacteria, Actinobacteria, Gemmatimonadetes, Verrucomicrobia, Thaumarchaeota, Firmicutes*, and *Nitrospirae*. In this study, it is pointed out that *Actinobacteria* are tolerant to Cd, whereas *Proteobacteria, Verrucomicrobia*, and *Nitrospirae* are sensitive [48].

Sun et al. [49] found that the biodiversity and composition of the soil microbial community were affected in Cd-contaminated soil, with bacterial diversity impacted more than fungal diversity. The abundance of the soil microbial community decreased. At the same time, the composition changed at the phylum level, specifically in biomarkers for bacteria such as *Saccharibacteria* and *Gemmatimonadetes*, as well as in *Arenimonas, Xanthomonadales, Nitrosomonadaceae, Methylophilales, Caulobacteriales, Aeromicrobium, Chitinophagaceae, Acidimicrobiales, Nocardioidaceae, Propionibacteriales, Frankiales*, and *Gemmatimonadaceae*, which were found to be positively correlated with the total and available Cd [49]. Khan et al. describe an overview of Cd toxicity against living organisms and microbial resistance mechanisms, emphasizing the efflux systems, antioxidant profiling, and Cd eradication potential exhibited by microorganisms when exposed to Cd^{2+}. Cd resistance and bioremediation potential make these microorganisms a good bioresource for green chemistry to exterminate environmental Cd^{2+} [50].

Heavy metals alter the soil microbial community composition, and the microorganisms that adapt to this stress increase in abundance [51]. Usually, the highest bacterial diversity is detected in severely contaminated soils [52]. As reported by authors, phyla *Proteobacteria* and *Acidobacteria* are abundant in soils contaminated by Cd and other heavy metals. The prevalence of *Actinobacteria, Acidobacteria, Proteobacteria*, and *Chloroflexi* in heavy metal-polluted soils was reported by Hemmat-Jou et al. [53]. Besides, the soil pH influences the heavy metals' mobility and their toxic effects on bacterial communities. Acidic pH turns metals more bioavailable than neutral pH.

Using minimal bacteriological media with heavy metals added and no carbon source, researchers are able to grow, isolate, and further study the heavy metal-tolerant bacteria. To determine the bacterial sensitivity to heavy metals, growth of the bacteria is then assessed based on different criteria such as turbidity, biomass, and enzyme activities [54].

Several studies have shown that some lactic acid bacteria (LAB), including Lactobacillus plantarum, *Lactobacillus rhamnosus, Bifidobacterium breve*, and *Bifidobacterium lactis*, can bind and remove heavy metals such as Cd and Pb in vitro [55]. There is also substantial evidence that probiotic LAB have antioxidative properties that may be effective against Cd-induced oxidative stress in humans. Based on these unique characteristics, daily LAB consumption may be a preventive dietary strategy for people exposed to Cd [55]. Table 2 shows Cd removal potential by some bacterial and yeast strains.

Table 2. Cd^{2+} removal (%)/uptake (mM/g) potential in some bacterial and yeast strains [50].

Organism	Cd^{2+} Resistance (mM)	Cd^{2+} Removal (%) Uptake (mM/g)
Klebsiella pneumoniae	13.3	57.4
Escherichia coli P4	10.6	56
Salmonella enterica 43C	13.3	22
Bacillus sp.	–	50
Rhodobacter sphaeroides	–	30.7
Microbacterium oxydans CM3	–	43
Candida tropicalis	25.0	92
Pichia hampshirensis 4Aer	24.0	28
Candida tropicalis 3Aer	25.1	31
Trametes versicolor	5.0	0.300
Trichosporon ashii	10	78
Pichia kudriavzevii	15	61

3.2. Cd Accumulation and Toxicity on Fungi

The primary impact of Cd's toxicity is mainly on fungi's growth and replication. Fungi are particularly affected, as some fungal species may be eradicated following Cd exposure in soil. There is a selection pressure for resistant strains after low-level exposure to Cd in soil.

Soil fungi have an essential role in detoxifying and improving Cd-contaminated soils. The fungal cell wall contains polysaccharides and chitin and effectively helps control Cd tolerance and its uptake while acting as a barrier against Cd^{2+} entry into plant cells. Functional groups like carboxylic and hydroxylic groups and amino acids in the fungal cell wall carry a negative charge, enabling the cell wall to attract positively charged metallic ions. Research has shown that species such as *Absidia cylindrospora*, *Suillus luteus* (ectomycorrhizal fungi), and the group of *Neotyphodium endophytes* have significant potential for Cd tolerance and can be used effectively in remediating Cd-contaminated soils [56].

Heavy metals like Cd are known to be some of the most hazardous pollutants. In research conducted by Zheng et al., a filamentous fungus strain YZ1 was found in soil from wheat farmland. This strain, which belongs to *Purpureocillium* sp. based on its appearance and genetic evidence, had a minimum inhibitory concentration of 1 mM Cd and could survive in 100 mM Cd, reaching maximum biomass at 0.4 mM Cd. The strain YZ1 can be a great choice for cleaning up Cd-contaminated farmland in wheat-producing areas [57].

In another study by Fazli et al., seven fungi that were highly tolerant to Cd were tested: *Aspergillus versicolor, Aspergillus fumigatus, Paecilomyces* sp.9, *Paecilomyces* sp.G, *Terichoderma* sp., *Microsporum* sp., and *Cladosporium* sp. [58]. The findings revealed that the fungi displayed various resistance mechanisms against Cd and could sequester Cd from liquid environments. *Aspergillus versicolor* showed a remarkable difference in detoxification behavior compared to the other isolated fungi. It demonstrated a strong ability to grow actively in the presence of Cd and reduce the Cd concentration to less toxic levels. The introduction of *Aspergillus versicolor* as a scavenger organism marks the initial step in the emergence of this fungus in bioremediation science [58].

3.3. Cd Accumulation and Toxicity on Edible Mushrooms

The accumulation of heavy metals in the fruiting bodies of mushrooms is extensively studied. The mean content of Cd in analyzed mushrooms ranges from 0.370 to 2.151 mg/kg d.w., while Pb is found at the level of 0.243–0.424 mg/kg d.w. Heavy metals are believed to cause pronounced toxicological harm to human health when contaminated mushrooms, even at low metal concentrations, are consumed [59]. Furthermore, consuming mushrooms contaminated with heavy metals can lead to damage to the kidneys and heart, as well as the impairment of the digestive, immunological, skeletal, and nervous systems [59]. Mushrooms can gather high amounts of minerals, even when grown in soils with low metal content. This is attributed to the species' genetic characteristics, including many transport genes and binding ligands [60].

Mushrooms can absorb specific forms of heavy metals, such as Cd^{2+}, Cd^{6+}, Hg^{2+}, As^{5+}, etc., into their fruiting bodies. In connection, metals' intracellular speciation and uptake are typically regulated by metallothioneins and GT complexes that are directly linked to fungal physiology [61]. Species with the ability to accumulate Cd include *Agaricus bisporus, A. campestris, A. macrosporus, Armillaria mellea, Amanita muscaria* and *A. allies, Boletus edulis, Cantharellus cibarius, Cystoderma carcharias, Macrolepiota procera, Xerocomus badius,* and *Tricholoma matsutake*. The concentration of Cd in edible mushrooms can go up to 1.3925 mg/kg and readily accumulate in more significant quantities within the caps.

The highest allowable levels of Cd for *Agaricus bisporus* (common mushroom), *Pleurotus ostreatus* (oyster mushroom), and *Lentinula edodes* (shiitake mushroom) are set at 0.20 mg/kg. For other mushroom species, it is 1.0 mg/kg. The mentioned maximum levels are applicable after washing the mushrooms and separating the edible part [62].

As the number of wild mushroom consumers continues to rise, monitoring the presence of toxic elements and their potential risks has become important. In this regard, the elevated Cd levels in wild edible mushrooms can pose health risks to consumers, especially as several species of *Tricholoma* mushrooms are consumed in fresh or processed form [61].

3.4. Cd Accumulation and Toxicity on Plants

In plants, Cd predominantly accumulates in the roots, with lesser amounts in the leaves [63]. Cd exhibits toxicity to a broad range of plants, but its harmful effects are mitigated by sediment, high concentrations of dissolved salts, or organic matter. Cd adversely affects plant growth in experimental settings, although no field effects have been reported. Plants more readily take the metal from nutrient solutions than from soil, with most studies demonstrating its effects in nutrient solution cultures. Cd has been shown to impact stomatal opening, transpiration, and photosynthesis in plants grown in nutrient solutions [63].

Chlorosis and stunted growth in plants are easily identifiable indicators of Cd poisoning. Higher toxicity slows plant growth and causes necrosis [64]. Cd toxicity harms plants by limiting carbon fixation and reducing chlorophyll concentration and photosynthetic activity [65]. Cd accumulation in plants can result in various physiological, biochemical, and structural alterations. Its accumulation modifies mineral nutrient intake, slows stomatal opening by interfering with plant water balance, disrupts Calvin cycle enzymes, photosynthesis, and carbohydrate metabolism, changes antioxidant metabolism, and reduces agricultural output [66].

Exposure to Cd in the soil causes osmotic stress in plants by reducing leaf-relative water content, stomatal conductance, and transpiration, resulting in physiological damage [67]. Cd poisoning generates an overproduction of reactive oxygen species (ROS), which damages plant membranes and destroys cell proteins and organelles [65].

As reported by Huybrechts et al., exposing *Trigonella foenum-graecum* seeds to Cr, Pb, and Cd solutions revealed that Cd at 10 mg L^{-1} showed the most significant inhibitory effect on germination, the highest concentration tested. Similarly, *Triticum aestivum* required less Cd than Pb to hinder seed germination. However, it is important to note that there can be significant variations between plant species [64].

Cd is recognized for inhibiting seed germination through various mechanisms [68]. In *Vigna unguiculata* seeds, the inhibitory effect of Cd was suggested to be caused by impaired water uptake, limiting the water availability for the developing embryo. In addition to a restricted water supply, inhibition of starch mobilization from the endosperm, combined with an impaired translocation of soluble sugars to the embryonic axis, can further starve the embryonic axis. A decrease in hydrolyzing enzymes, including α-amylase, proteases, and acid phosphatases, in *Sorghum bicolor* seeds was proposed to be responsible for the reduced storage mobilization. Calcium is crucial for amylase activity, and replacing the chemically similar Cd ion could disrupt normal enzyme functioning. Furthermore, radish seeds experienced direct competition for Ca-calmodulin binding sites between Ca and Cd ions [64].

Cd exposure is widely known to cause DNA damage. The mechanisms behind this Cd-induced DNA damage involve the ROS-induced formation of 8-hydroxyguanosine and the inhibition of DNA repair systems. Table 3 describes Cd's effect on different plant species.

The adverse effect of Cd on the cell cycle is presented in Table 4.

The structure that provides the primary defense for a plant cell against pathogen attacks and adverse environmental conditions such as drought and metals is the cell wall. Preventing excess Cd from entering the cytoplasm is crucial because it can cause damage to macromolecules, proteins, and DNA due to Cd-induced oxidative stress. Roots in direct contact with Cd from the soil have cell walls that play a crucial role in this process. Research indicates that most plant species store the majority of Cd within the cell walls of roots [64]. When the capacity of the cell wall is surpassed, Cd may form complexes with PCs and then can finally be sequestered *within vacuoles* by transferring ABC transporters through the tonoplast.

Table 3. Plant species show varying responses to Cd-induced DNA damage. This damage includes DNA strand breaks, chromosomal aberrations, and micronuclei formation. Cd also impacts the expression of DNA repair genes and causes alterations in amplified fragment length polymorphism (AFLP), inter-simple sequence repeat (ISSR), random amplified polymorphic DNA (RAPD), sequence-related amplified polymorphism (SRAP), and simple sequence repeat (SSR) profiles, leading to a reduction in genomic template stability (GTS). The symbols ↑ and ↓ indicate increases and decreases, respectively [64].

Species	Organ	Cd Concentration	Exposure Duration	Effect
Allium cepa	Root tip	50–200 μM	2 h + 24 h recovery	Micronucleus formation Chromosomal aberrations
	Root tip	25 μM	48 h	
	Root tip	25 μM	48 h	% tail DNA ↑
Arabidopsis thaliana	Root tip	0.125–2.5 mg L^{-1}	5 d	Altered expression DNA repair genes
	Root	1.25–4 mg L^{-1}	5 d	Altered RAPD profile Altered expression DNA repair genes
	Leaf	0.5–5 mg L^{-1}	16 d	Altered AFLP profile
	Leaf	0.25–8 mg L^{-1}	15 d	Microsatellite instability Altered RAPD profile
	Leaf	5 μM	72 h	Altered expression DNA repair genes
Brassica chinensis	Leaf	15–120 mg kg^{-1} soil	30 d	Altered RAPD profile
Brassica oleracea	Root	2.5–20 mg kg^{-1} soil	3–56 d	Altered % tail intensity
Capsicum annuum	Root tip	20–100 ppm	24 h	Chromosomal aberrations
	Leaf	20–100 ppm	24 h	Altered RAPD profile
Hordeum vulgare	Root tip	75–225 μM	7 d	Altered RAPD profile (GTS ↓)
	Leaf	5 μM	15 d	DNA damage ↑
Ipomoea aquatica	Entire seedling	15–120 mg kg^{-1} soil	21 d	Altered RAPD profile (GTS ↓)
Lactuca sativa	Root tip	25 μM	48 h	Chromosomal aberrations Micronucleus formation % DNA damage ↑
Lathyrus sativus	Root tip	5–50 μM	3–7 d	Chromosomal aberrations Micronucleus formation
Leucaena leucocephala	Leaf	50 mg L^{-1}	15 d	Altered RAPD profile
Nicotiana tabacum	Root and leaf	10–15 μM	7 d	% tail DNA ↑
Oryza sativa	Root tip	50–200 μM	48–96 h	Altered SRAP profile (GTS ↓)
Sphagnum palustre	Shoot	0.1–10 μM	24–48 h	Altered ISSR profile (GTS ↓)

For a plant to go through its life cycle, it needs to start the reproductive phase, which in higher plants involves the development of flowers, pollination, and fertilization, followed by seed production. When *Brassica campestris* plants were exposed to Cd during the flowering stage, the content of glutathione (GSH) and ascorbate (AsA) decreased the most, suggesting that the reproductive phase was highly vulnerable [64]. Cd has been demonstrated to harm pollen germination and disrupt pollen tube morphology in various plant species [69].

Table 4. An analysis of the effects of Cd on cell cycle-related parameters organized by plant species. Exposure to Cd decreases the mitotic index (i.e., the ratio of cells undergoing mitosis to the total number of cells) and changes in nuclear ploidy levels. It influences the expression of cell cycle-related genes in various plant species. The symbols ↑ and ↓ represent increases and decreases, respectively [64].

Species	Organ	Cd Concentration	Exposure Duration	Effect
Allium cepa	Root tip	50–200 μM	2 h + 24 h recovery	Mitotic index ↓
	Root tip	25 μM	48 h	Mitotic index ↓
Arabidopsis thaliana	Root tip	0.125–2.5 mg L^{-1}	5 d	2C ↓, 4C ↑, 8C ↑ Altered cell cycle phase distribution Altered expression cell cycle-related genes
	Root	1.25–4 mg L^{-1}	5 d	2C ↓, 4C ↑ Altered expression of cell cycle-related genes
	Leaf	5 μM	3–12 d	Endoreduplication factor ↓ Epidermal cell number and cell surface area ↓ Altered expression of cell-cycle related genes
Capsicum annuum	Root tip	20–100 ppm	24 h	Mitotic index ↓
Lactuca sativa	Root tip	25 μM	48 h	Mitotic index ↓
Lathyrus sativus	Root tip	5–50 μM	3–7 d	Mitotic index ↓
Oryza sativa	Root	200 μM	7 d	Cortex cell length in the elongation zone ↓ Cortex cell number in the elongation zone ↓ Altered expression of cell cycle-related genes
Sorghum bicolor	Root tip	50–200 μM	5 d	Inhibition of S phase progression

3.5. Cd Accumulation and Toxicity on Animals

Cd toxicity causes a wide range of health issues, including some of the deadliest diseases, such as heart disease [70,71], kidney disease [72,73], liver disease [72], cancer, and diabetes [70]. Practically every system in an animal's body can be harmed by Cd. Over an extended period, Cd accumulates in the kidney and liver [74]. Farm animals can come into contact with contaminated water, soil, vegetation, and car and industrial emissions [75,76]. Foods of all kinds contain a lot of different types of Cd. Foods made from grains, leafy vegetables like spinach, and mainstays like potatoes have relatively high levels of Cd, ranging from 30 to 150 ppb [77]. Sunflower, soybean, and peanut seeds all have naturally high levels of Cd, seemingly with no adverse health effects. Fish and meat typically have lower levels of Cd, ranging from 5 to 40 ppb [78]. Offal organs where Cd accumulates, like the kidney and liver, can contain remarkably high Cd levels—up to 1000 ppb [78].

Numerous organisms readily accumulate Cd, notably mollusks, where bioconcentration factors can reach several thousand. Soil invertebrates also exhibit significant Cd accumulation. In contrast, most organisms display low to moderate concentration factors, typically under 100 [79]. Cd in tissues is often bound to proteins, including specific heavy-metal-binding proteins known as metallothioneins, which have been isolated from organisms exposed to Cd [80]. The highest Cd concentrations are found in the kidney, gills, and liver (or their equivalents). The primary route of Cd elimination in organisms is likely through the kidney, although in crustaceans, substantial amounts can also be shed via the exoskeleton.

The acute toxicity of Cd to aquatic organisms varies considerably, even among closely related species, and is influenced by the free ionic concentration of the metal. Cd interferes with calcium metabolism in animals, causing hypocalcemia in fish by inhibiting calcium uptake from the water. However, high calcium concentrations in the water can protect fish from Cd uptake by competing at the uptake sites [81]. Zinc exacerbates Cd toxicity in aquatic invertebrates. Sublethal effects on aquatic invertebrates include impaired growth and reproduction and structural damage to gills. Fish, particularly salmonids, show variable sensitivity to Cd, with sublethal effects, including spinal malformations [82]. Embryos and early larvae are the most susceptible life stages, while eggs are the least affected. There is no consistent interaction between cadmium and zinc in fish.

In a recent study by Djedjibegovic et al., Cd content (mean concentration) was tested in seafood samples in Bosnia and Herzegovina. The team concluded that mercury and cadmium were detected in all analyzed samples (100%), while lead was detected in 33 samples (89.2%). Metals content was in the order Hg > Cd > Pb in most of the species, except blue mussel (Pb > Cd > Hg) and Indian white prawn (Hg > Pb > Cd). Cadmium content was close to the corresponding MRL in two samples of Patagonian squid (0.918 and 0.896 mg kg^{-1}). It was also relatively high in the other three samples of the same species (0.591, 0.425, and 0.391 mg kg^{-1}) [83].

Exposure to Cd at relatively low doses impairs sperm motility and morphology and can reduce male fertility in rats [84]. In a study conducted in the Czech Republic by Drapal et al., higher average levels of Cd were found in the liver (0.10 mg/kg) and kidney (0.62 mg/kg) of cattle over 2 years old compared to lower levels in the liver (0.06 mg/kg) and kidney (0.24 mg/kg) of cattle under 2 years old. Ruminants are exposed to Cd contamination by consuming pasture during the summer and preserved feed, including cereals, during the winter [85,86]. According to Chirinos-Peinado et al., Cd and Pb can accumulate in fresh cow's milk [87]. Fay et al. discussed that Cd nephrotoxicity is associated with altered microRNA expression in the rat renal cortex [88].

3.6. Cd Accumulation and Toxicity in Humans

Cd poses a significant health danger to people, even at low concentrations. The body's ability to adapt to Cd exposure is limited due to its inability to undergo metabolic degradation and poor excretion [14]. Cd toxicity affects various animal organs, including the liver, kidney, lungs, testes, prostate, heart, skeletal system, neurological system, and immune system. Prolonged exposure to Cd can accumulate in the body and cause illnesses primarily affecting the lungs and kidneys [89]. It is believed that chronic lung diseases, high blood pressure, cancer, leukemia, genetic toxicity, damage to human organs, abdominal pain, burning sensations, nausea, vomiting, and various cancers may be related to slow poisoning by Cd in small doses. Additionally, Cd exposure may be associated with diseases related to the central nervous system, cognitive and behavioral functions, chronic diseases, teratogenic effects, cardiovascular disease, lung function abnormalities, and damage to the kidneys [90].

Acute Cd poisoning symptoms typically develop after 24 h and include shortness of breath, weakness, and fever. Cd poisoning can lead to pulmonary edema, pneumonia, and, in severe cases, respiratory failure, and death. Cadmium's distribution in the body is determined by its chemical form. Exposure to Cd in the form of inorganic salts (e.g., $CdCl_2$) leads to a higher accumulation of Cd^{2+} ions in the liver, kidneys, and bones than Cd combined with metallothionein. $CdCl_2$ primarily accumulates in the liver, and CdMT accumulates in the kidney [91]. Cd is deposited in various organs, including the liver, kidney, testis, spleen, heart, lungs, thymus, salivary glands, epididymis, and prostate. However, because of their high MT content, the liver and kidney store around 50% of the accumulated Cd [92]. Cd can accumulate in the pancreas, lungs, central nervous system, and testes in men. Cd particles are carried through the primary olfactory neurons [93].

3.6.1. Cd Distribution in the Human Body

Cd, a subject of increasing scientific and medical interest due to its detrimental effects on health, presents a pressing need for a comprehensive understanding of its distribution in human tissues and organs. This knowledge forms the basis for crucial in vitro and in vivo animal studies [94]. The specific amounts of Cd in various cell types and tissue layers, in particular, have not been thoroughly explored. This underscores the necessity for further investigation [95]. Only a handful of researchers have dealt with the total Cd content in the human body [94]. Figure 4 below presents the primary sources of Cd and its effects on the human body.

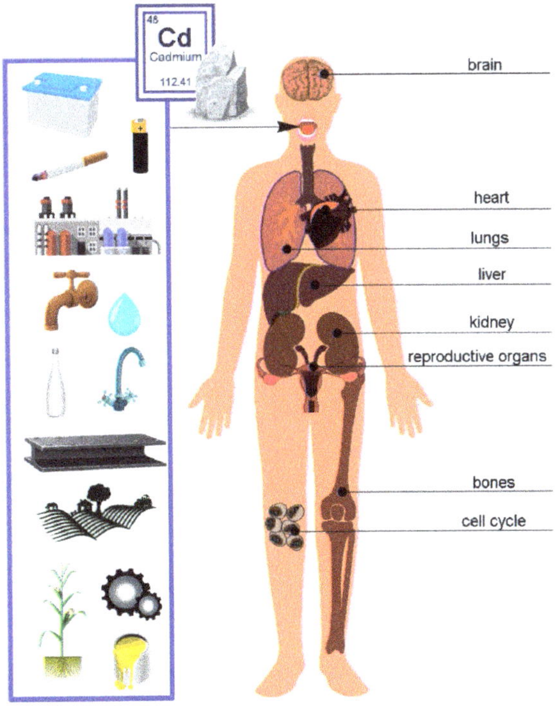

Figure 4. Sources of Cd and its most significant effects on different parts of the human body [90].

Ingestion

Cd tends to persist and accumulate in soil, eventually entering plant metabolism. Cd enters the food chain as it accumulates in edible plant parts such as fruits and seeds [96]. This accumulation rises when soil pH decreases; therefore, acid rain has the effect of raising Cd concentrations in plants. Cereals and bread account for 34% of daily Cd intake in Western countries, followed by leafy vegetables, particularly spinach, among adults (20%), potatoes (11%), legumes and nuts (7%), stem/root vegetables (6%), and fruits (5%) [97]. Mei et al. studied the accumulation and molecular mechanisms of Cd in cereal crops and tobacco [98]. Wheat can also accumulate high levels of Cd in the grain [99]. In Eastern countries, fish and shellfish are the most common Cd sources, followed by grains and vegetables, particularly rice [100]. Rice is one of the most important food crops worldwide, feeding over 5 billion people. Thus, Cd pollution in rice has attracted great attention [101]. Although dietary Cd exposure poses a minor health risk in most Eastern countries, it remains a concern for specific subgroups. Cd is dispersed over the earth, and there are locations with extremely high concentrations of Cd in the soil. Crop uptake of Cd in these locations can result in high food exposures for those living nearby. For example, in Japan's Jinzu and Kakehashi river basins, soil is extensively contaminated with Cd from industrial waste [102,103]. Locals who habitually ingested rice cooked with Cd-contaminated water experienced a severe kidney and bone illness known as "itai-itai" sickness, which was marked by bone distortion and many fractures, particularly in women [104]. The decrease in Cd intake over the last 50 years could be ascribed to less sewage sludge escaping into agricultural soil due to improved control and environmental awareness in industrialized countries. These processes primarily transferred Cd from plants into the food chain, increasing human exposure to the metal. Some aquatic creatures can also be significantly impacted by Cd deposition at levels above the regulatory limits [82].

Cd intake from meals in humans ranges from 8 to 25 μg per day, with only 0.5 to 1.0 μg retained in the body. Factors affecting this form of absorption include dose, exposure period, dietary components, nutritional status, age, and gender [93]. As reported by Raikwar et al., the safe recommended intake of Cd is 15–50 μg/day for adults and 2–25 μg/day for children [74].

Inhalation

Cd air levels can be hundreds of times higher in the workplace than in the broader environment [105]. The Occupational Safety and Health Administration (OSHA) sets the permissible exposure limit (PEL) for Cd fumes or Cd oxide in the workplace at 0.1 mg/m^3. However, Cd concentrations in ambient air range from 1×10^{-6} mg/m^3 in non-industrialized areas to 4×10^{-5} mg/m^3 in urban areas. Non-occupational Cd exposure from the air is unlikely to cause harmful health effects. Cd air levels are typically insufficient to create health issues in the general population. Even in places with substantial industrial Cd emissions, the average atmospheric concentration does not exceed 35 ng Cd/m^3 of air [106]. Tobacco smokers have four to five times greater levels of Cd in their blood and two to three times higher amounts in their kidneys than nonsmokers (Figure 5) [107,108]. According to Gray et al., the liquids used in electronic cigarettes also contain Cd with a concentration of 0.108 μg/g [109].

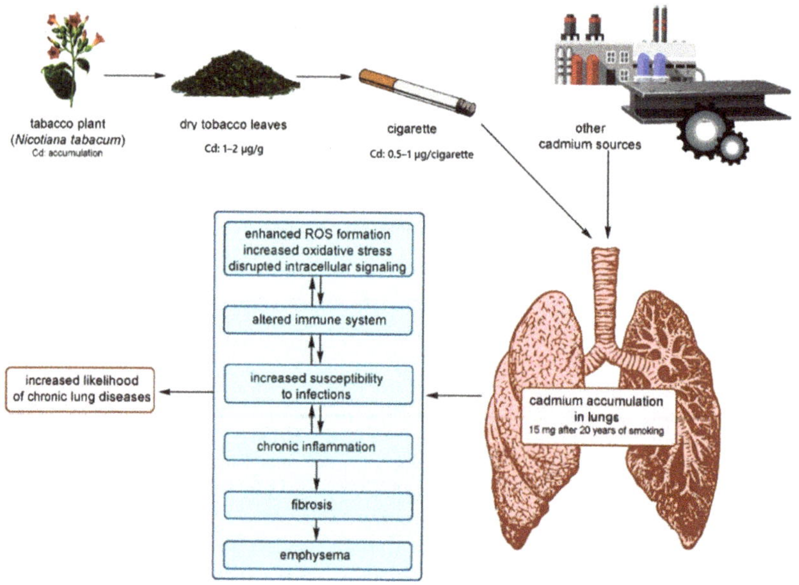

Figure 5. Cd exposure leads to the development of smoking-related lung diseases [90].

Permeation

The skin absorbs tiny amounts of Cd. Hence, it is not considered a critical route of exposure. However, recent research has highlighted the environmental significance of photosensitive CdS and CdSe pigments and nano semiconductors, whose oxidized products (cadmium sulfate ($CdSO_4$) and cadmium selenite ($CdSeO_4$)) are significantly more soluble and bioavailable, and thus potentially more dangerous [110]. An in vitro investigation employing human full-thickness skin as a model to describe the impact of Cd exposure on skin revealed that the metal only enters the epidermis; previously, it was demonstrated that its solubility into the *stratum corneum* layer is a rate-limiting process (Figure 6) [92].

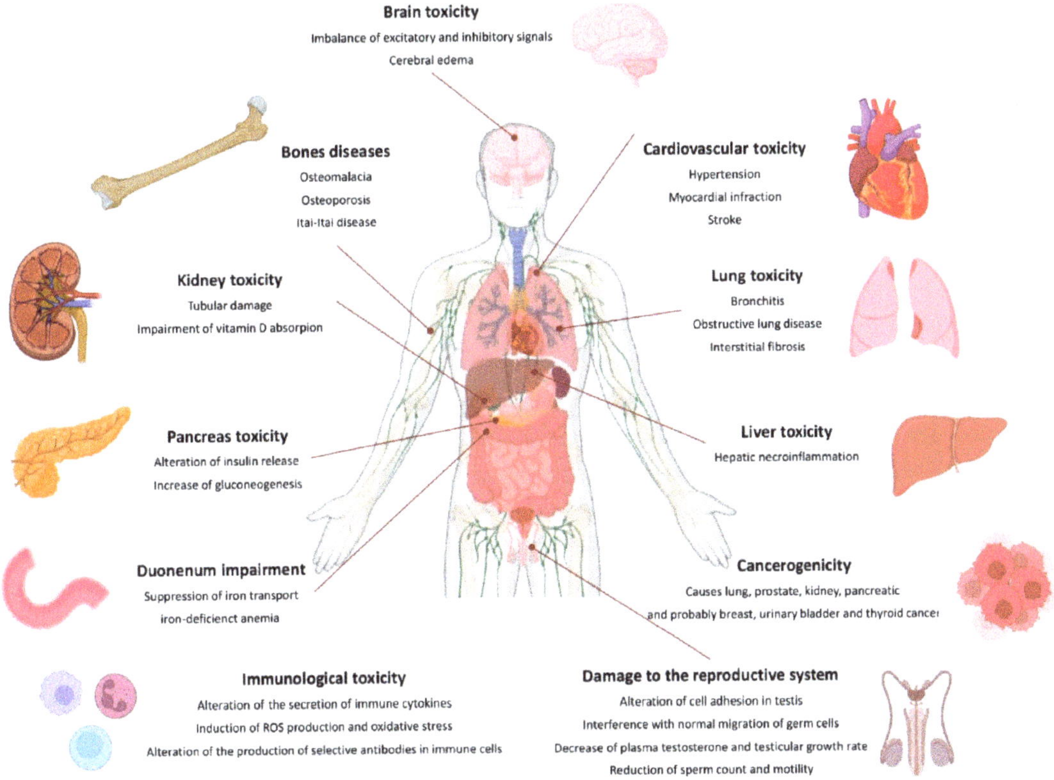

Figure 6. Primary outcomes in health effects following chronic Cd exposure [1].

3.6.2. Harmful Effects on Human Organs and Systems

Cd can act through molecular mimicry, which occurs when metals attach to nucleophilic sites on specific biological molecules, forming complexes that can structurally and/or functionally resemble endogenous substrates that typically bind to the active sites of carrier proteins, channels, structural proteins, enzymes, and/or transcription factors. In principle, ionic mimicry shares similarities with molecular mimicry. Generally, the term ionic mimicry refers to the capacity of an unbound, cationic species of metal to act as a structural and/or functional counterpart or mimic of another (usually an essential) element at the site of a carrier protein, ion channel, enzyme, structural protein, transcription factor, and/or metal-binding protein. For instance, considerable evidence has been gathered demonstrating that the cationic forms of certain toxic metals, such as Cd, can utilize ion channels (especially Ca^{2+} channels) and specific membrane transporters, like the divalent metal transporter 1 (DMT1/DCT1/Nramp1), to enter target cells in mammalian organisms [111]. Cd^{2+} can interact with membrane transporters responsible for the uptake of essential metals like Ca^{2+}, Fe, and Zn, allowing it to enter target cells in organs negatively impacted by this metal. This process occurs through ionic mimicry, where Cd^{2+} imitates the divalent cationic forms of these essential metals at the binding sites of various carrier proteins and/or channels that facilitate their transport [112].

Effects on the Blood and Circulatory System

Smokers have higher levels of Cd in their blood and are more likely to develop atherosclerosis, specifically peripheral vascular disease, than non-smokers [113]. Research suggests that low-dose Cd exposure increases the incidence of peripheral arterial disease (PAD). This element also has a detrimental effect on the cardiovascular system [114]. Cd's harmful effects on the vasculature of many organs, particularly endothelial cells, have been demonstrated through experimental research. Cd damages endothelium and smooth muscle cells, leading to the production of atherosclerotic plaques [115]. This is supported by epidemiological and clinical research. Cd can also influence homocysteine metabolism. High levels of homocysteine can lead to heart disease, stroke, peripheral vascular disease, and cognitive impairment [116].

Effects on the Reproductive System

Cd can impair reproductive functioning. Exposure to Cd poisoning predominantly impacts testicular function [117]. The toxic effects of Cd in the testis include damage to the vascular endothelium. Cells experience morphological and functional changes, including inhibition of testosterone synthesis and impaired spermatogenesis due to oxidative stress, impaired antioxidant defense mechanisms, and inflammatory response severity. Cd can disrupt prostate function, affecting hormonal activity, secretion, and fertility in men [93]. Exposure to Cd harms human male reproductive organs/systems, impairing spermatogenesis, semen quality, particularly sperm motility, and hormone synthesis/release [118]. According to experimental and human investigations, it also disrupts female fertility, reproductive hormone balance, and menstrual cycles [119]. Based on the available data, it is possible to conclude that low-dose Cd exposure harms both male and female reproduction and impacts pregnancy or its outcome [120]. Furthermore, maternal prenatal Cd exposure may have a distinct effect on male and female offspring, particularly female offspring. As a result, measures must be taken to limit exposure to Cd [121]. The testicles are the primary target of Cd-induced acute toxicity [122]. The effect is quick. The testicles begin to shrink, followed by inflammation, edema, acute bleeding, and necrosis in less than 24 h after a single dose of cadmium [123].

Even a signal injection of Cd^{2+} can cause acute testicular necrosis, as reported by Bridges and Zalups [111]. In testes, Cd^{2+} mimics Zn^{2+} using a Zn^{2+} transporter, which contains many amino acids. In the same paper, it is shown that Cd^{2+} uptake occurs via passive and active mechanisms of interstitial cells of rat testes. Fe^{2+} is also involved in the Cd^{2+} uptake in the testes via the Fe^{2+} transporter DMT1, which is identified in the Sertoli cells of the testes.

Cd^{2+} can also mimic estrogen (estradiol). Twenty-four-hour exposure to $CdCl_2$ leads to the same physiological effect when treated with estradiol, both activating the estrogen receptor proven in a breast cancer cell line (MFC-7). As Cd^{2+} does not exist in an unbound state in physiological solutions, data obtained from studies in which cells were treated with $CdCl_2$ may not accurately represent physiological events in vivo [111].

Effects on the Respiratory System

Inhalation produces respiratory distress and damages the respiratory tract [124]. High Cd concentrations in contaminated air have been linked to conditions like emphysema, anosmia, and chronic rhinitis. Lampe et al. [125] studied the impact of Cd exposure on lung function among 96 males with one to three lung function tests between 1994 and 2002. Researchers discovered that smoking was linked to lower forced expiratory volume in 1 s (a measure of lung function) and higher urine cadmium levels [125]. Inhaling Cd can cause respiratory distress syndrome [93].

Effects on the Kidney System and Bones

Long-term exposure to high doses of Cd results in itai-itai disease, primarily affecting women. This disease is marked by severely compromised tubular and glomerular function, generalized osteomalacia, and osteoporosis, leading to multiple bone fractures [126]. Prolonged exposure to low doses of Cd has been associated with tubular damage, resulting in reduced reabsorptive capacity for nutrients, vitamins, and minerals; nephropathy; and proteinuria [77,107]. Non-absorbed molecules include zinc or copper bound to metallothionein (MT), glucose, amino acids, phosphate, calcium, and low-molecular-weight (LMW) proteins like 2-microglobulin (2-M) and 1-microglobulin (1-M), also known as protein HC, retinol-binding protein (RBP), and uric acid, resembling Fanconi syndrome, a genetic disorder of renal tubular transport. Urinary markers for Cd exposure include Cd itself, LMW proteins (2-M, 1-M), and enzymes of renal tubular origin, such as *N*-acetyl glucosaminidase (NAG). Generally, urinary Cd levels indicate long-term body burden before kidney damage develops, while blood Cd levels indicate recent exposure [127]. Cd disrupts the metabolism of calcium, magnesium, iron, zinc, and copper in cells, leading to demineralization, osteomalacia, osteoporosis, and bone disorders, necessitating the replacement of these ions. Cd's competitive displacement of calcium ions weakens bone structure, often causing fractures, especially in children and postmenopausal women. Cd also inhibits the activity of 1-hydroxycholecalciferol hydroxylase, an enzyme essential for converting 25(OH)D3 to the active form of vitamin D3, 1,25$(OH)_2$D3, in the kidney [91]. This active form is crucial for calcium absorption in the intestine [93]. Research of 31- to 60-year-old men from Pakistan's growing industrial nation found that diabetic males (N = 196) had considerably higher Cd levels in their blood and urine than non-diabetic males (N = 238). These disparities were observed in both smokers (N = 209) and non-smokers (N = 225) [128]. Although Cd is known for its renal toxicity, the dose-response relationship at low environmental exposure levels remains unclear. The Centers for Disease Prevention and Control (CDC) and European health agencies acknowledge this uncertainty and the potential public health impact of low-level Cd exposure, urging further research.

The Avonmouth lead/zinc smelter in southwest England closed in early 2003 and was the UK's largest source of atmospheric Cd emissions. It emitted 978 kg of Cd in its final year, accounting for nearly 30% of the UK's point source emissions. The site had produced zinc for over 70 years, emitting other nephrotoxic metals like lead, mercury, and arsenic. Given Cd's high nephrotoxicity and long biological half-life, this study focused on Cd exposure. With approximately 50,000 people living within 5 km of the smelter, there is concern about increased Cd exposure from inhalation contaminated air and ingesting homegrown vegetables and house dust. A total of 180 volunteers (74 men and 106 women) were studied for health problems caused by Cd. Among the participants, 109 (40 men and 69 women; 61%) were never-smokers, 36 (19 men and 17 women; 20%) were former smokers, and 32 (13 men and 19 women; 18%) were current smokers. Results obtained by urine analyses showed early kidney damage [129].

The kidney is one of the main organs that suffer negatively in humans after long-term oral or inhalation exposure to Cd^{2+}. The harmful impacts of Cd^{2+} on the kidneys became evident when factory workers involved in the production of nickel-cadmium batteries were exposed to cadmium oxide dust and cadmium vapors. The kidney function of these workers experienced notable changes, leading to proteinuria and a decreased glomerular filtration rate [111].

The majority of Cd^{2+} found in the kidney is concentrated in the epithelial cells of the proximal tubule. Due to Cd^{2+}'s strong binding affinity for thiol-containing biomolecules like GSH and Cys, the forms of Cd^{2+} likely presented to the luminal membrane of proximal tubular cells are complexes of these molecules, such as G-S-Cd-S-G or Cys-S-Cd-S-Cys. Furthermore, given that γ-glutamyltransferase and cysteinylglycinase are highly abundant in the luminal plasma membrane of proximal tubular cells, it seems unlikely that G-S-Cd-S-G is absorbed as an intact complex [111]. In vivo research in rats has shown that after administering $CdCl_2$ subcutaneously with excess Cys, increased amounts of Cd accumulate in the epithelial cells of the proximal tubule. At least one mechanism for renal uptake of Cd^{2+} exists at the luminal side and another at the basolateral side. In addition to other pathways, Cd^{2+} might gain entry into proximal tubular cells via Ca^{2+} channels, though this theory lacks definitive evidence. Nonetheless, studies have indicated that Cd^{2+} can use Ca^{2+} channels in isolated cells from different organs, such as the liver and intestine [111].

Effects of the Liver and Intestines

Following oral intake, Cd^{2+} is absorbed by the intestines and transported to the liver through portal blood, where hepatocytes readily uptake it. After administering Cd^{2+} intravenously, up to 60% of a non-toxic dose (5 µmol/kg) can be found in rat livers within an hour. In studies using low doses of $CdCl_2$, G-S-Cd-S-G, or Cys-S-Cd-S-Cys, less than 1% remained in the bloodstream after one hour, demonstrating the swift extraction by the liver and other tissues [111]. Approximately 50% of the Cd^{2+} still present in the blood is associated with cellular components, mainly within erythrocytes, likely via an anion exchanger. One membrane transporter that is probably responsible for the sinusoidal uptake of the cationic version of Cd^{2+} is DMT1. This transporter has been isolated and identified as a proton-coupled iron transporter. In liver cells, DMT1 is situated in the sinusoidal membrane [111].

The uptake of Cd^{2+} by hepatocytes may involve calcium channels. Given that the ionic radius of Cd^{2+} closely resembles that of Ca^{2+}, Cd^{2+} can likely imitate Ca^{2+} at calcium channels to enter hepatocytes. In fact, in vitro research using primary cultures of rat hepatocytes and cultured immortalized hepatocytes shows that Cd^{2+} may be transported via calcium channels. These studies found that the uptake of Cd^{2+} into hepatocytes was significantly blocked by Ca^{2+} channel antagonists, such as diltiazem and verapamil. Cd^{2+} entered these cells through voltage-gated L-type calcium channels. Moreover, additional experiments where hepatocytes were treated with Ca^{2+} channel antagonists indicated that around one-third of the Cd^{2+} that entered the cells did so through calcium channels [111].

The duodenum and the initial portion of the jejunum account for most of the absorption of ingested Cd^{2+}. The duodenum is a key area for the absorption of Fe^{2+}. The similarity in ionic radii between Cd^{2+} (0.95 Å) and Fe^{2+} (0.55 Å) implies that these two cations might share some transport mechanisms. In vivo studies in rats reveal that Cd^{2+} disrupts the intestinal uptake of Fe^{2+}, indicating that these two metal ions could use similar transport pathways. DMT1, found in the luminal plasma membrane of enterocytes, seems to play a crucial role in the intestinal absorption of Fe^{2+}. In another research, rats were fed a diet deficient in Fe or a diet supplemented with Fe, following which they were given an oral dose of $CdCl_2$. The mRNA levels of DMT1 in the small intestine of the rats fed the Fe-deficient diet were 15-fold greater than those in the ones fed the supplemented diet [111]. The content of Cd^{2+} in the small intestines of animals with depleted Fe stores was significantly greater than that in rats fed a supplemented diet. Thus, it can be concluded that the Fe status significantly affects the expression of DMT1. Ca^{2+} channels in the luminal plasma membrane can serve as a way for Cd^{2+} to access the cytosolic compartment of enterocytes. The paper by Bridges and Zalups describes it as a two-step mechanism of Cd^{2+} transport across the luminal plasma membrane of enterocytes. The first is the binding of Cd^{2+} to the plasma membrane. The binding was susceptible to EDTA but insensitive to temperature, whereas the second step was sensitive to temperature and insensitive to EDTA. The same paper highlighted that duodenum was rich in amino acids and peptides that can absorb significant amounts of Cd^{2+}. In contrast, Cd^{2+} can evacuate from the basolateral membrane of the enterocytes via Zn^{2+} transporter (ZnT-1), which was demonstrated by in vivo and in vitro studies. Based on these statements, it can be concluded that Cd^{2+} is a competitive inhibitor of Ca, Fe, and Zn [111].

Effects on the Central Nervous System

Accumulating evidence over the years suggests that Cd may be a potential cause of neurodegenerative diseases. This could be related to the excessive production of free radicals, which damage organs depending on their defense mechanisms [130]. Since Cd is a toxic agent affecting various cell types, this study aimed to investigate Cd's effects and consequences on different organs in mice. To test the hypothesis of concentration-dependent reactive oxygen species (ROS) generation and DNA damage, Agnihotri et al. [130] assessed the serum of 4–5-week-old male Swiss albino mice, which were treated with cadmium chloride ($CdCl_2$), added to their drinking water for 30 days. The expression of Bcl-2-associated X protein (Bax), an apoptotic marker, was twice as high in the brain compared to the liver at an exposure level of 0.5 mg L^{-1} $CdCl_2$ [131]. Additionally, correlation and linkage data analysis of the antioxidant defense system showed rapid changes in the brain compared to other organs in the study. The results indicate that even at low doses, Cd impairs the brain due to its lipid peroxidase sensitivity, which promotes Cd-induced oxidative injury in the brain [132]. Human studies indicate that exposure to Cd can lead to aberrant behavior and lower IQ in children and adults. The blood-brain barrier is susceptible to Cd. Cd has direct harmful effects when it crosses the blood-brain barrier, which can happen in young children or with certain medical conditions. The choroid plexus epithelium can accumulate significant levels of Cd, which reduces concentrations in other body regions [133].

Cd toxicity can cause cell injury, cell death, and organ failure through various chemical pathways, the most common of which is oxidative stress, in numerous body compartments and tissues, including the central nervous system [134]. In this context, many in vivo and in vitro research studies have provided evidence showing Cd-induced neurotoxicity in the CNS and Cd's involvement in many compartments and cells of the CNS [135]. Cd-treated embryos developed a smaller head with unclear boundaries between the brain subdivisions, particularly in the mid-hindbrain region. Embryos display normal anterior-to-posterior regionalization; however, the commitment of neural progenitor cells was affected by Cd [81]. Cd has been shown to produce free radicals in the brain, potentially damaging neurons and oligodendrocytes (OLG). OLGs are the glial cells which myelinate axons in the CNS. An early study reported that Cd toxicity affected CNS white matter [136], and one laboratory demonstrated that OLGs are direct targets of this structure [137]. Experimental studies have shown that Cd can also be a potent neurotoxicant for the peripheral nervous system [138]. Moreover, Cd has a half-life of more than 15 years in humans. Elderly workers may be more susceptible to an increased Cd body burden and may develop peripheral polyneuropathy (PNP) over time [139]. There is significant evidence suggesting that Cd^{2+} can be absorbed by the distal segments of the nephron. Additionally, Ca^{2+} channels could offer a pathway for the uptake of luminal Cd^{2+} in the cells that line the distal nephron [111].

Effects on the Immune System

Chronic Cd exposure can negatively impact the immune system [140]. Cd targets several cells, including T cells, macrophages, B cells, and natural killer cells. Cd's direct immunotoxicity alters immunological responses, including cell-mediated and humoral immunity. Cd toxicity may cause anemia and eosinophilia, according to Sarkar et al. [93]. The effect of Cd on the immune system has been investigated utilizing experimental animals, particularly rodents. Cd can interact directly with immune system cells, causing modifications in their status and functionality that can be evaluated in vitro. Direct consequences include cell death and signaling pathways interference, altered cytokine production, cell surface marker expression, cell activation, and differentiation [141]. It has been demonstrated that Cd exposure impairs immune cells' ability to operate normally. When administered orally to mice, Cd increases vulnerability to herpes simplex virus type 2, reduces T and B cell production, and enhances macrophage phagocytosis [142]. It has been shown that school-age children exposed to Cd experience a drop in sensitivity and IgG antibody titers. Stress is thought to be the cause of the induction of apoptosis, as is the case with many other immunotoxic drugs. Cd also has an impact on cytokine production [143].

4. Molecular Mechanisms of Cd Toxicity

Cd affects cellular proliferation, differentiation, and apoptosis. The International Agency for Research on Cancer (IARC) classed it as a proven carcinogen, a member of Group No. 1. However, it has a low genotoxic potential. Cd's indirect effects cause reactive oxygen species (ROS) production and DNA damage [144,145]. In vitro studies indicate that Cd has several activities that still need to be fully understood. Chronic heavy metal exposure leads to increased expression of stress proteins (such as heat shock protein 70 and metallothioneins), which can cause apoptosis, growth inhibition, proliferation, or carcinogenicity in animal cells, depending on factors like amount, timing, cell line, and presence of other chemicals. Cd carcinogenesis is primarily caused by oxidative stress, DNA repair inhibition, and altered apoptosis rates [93,146]. A recent study describes the effects of Cd on signaling through Ca^{2+}, NO, c-AMP, nuclear factor kappa-light-chain-enhancer of activated B cells (NF-κB), and developmental pathways such as Wnt signaling, as well as kinases [132]. There are well-established effects of Cd on kinase activation and downstream events of immediate early response oncogene induction, as these events are likely involved in cancer promotion and progression and cell survival [147]. Mitochondria regulate cell homeostasis, proliferation, motility, senescence, and death. Cell and tissue aging, as well as numerous illnesses, such as Alzheimer's disease, Parkinson's disease, Huntington's disease, and even cancer, are associated with mitochondrial dysfunction [148].

Cd can change the expression of various genes, including immediate early response genes, stress response genes, transcriptional factors, and translational factors. It activates the c-jun N-terminal kinase (JNK) pathway, leading to the over-expression of genes responsible for the synthesis of metallothionines and heat shock proteins. Cd also affects transcription factors such as metal-regulatory transcription factor, nuclear factor-κB, and NF-E2-related factor 2, as well as translational factors like TIF3 and TEF-1δ. These changes can provoke the development and progression of tumors [149].

The mechanisms by which Cd disrupts gene expression include changes in Ca intracellular level, ROS generation, effect on cell kinases, and DNA methylation. Cadmium increases intracellular Ca content, affecting gene expression directly by binding to target sites in different genes or indirectly through activation of kinases. Additionally, it mimics Ca and activates Ca-dependent genes. Changes in Ca level lead to ROS production and an increase in gene expression. Cadmium activates cellular protein kinases, leading to increased phosphorylation of transcription factors and an increase in certain gene expression [149]. Additionally, cadmium exposure inhibits DNA-methyltransferase-1 (DNMT1) and decreases DNA methylation, which can be associated with cell transformation and hyperproliferation. Cadmium exposure inhibits DNA repair and damages the genome, leading to potential cell transformation. It interferes with repair processes and inhibits DNA repair genes expression, transcription factor activity, and protein function. Additionally, cadmium can replace zinc in proteins, leading to nonfunctional enzymes. Zinc supplementation can help correct some of the DNA damage [149].

Autophagy, a process of self-degradation that plays a crucial role in eliminating proteins and clearing damaged organelles, is increasingly acknowledged to be involved in Cd toxicity. Cd can function as both a protector and a promoter of cell death [150]. The conflicting impact on cell fate is determined by the appropriate level of autophagy needed to sustain cell survival. Cd exposure disrupts normal cellular autophagy, and both excessive autophagy and its absence can lead to cell death [148]. The effects of Cd on the autophagy process are observed as either stimulation or disruption, likely based on gene expression [148]. Autophagy following Cd exposure appears to either suppress or trigger apoptosis, as the increased accumulation of ROS can activate both autophagy and apoptosis. Moreover, Cd-induced elevation of intracellular Ca leads to ROS induction, initiating cell apoptosis due to the interaction between Ca signaling and ROS in normal and pathological conditions (Figure 7) [149].

Cd triggers cell death by modifying the behavior of protein kinases such as mitogen-activated protein kinase (MAPK). Cd enhances the activity of p38-mitogen-activated protein kinase (p38 MAPK), leading to increased expression of genes related to inflammation and cell death, ultimately resulting in tissue necrosis and kidney damage in rats. Furthermore, recent findings show that Cd prompts cell death in TM3 Leydig cells by generating reactive oxygen species (ROS) and promoting phosphorylation via the JNK pathway. Consequently, this causes a reduction in the levels of the anti-cell death protein Bcl-2, followed by the activation of caspase-3 and cell demise [149].

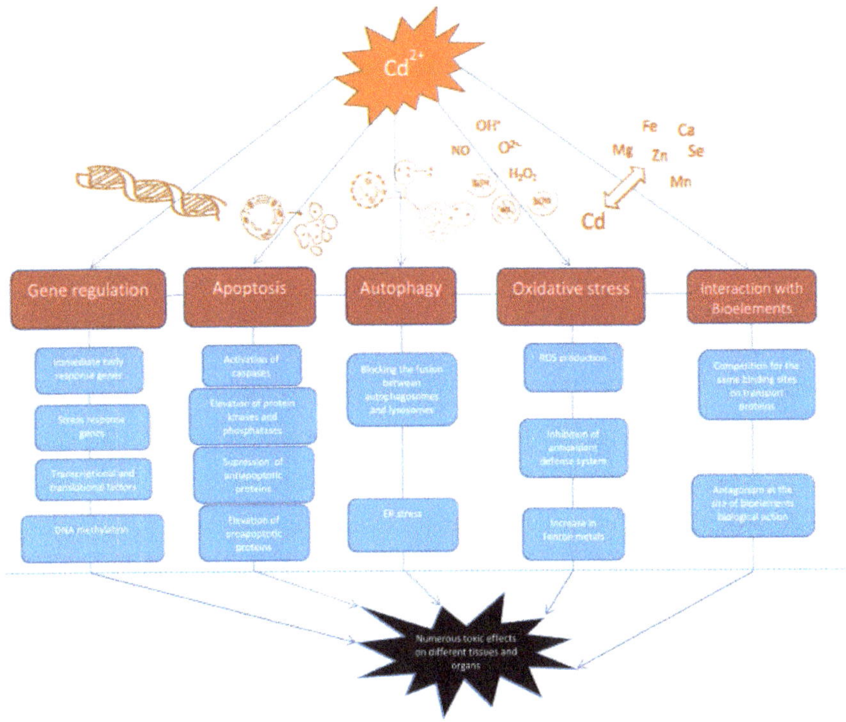

Figure 7. General mechanisms and specific molecular pathways of Cd toxicity. The figure shows general mechanisms (gene regulation, apoptosis, autophagy, oxidative stress, and interaction with bioelements) along with the specific molecular pathways of Cd toxicity [149].

Exposure to Cd impacts the functioning of glutamate, acetylcholine, GABA (gamma-aminobutyric acid), and dopamine neurotransmitter receptors in the brain. N-methyl-D-aspartate receptor (NMDAR) voltage-dependent calcium channels facilitate neuronal uptake of Cd, which leads to increased Cd influx following stimulation with glutamate or N-methyl-D-aspartate (NMDA) and glycine. Additionally, Cd interacts with muscarinic acetylcholine receptors, leading to cell death in primary cholinergic neurons from the basal forebrain by suppressing the muscarinic receptor M1. A subsequent study by the same authors demonstrated that oxidative stress caused Cd-induced muscarinic receptor disruption [150].

Early research suggests that Cd inhibits the neuronal $GABA_A$ receptor channel complex through a binding site different from the recognition sites for GABA and other drugs. Moreover, exposure to Cd alters the expression of $GABA_A$ receptors in animal studies. Specifically, altered protein expression levels of $GABA_AR\alpha5$ and $GABA_AR\delta$ were observed in the hippocampus of mice offspring following Cd exposure during pregnancy and lactation, indicating that $GABA_AR\alpha5$ is more susceptible to environmental pollutants during puberty and young adulthood. Conversely, $GABA_AR\delta$ may reflect the accumulation of environmental contaminants in adulthood [150].

Cd-induced neurotoxicity causes impairments in movement due to Cd's specific impact on DA receptors. Cd exposure reduces the production of mRNA and proteins associated with dopamine (DA)-D2 receptors in the stratum of rat brains. In contrast, the levels of expression for DA-D1 receptors remain unchanged. Additionally, experiments using molecular docking have shown that Cd may directly attach itself to the competitive site of dopamine on DA-D2 receptors (Figure 8) [150].

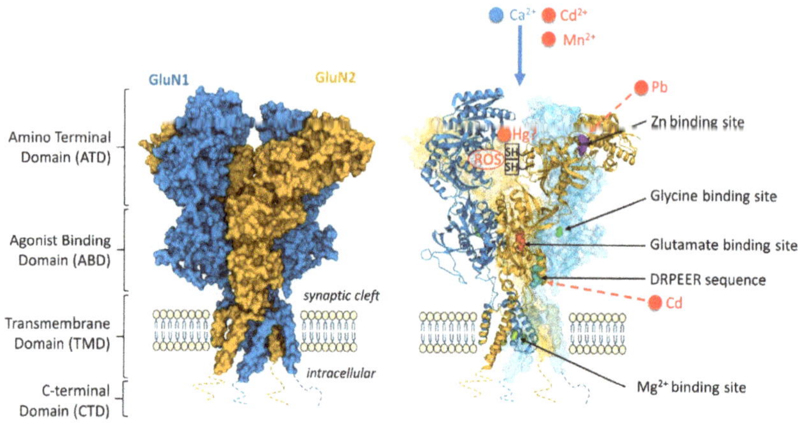

Figure 8. Interaction sites of environmental metal toxicants with the NMDA receptor are as follows: Cd^{2+} facilitates neuronal uptake after receptor stimulation. Similarly, Mn^{2+} can enter the plasma membrane through the NMDAR. Cd can directly attach to the DRPEER sequence in the extracellular domain (in the agonist binding domain (ABD)/transmembrane domain (TMD) linker) of the GluN1 subunit, inhibiting NMDA-mediated current. Pb competes with zinc for the zinc-binding site of the GluN2 subunit, thereby affecting receptor function. As for Hg, there are indications of interactions with cysteine -SH groups that regulate NMDAR activity, although this suggestion still requires experimental evidence. In most instances, metals may induce ROS, which can interact with the –SH groups of the NMDAR [150,151].

$CdCl_2$ leads to the disassembly of the cytoskeleton in various cultured neuronal cells, affecting both the actin and microtubule networks. In primary rat cortical neurons, Cd causes the destruction of microtubules and reduces acetylated tubulin levels. Furthermore, Cd down-regulates the gene expression of microtubule dynamics and microtubule motor-based proteins in a neuronal human cellular model [150,152].

The toxic effects of Cd on human health are widespread and are caused by various biochemical and molecular mechanisms. The main ways in which Cd causes harm include inducing oxidative stress, disrupting Ca^{2+} signaling, interfering with cellular signaling pathways, and making epigenetic modifications. Cd interacts with cellular components such as mitochondria and DNA and causes extensive damage at both cellular and tissue levels [114]. Cd induces oxidative stress, which is a crucial mechanism behind its toxicity, and thus disrupts the balance between oxidants and antioxidants, leading to cellular damage and apoptosis. Furthermore, Cd negatively impacts signaling pathways such as mitogen-activated protein kinase (MAPK), nuclear factor kappa-light-chain-enhancer of activated B cells (NF-κB), and tumor protein 53 (p53) pathways. Cd's interference with these pathways contributes to pathological conditions and carcinogenesis. The epigenetic effects of Cd include DNA methylation and histone modifications, and it causes long-term impact on gene expression and disease manifestation (Figure 9) [114].

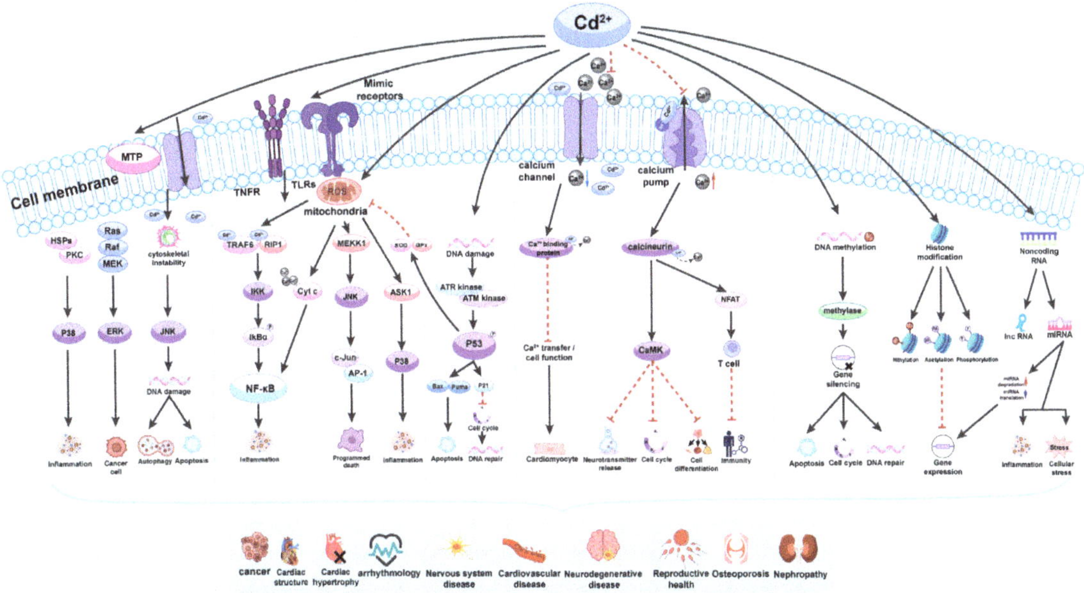

Figure 9. Cd's effect on MAPK pathways disrupts cellular signaling and health. The figure demonstrates how Cd^{2+} disrupts the MAPK signaling pathways, highlighting direct and indirect impacts on cellular function. This leads to abnormal cell responses and an increased risk of disease. Cd's role in activating NF-κB signaling is a pathway to increased inflammation. Cd^{2+} activates the NF-κB pathway and interacts with ROS, receptors, and kinases, resulting in heightened inflammation and potential chronic conditions. Cd^{2+} also affects the p53 pathway, which leads to DNA damage responses, including apoptosis and cell-cycle control. The role of p53 is in protecting cellular health. Cd^{2+} disrupts Ca^{2+} homeostasis by mimicking and interfering with Ca^{2+} in cells. The figure depicts the effects of Cd on calcium channels and pumps and the subsequent risks to cell function and activity. The figure also demonstrates the pathways through which Cd^{2+} influences epigenetic processes, impacting DNA methylation, histone modification, and non-coding RNA activity. These changes lead to significant alterations in gene expression, affecting cellular functions and contributing to disease progression, particularly in cancer and other severe health conditions [114].

5. Methods for Cd Removal

5.1. From the Environment

5.1.1. Chemical Precipitation

Chemical precipitation is the most efficient and widely used technology for removing Cd^{2+} from wastewater because it is simple, inexpensive, and easy to apply. In this procedure, substances react with Cd^{2+} to produce insoluble precipitates [153]. Precipitants such as hydroxides and sulfides are often employed to precipitate Cd^{2+}, resulting in insoluble $Cd(OH)_2$ and CdS residues, respectively. Simple sedimentation or filtration procedures could remove insoluble precipitates from water before they are released into the environment or reused [50].

5.1.2. Adsorption

Adsorption is another excellent way to remove Cd^{2+} from wastewater. Physiological interactions transfer metal ions from the aqueous to the solid phase [154]. Mineral clay, mineral ores, agricultural wastes, synthetic polymers, and industrial leftovers are all examples of low-cost absorbent materials [41].

5.1.3. Ion Exchange

This approach uses the ion exchange principle to adsorb Cd^{2+} on solid surfaces. Metal ions in solution are replaced by ions bonded to the surface of an insoluble matrix, commonly known as an ion exchanger, and covalently bound to negative or positive functional groups. As a result, these functional groups are connected to oppositely charged ions, which are then replaced by the same charged ions found in the solution due to their higher affinity for the functional groups. Many inexpensive materials, such as waste iron, fly ash, resins, zeolites, and silicate minerals, have been widely used as ion exchangers [50]. pH and temperature are the most important parameters influencing ion exchangers' uptake of metal ions.

5.1.4. Membrane Filtration

Membrane filtration is a practical approach for removing hazardous metal ions from polluted water. Several methods use membranes to remove metal ions from effluent, including nanofiltration, ultrafiltration, electrodialysis, and reverse osmosis. Generally, metal-containing wastewater is passed over membranes with varying pore diameters depending on the nature and kind of metal ions to be extracted using selective pressure or electric current (for example, in electrodialysis). The membrane's pore size is smaller than the metal ion, preventing metal ions from passing through and decontaminating the effluent.

All conventional treatment methods have disadvantages, such as waste byproducts, sludge production, membrane fouling, low selectivity and efficiency, and high capital and operating costs [155]. Thus, using living and dead microbial biomass to remove heavy metal ions from the environment has gained popularity due to its benefits, such as cost-effectiveness, efficiency, and environmental friendliness [156].

5.2. From the Human Body

Nano selenium-enriched probiotics have been investigated recently for their potential in addressing Cd liver toxicity [157]. The presence of Cd and other heavy metals like Pb in the gut and food-derived microbes can actively or passively affect the bioavailability of these toxins within the gut [158]. Probiotic treatment can alleviate Cd-induced cytotoxicity, reduce oxidative stress and inflammation, restore tight-junction integrity, and lower gut permeability in intestinal epithelial cells and animals. This safeguarding of the gut barrier leads to higher levels of fecal Cd and reduced Cd accumulation in mouse tissues, indicating probiotics' ability to inhibit intestinal Cd absorption [159]. Capriglione et al. studied the impact of $CdCl_2$ on immortalized, non-tumorigenic thyroid cells Nthy-ori-3-1 and the protective influence of quercetin against $CdCl_2$-induced damage [160]. Nutritional trace metals such as Zn and Se have the ability to alleviate Cd-induced mitochondrial toxicity. Furthermore, Se supplementation can reduce Cd-induced oxidative stress and the mitochondrial apoptosis pathway [161]. A study reported by Smereczanski et al. found that an extract from *Aronia melanocarpa* L. berries has a protective effect against Cd in a rat model. The co-administration of that extract significantly reduces the harmful effect of Cd-induced kidney damage [162]. Resveratrol has also been found to have a protective effect against hepatotoxicity, according to Al-Baqami et al. [163]. Magnesium chloride was reported to diminish Cd toxicity via drinking water with a concentration of 500 mg Mg/L [164]. Asperuloside, an iridoid monoterpenoid glycoside found in many medicinal plants, is described as a substance attenuating Cd-induced toxicity [165].

6. Cd Resistance Mechanisms

Some microbes, such as bacteria, absorb heavy metals such as Cd^{2+} and necessary metal ions, helping to eliminate hazardous metal ions from the aquatic environment [158]. Heavy metal ions generate ROS and thus disorder bacterial metabolism by damaging DNA, RNA, and proteins. Hyperaccumulation of Cd^{2+} may cause bacterial respiratory proteins to disorganize, disrupting the physiological functioning of the cell. To battle the fatal effects of heavy metals, bacteria have evolved various metal resistance methods, including biosorption, efflux transport, intracellular and extracellular sequestration, transformation, and physiological adaptations [166].

Organisms have mechanisms to detoxify Cd, including chelation, compartmentalization, and efflux. In yeast *S. cerevisiae*, yeast cadmium factor 1 (YCF1) provides resistance to Cd by sequestering glutathione-conjugated cadmium, bis(glutathionato)Cd [167]. Efflux and sequestration of heavy metals via metallothioneins and other molecules containing thiol groups are well-known metal resistance mechanisms in bacteria against Cd^{2+} and are briefly described below [50].

ABCC proteins in humans are linked to the detoxification of chemicals, their distribution, and regular cell functions. In Saccharomyces cerevisiae, the ABCC protein ScYCF1 has the ability to provide resistance to Cd. In plant systems, ABCCs are known to be involved in detoxifying and sequestrating potentially harmful elements, as well as transporting chlorophyll breakdown products and regulating ion channels [168]. For instance, in *Arabidopsis*, AtABCC1 facilitates the export of glutathione-S (GS) conjugates from the cytosol. Furthermore, the AtABCC1 and AtABCC2 genes have been associated with the vascular sequestration of phytochelatin (PC)-Cd(II) and PC-Hg(II), while the AtABCC3 gene encodes a transporter for PC-Cd complexes. The wheat ABCC protein, TaABCC13, is crucial for the glutathione-mediated detoxification process, and OsABCC1 in rice lowers As levels in grains by sequestering As in the vacuoles of phloem companion cells within diffuse vascular bundles. These observations underscore the importance of ABCC members in understanding the transport and detoxification of potentially toxic elements [168]

6.1. Biosorption

Bacteria and yeasts have the intrinsic ability to bioabsorb metal ions due to their unique cell envelope, which limits cellular intake of harmful metal ions to preserve homeostasis [169]. When dealing with microbial organisms such as yeast or bacteria, the cell wall is the first structure interacting with metal ions. Several studies have indicated the existence of critical functional groups on biomass/biomaterial surfaces, such as hydroxyl, thiol, carboxyl, and amino groups, which play a crucial role in metal ion biosorption [170]. Nonetheless, the specific process of Cd^{2+} biosorption is unknown and varies depending on factors such as biomass type, heavy metal properties, co-metal ion presence, pH, and medium temperature. The most influential factor is the composition of biomaterials' surfaces [50].

Understanding biosorption's mechanism requires a better understanding of the cell surface's structure and chemical makeup. Gram-negative bacteria's cell wall comprises peptidoglycan, phospholipids, and lipopolysaccharides. The lipopolysaccharides have a negative charge, contributing to the cell wall's anionic character. The cell wall of Gram-positive bacteria contains about equal amounts of peptidoglycan and teichoic acids (TAs). While both carry an anionic charge at neutral pH, TAs (anionic, phosphate-based linear polymers) are critical in keeping a net negative charge on the bacterial surface [50].

6.2. Efflux Transport Systems

Heavy metals enter microbial cells via the critical metal ion absorption pathways earlier described in this review. Cd uses manganese and magnesium transport systems in Gram-positive and Gram-negative bacteria. Cd hyperaccumulation causes the cell to produce efflux systems for its removal to maintain homeostasis. The efflux systems may be chromosomal or plasmid controlled. Three distinct efflux systems, notably resistance nodulation cell division (RND), P-type ATPases, and cation diffusion facilitator (CDF), have been discovered in bacteria to remove heavy metal divalent cations such as Cd^{2+} from the cells [50].

7. Conclusions

Cd's toxic effects, environmental impact, sources, health effects, and biological impacts make it a pressing issue. This review highlights its toxic nature, widespread pollution, entry into living organisms' systems, including the human body, through ingestion, inhalation, and permeation, and its accumulation in organs, leading to severe health issues. Cd exposure leads to extensive environmental and health harm. Cd is highly toxic and can be absorbed by plants and crops from the soil, then enters the animal body, leading to potential exposure for humans through the food chain. It accumulates in various organs, particularly the kidneys and liver, and is known to cause severe health problems, including renal dysfunction, bone diseases, cardiovascular problems, and many others. On a cellular level, Cd disrupts numerous biological processes, inducing oxidative stress generation and DNA damage, and is classified as a carcinogen by the International Agency for Research on Cancer (IARC). Preventing the use of products containing Cd is essential to minimize its adverse effects on humans and other living beings. The current Cd usage trend will lead to more severe consequences if it continues. To avoid Cd toxicity, restoring Cd-contaminated sites and appropriately disposing of materials containing Cd is essential.

Author Contributions: Conceptualization, supervision, writing—review and editing, data curation, project administration, funding Acquisition: G.S.; writing—original draft, data curation, visualization: S.D. and V.M.; data curation, visualization: S.D. and V.M.; supervision, writing—review and editing, data curation: G.S. All authors have read and agreed to the published version of the manuscript.

Funding: APC was funded by Grant NBU-FBE-UP-012/2024, Faculty of Basic Education and Grant NBU-CFSD-1916/2024, Central Fund for Strategic Development—New Bulgarian University, Sofia, Bulgaria.

Institutional Review Board Statement: Not applicable.

Informed Consent Statement: Not applicable.

Data Availability Statement: The data used during the current study are available from the corresponding author upon request.

Acknowledgments: This work was conducted at UPIZ Educational and Research Laboratory of Biology-MF-NBU, New Bulgarian University, Sofia, Bulgaria.

Conflicts of Interest: The authors declare no conflicts of interest.

References

1. Peana, M.; Pelucelli, A.; Chasapis, C.T.; Perlepes, S.P.; Bekiari, V.; Medici, S.; Zoroddu, M.A. Biological Effects of Human Exposure to Environmental Cadmium. *Biomolecules* **2023**, *13*, 36. [CrossRef] [PubMed]
2. Thomas, O. *Llewellyn IC 9380 Cadmium (Materials Flow)*; The National Institute for Occupational Safety and Health (NIOSH): Washington, DC, USA, 1994.
3. Goering, P.L.; Waalkes, M.P.; Klaassen, C.D. Toxicology of Cadmium. In *Toxicology of Metals*; Springer: Berlin/Heidelberg, Germany, 1995; pp. 189–214. [CrossRef]
4. Genchi, G.; Sinicropi, M.S.; Lauria, G.; Carocci, A.; Catalano, A. The Effects of Cadmium Toxicity. *Int. J. Environ. Res. Public Health* **2020**, *17*, 3782. [CrossRef] [PubMed]
5. Zhang, Y.; Feng, J.; Gao, Y.; Liu, X.; Qu, L.; Zhu, L. Physiologically Based Toxicokinetic and Toxicodynamic (PBTK-TD) Modelling of Cd and Pb Exposure in Adult Zebrafish Danio Rerio: Accumulation and Toxicity. *Environ. Pollut.* **2019**, *249*, 959–968. [CrossRef] [PubMed]
6. Fatima, G.; Raza, A.M.; Hadi, N.; Nigam, N.; Mahdi, A.A. Cadmium in Human Diseases: It's More than Just a Mere Metal. *Indian J. Clin. Biochem.* **2019**, *34*, 371–378. [CrossRef] [PubMed]
7. Nordberg, G.F. Historical Perspectives on Cadmium Toxicology. *Toxicol. Appl. Pharmacol.* **2009**, *238*, 192–200. [CrossRef]
8. Chen, M.; Ding, S.; Li, C.; Tang, Y.; Fan, X.; Xu, H.; Tsang, D.C.; Zhang, C. High Cadmium Pollution from Sediments in a Eutrophic Lake Caused by Dissolved Organic Matter Complexation and Reduction of Manganese Oxide. *Water Res.* **2021**, *190*, 116711. [CrossRef]
9. Sienko, M.J. *Chemistry, Principles and Applications*; McGraw-Hill: New York, NY, USA, 1979.
10. Friberg, L.T.; Elinder, G.G.; Kjellstrom, T.; Nordberg, G.F. *Cadmium and Health*; CRC Press: Boca Raton, FL, USA, 2019.
11. Lindén, K.; Andersson, A.; Os, A. Cadmium in Organic and Conventional Pig Production. *Arch. Environ. Contam. Toxicol.* **2001**, *40*, 425–431. [CrossRef]

12. Suciu, N.A.; Devivo, R.; Rizzati, N.; Capri, E. Cd Content in Phosphate Fertilizer: Which Potential Risk for the Environment and Human Health? *Curr. Opin. Environ. Sci. Health* **2022**, *30*, 100392. [CrossRef]
13. Alloway, B.J.; Steinnes, E. Anthropogenic Additions of Cadmium to Soils. In *Cadmium in Soils and Plants*; Springer: Berlin/Heidelberg, Germany, 1999; pp. 97–123. [CrossRef]
14. Satchanska, G. *Toxicity of Heavy Metals and Pesticides on Living World*; Monograph; NBU Publishing House: Sofia, Bulgaria, 2024; ISBN 978-619-233-322-5.
15. Yuan, Z.; Luo, T.; Liu, X.; Hua, H.; Zhuang, Y.; Zhang, X.; Zhang, L.; Zhang, Y.; Xu, W.; Ren, J. Tracing Anthropogenic Cadmium Emissions: From Sources to Pollution. *Sci. Total Environ.* **2019**, *676*, 87–96. [CrossRef]
16. Wu, W.; Zhang, K.; Jiang, S.; Liu, D.; Zhou, H.; Zhong, R.; Zeng, Q.; Cheng, L.; Miao, X.; Tong, Y.; et al. Association of Co-Exposure to Heavy Metals with Renal Function in a Hypertensive Population. *Environ. Int.* **2018**, *112*, 198–206. [CrossRef] [PubMed]
17. Goncharuk, E.A.; Zagoskina, N.V. Heavy Metals, Their Phytotoxicity, and the Role of Phenolic Antioxidants in Plant Stress Responses with Focus on Cadmium: Review. *Molecules* **2023**, *28*, 3921. [CrossRef]
18. Wen, H.; Zhang, Y.; Cloquet, C.; Zhu, C.; Fan, H.; Luo, C. Tracing Sources of Pollution in Soils from the Jinding Pb–Zn Mining District in China Using Cadmium and Lead Isotopes. *Appl. Geochem.* **2015**, *52*, 147–154. [CrossRef]
19. *GB 15618-2018*; Soil Environmental Quality—Risk Control Standard for Soil Contamination of Agricultural Land. State Administration for Market Regulation: Beijing, China, 2018. (In Chinese)
20. Jiang, Y.; Guo, X. Multivariate and Geostatistical Analyses of Heavy Metal Pollution from Different Sources among Farmlands in the Poyang Lake Region, China. *J. Soils Sediments* **2019**, *19*, 2472–2484. [CrossRef]
21. Kubier, A.; Wilkin, R.T.; Pichler, T. Cadmium in Soils and Groundwater: A Review. *Appl. Geochem.* **2019**, *108*, 104388. [CrossRef]
22. Mikhailenko, A.V.; Ruban, D.A.; Ermolaev, V.A.; van Loon, A.J. Cadmium Pollution in the Tourism Environment: A Literature Review. *Geosciences* **2020**, *10*, 242. [CrossRef]
23. Hayat, M.T.; Nauman, M.; Nazir, N.; Ali, S.; Bangash, N. Environmental Hazards of Cadmium: Past, Present, and Future. In *Cadmium Toxicity and Tolerance in Plants*; Academic Press: Cambridge, MA, USA, 2019; pp. 163–183. [CrossRef]
24. European Commission. Europa.eu. Available online: https://esdac.jrc.ec.europa.eu/search/node/cadmium (accessed on 21 August 2024).
25. Ballabio, C.; Jones, A.; Panagos, P. Cadmium in Topsoils of the European Union—An Analysis Based on LUCAS Topsoil Database. *Sci. Total Environ.* **2024**, *912*, 168710. [CrossRef] [PubMed]
26. Directive 2000/60/EC of the European Parliament and of the Council of 23 October 2000 establishing a framework for Community action in the field of water policy. *Off. J. L* **2000**, *327*, 0001–0073.
27. European Union. *Directive 2008/50/EC of the European Parliament and of the Council of 21 May 2008 on Ambient Air Quality and Cleaner Air for Europe*; European Union: Maastricht, The Netherlands, 2008.
28. Wu, Y.; Yang, X.; Wang, H.; Jia, G.; Wang, T. Relationship between Ambient PM2.5 Exposure and Blood Cadmium Level in Children under 14 Years in Beijing, China. *J. Hazard. Mater.* **2021**, *403*, 123871. [CrossRef]
29. Scaccabarozzi, D.; Castillo, L.; Aromatisi, A.; Milne, L.; Búllon Castillo, A.; Muñoz-Rojas, M. Soil, Site, and Management Factors Affecting Cadmium Concentrations in Cacao-Growing Soils. *Agronomy* **2020**, *10*, 806. [CrossRef]
30. Satchanska, G.; Ivanova, I.; Groudeva, V.; Pentcheva, E.N.; Kerestedjian, T.; Golovinsky, E. Culture dependent approach for determining microbial diversity in soils from KCM/AGRIA region. *Comptes Rendue Acad. Bulg. Sci.* **2005**, *58*, 409–416.
31. Satchanska, G.; Selenska-Pobell, S. Bacterial diversity in the uranium mill-tailings Gittersee as estimated via a 16S rDNA approach. *Comptes Rendue Acad. Bulg. Sci.* **2005**, *58*, 1105–1112.
32. Satchanska, G.; Golovinsky, E.; Selenska-Pobell, S. Bacterial diversity in a soil sample from a uranium mining waste pile as estimated via a culture-independent 16S rDNA approach. *Comptes Rendue Acad. Bulg. Sci.* **2004**, *57*, 75–82.
33. Selenska-Pobell, S.; Kampf, G.; Flemming, K.; Radeva, G.; Satchanska, G. Bacterial diversity in soil samples from two uranium waste piles as determined by rep-APD, RISA and 16S rDNA retrieval. *Antonie Leeuwenhoek* **2001**, *79*, 149–161. [CrossRef] [PubMed]
34. Zulfiqar, U.; Haider, F.U.; Maqsood, M.F.; Mohy-Ud-Din, W.; Shabaan, M.; Ahmad, M.; Kaleem, M.; Ishfaq, M.; Aslam, Z.; Shahzad, B. Recent Advances in Microbial-Assisted Remediation of Cadmium-Contaminated Soil. *Plants* **2023**, *12*, 3147. [CrossRef]
35. Surenbaatar, U.; Lee, S.; Kwon, J.-Y.; Lim, H.; Kim, J.-J.; Kim, Y.-H.; Hong, Y.-S. Bioaccumulation of Lead, Cadmium, and Arsenic in a Mining Area and Its Associated Health Effects. *Toxics* **2023**, *11*, 519. [CrossRef]
36. Satchanska, G. *Bioindicatory and Biotransformation Activity of Microorganisms Obtained from Xenobiotics Polluted Soils and Waters*; Edd. PSSE: Sofia, Bulgaria, 2022; 146p, ISBN 978-954-749-126-7.
37. Gómez, M.; Grimes, S.; Qian, Y.; Feng, Y.; Fowler, G. Critical and Strategic Metals in Mobile Phones: A Detailed Characterisation of Multigenerational Waste Mobile Phones and the Economic Drivers for Recovery of Metal Value. *J. Clean. Prod.* **2023**, *419*, 138099. [CrossRef]
38. Igweze, Z.N.; Ekhator, O.C.; Orisakwe, O.E. A Pediatric Health Risk Assessment of Children's Toys Imported from China into Nigeria. *Heliyon* **2020**, *6*, e03732. [CrossRef]
39. Bielak, E.; Marcinkowska, E. Heavy Metals in Leathers, Artificial Leathers, and Textiles in the Context of Quality and Safety of Use. *Sci. Rep.* **2022**, *12*, 5061. [CrossRef]
40. Kern, M.S.; Boron, M.L.; Weidenhamer, J.D. Buyer Beware: Inexpensive, High Cadmium Jewelry Can Pose Severe Health Risks. *Sci. Total Environ.* **2020**, *764*, 142926. [CrossRef]

41. Satchanska, G.; Pentcheva, E.N.; Atanasova, R.; Groudeva, V.; Trifonova, R.; Golovinsky, E. Microbial Diversity in Heavy-Metal Polluted Waters. *Biotechnol. Biotechnol. Equip.* **2005**, *19*, 61–67. [CrossRef]
42. Cordoba-Novoa, H.A.; Cáceres-Zambrano, J.; Torres-Rojas, E. Isolation of Native Cadmium-Tolerant Bacteria and Fungi from Cacao (*Theobroma cacao* L.)—Cultivated Soils in Central Colombia. *Heliyon* **2023**, *9*, e22489. [CrossRef] [PubMed]
43. Chellaiah, E.R. Cadmium (Heavy Metals) Bioremediation by *Pseudomonas aeruginosa*: A Minireview. *Appl. Water Sci.* **2018**, *8*, 154. [CrossRef]
44. Mergeay, M.; Monchy, S.; Vallaeys, T.; Auquier, V.; Benotmane, A.; Bertin, P.; Taghavi, S.; Dunn, J.; van der Lelie, D.; Wattiez, R. *Ralstonia metallidurans*, a Bacterium Specifically Adapted to Toxic Metals: Towards a Catalogue of Metal-Responsive Genes. *FEMS Microbiol. Rev.* **2003**, *27*, 385–410. [CrossRef] [PubMed]
45. Osman, G.E.H.; Abulreesh, H.H.; Elbanna, K.; Shaaban, M.R.; Samreen, S.; Ahmad, I. Recent Progress in Metal-Microbe Interactions: Prospects in Bioremediation. *J. Pure Appl. Microbiol.* **2019**, *13*, 13–26. [CrossRef]
46. Salam, L.B.; Obayori, O.S.; Ilori, M.O.; Amund, O.O. Effects of Cadmium Perturbation on the Microbial Community Structure and Heavy Metal Resistome of a Tropical Agricultural Soil. *Bioresour. Bioprocess.* **2020**, *7*, 25. [CrossRef]
47. Yu, X.; Zhao, J.; Liu, X.; Sun, L.; Tian, J.; Wu, N. Cadmium Pollution Impact on the Bacterial Community Structure of Arable Soil and the Isolation of the Cadmium Resistant Bacteria. *Front. Microbiol.* **2021**, *12*, 698834. [CrossRef]
48. Luo, L.; Xie, L.; Jin, D.; Mi, B.; Wang, D.H.; Li, X.F.; Dai, X.; Zou, X.; Zhang, Z.; Ma, Y.; et al. Bacterial Community Response to Cadmium Contamination of Agricultural Paddy Soil. *Appl. Soil Ecol.* **2019**, *139*, 100–106. [CrossRef]
49. Sun, H.; Shao, C.; Jin, Q.; Li, M.; Zhang, Z.; Liang, H.; Lei, H.; Qian, J.; Zhang, Y. Effects of Cadmium Contamination on Bacterial and Fungal Communities in Panax Ginseng-Growing Soil. *BMC Microbiol.* **2022**, *22*, 77. [CrossRef]
50. Khan, Z.; Elahi, A.; Bukhari, D.A.; Rehman, A. Cadmium Sources, Toxicity, Resistance and Removal by Microorganisms—A Potential Strategy for Cadmium Eradication. *J. Saudi Chem. Soc.* **2022**, *26*, 101569. [CrossRef]
51. Delgado-Baquerizo, M.; Oliverio, A.M.; Brewer, T.E.; Benavent-González, A.; Eldridge, D.J.; Bardgett, R.D.; Maestre, F.T.; Singh, B.K.; Fierer, N. A Global Atlas of the Dominant Bacteria Found in Soil. *Science* **2018**, *359*, 320–325. [CrossRef]
52. Li, X.; Meng, D.; Li, J.; Yin, H.; Liu, H.; Liu, X.; Cheng, C.; Xiao, Y.; Liu, Z.; Yan, M. Response of Soil Microbial Communities and Microbial Interactions to Long-Term Heavy Metal Contamination. *Environ. Pollut.* **2017**, *231*, 908–917. [CrossRef] [PubMed]
53. Hemmat-Jou, M.H.; Safari-Sinegani, A.A.; Mirzaie-Asl, A.; Tahmourespour, A. Analysis of Microbial Communities in Heavy Metals-Contaminated Soils Using the Metagenomic Approach. *Ecotoxicology* **2018**, *27*, 1281–1291. [CrossRef] [PubMed]
54. Rathnayake, I.V.N.; Megharaj, M.; Krishnamurti, G.S.R.; Bolan, N.S.; Naidu, R. Heavy Metal Toxicity to Bacteria—Are the Existing Growth Media Accurate Enough to Determine Heavy Metal Toxicity? *Chemosphere* **2013**, *90*, 1195–1200. [CrossRef] [PubMed]
55. Zhai, Q.; Yin, R.; Yu, L.; Zhao, J.; Chen, W.; Yu, R.; Liu, X.; Chen, Y.; Zhang, H.; Chen, W. Screening of Lactic Acid Bacteria with Potential Protective Effects against Cadmium Toxicity. *Food Control* **2015**, *54*, 23–30. [CrossRef]
56. Sidhu, G.P.S.; Bali, A.S.; Bhardwaj, R. Use of Fungi in Mitigating Cadmium Toxicity in Plants. In *Cadmium Toxicity and Tolerance in Plants*; Academic Press: Cambridge, MA, USA, 2019; pp. 397–426. [CrossRef]
57. Zheng, X.; Yang, S.; Chen, L.; Kimotho, R.N.; Chen, M.; Chen, J.; Zhang, J.; Li, X. A Newly-Isolated Cd-Loving *Purpureocillium* sp. Strain YZ1 Substantially Alleviates Cd Toxicity to Wheat. *Plant Soil* **2021**, *464*, 289–302. [CrossRef]
58. Fazli, M.M.; Soleimani, N.; Mehrasbi, M.; Darabian, S.; Mohammadi, J.; Ramazani, A. Highly Cadmium Tolerant Fungi: Their Tolerance and Removal Potential. *J. Environ. Health Sci. Eng.* **2015**, *13*, 19. [CrossRef]
59. Dowlati, M.; Sobhi, H.R.; Esrafili, A.; FarzadKia, M.; Yeganeh, M. Heavy Metals Content in Edible Mushrooms: A Systematic Review, Meta-Analysis and Health Risk Assessment. *Trends Food Sci. Technol.* **2021**, *109*, 527–535. [CrossRef]
60. Gałgowska, M.; Pietrzak-Fiećko, R. Cadmium and Lead Content in Selected Fungi from Poland and Their Edible Safety Assessment. *Molecules* **2021**, *26*, 7289. [CrossRef]
61. Širić, I.; Kumar, P.; Eid, E.M.; Bachheti, A.; Kos, I.; Bedeković, D.; Mioč, B.; Humar, M. Occurrence and Health Risk Assessment of Cadmium Accumulation in Three *Tricholoma* Mushroom Species Collected from Wild Habitats of Central and Coastal Croatia. *J. Fungi* **2022**, *8*, 685. [CrossRef]
62. Orywal, K.; Socha, K.; Nowakowski, P.; Zoń, W.; Kaczyński, P.; Mroczko, B.; Łozowicka, B.; Perkowski, M. Health Risk Assessment of Exposure to Toxic Elements Resulting from Consumption of Dried Wild-Grown Mushrooms Available for Sale. *PLoS ONE* **2021**, *16*, e0252834. [CrossRef]
63. Brima, E.I. Toxic Elements in Different Medicinal Plants and the Impact on Human Health. *Int. J. Environ. Res. Public Health* **2017**, *14*, 1209. [CrossRef]
64. Huybrechts, M.; Cuypers, A.; Deckers, J.; Iven, V.; Vandionant, S.; Jozefczak, M.; Hendrix, S. Cadmium and Plant Development: An Agony from Seed to Seed. *Int. J. Mol. Sci.* **2019**, *20*, 3971. [CrossRef] [PubMed]
65. Haider, F.U.; Liqun, C.; Coulter, J.A.; Cheema, S.A.; Wu, J.; Zhang, R.; Wenjun, M.; Farooq, M. Cadmium Toxicity in Plants: Impacts and Remediation Strategies. *Ecotoxicol. Environ. Saf.* **2021**, *211*, 111887. [CrossRef] [PubMed]
66. Nazar, R.; Iqbal, N.; Masood, A.; Khan, M.I.R.; Syeed, S.; Khan, N.A. Cadmium Toxicity in Plants and Role of Mineral Nutrients in Its Alleviation. *Am. J. Plant Sci.* **2012**, *03*, 1476–1489. [CrossRef]
67. Ismael, M.A.; Elyamine, A.M.; Moussa, M.G.; Cai, M.; Zhao, X.; Hu, C. Cadmium in Plants: Uptake, Toxicity, and Its Interactions with Selenium Fertilizers. *Metallomics* **2019**, *11*, 255–277. [CrossRef] [PubMed]
68. Shaari, N.E.M.; Tajudin, M.T.F.M.; Khandaker, M.M.; Majrashi, A.; Alenazi, M.M.; Abdullahi, U.A.; Mohd, K.S. Cadmium Toxicity Symptoms and Uptake Mechanism in Plants: A Review. *Braz. J. Biol.* **2024**, *84*, 252143. [CrossRef] [PubMed]

69. Pan, J.; Li, D.; Shu, Z.; Jiang, X.; Ma, W.; Wang, W.; Zhu, J.; Wang, Y. CsPDC-E1α, a Novel Pyruvate Dehydrogenase Complex E1α Subunit Gene from *Camellia Sinensis*, Is Induced during Cadmium Inhibiting Pollen Tube Growth. *Can. J. Plant Sci.* **2017**, *98*, 62–70. [CrossRef]
70. Nguyen, J.; Patel, A.; Gensburg, A.; Bokhari, R.; Lamar, P.; Edwards, J. Diabetogenic and Obesogenic Effects of Cadmium in Db/Db Mice and Rats at a Clinically Relevant Level of Exposure. *Toxics* **2022**, *10*, 107. [CrossRef]
71. Liu, Y.; Lin, X.; Hao, Z.; Yu, M.; Tang, Y.; Teng, X.; Sun, W.; Kang, L. Cadmium Exposure Caused Cardiotoxicity in Common Carps (*Cyprinus carpio* L.): MiR-9-5p, Oxidative Stress, Energetic Impairment, Mitochondrial Division/Fusion Imbalance, Inflammation, and Autophagy. *Fish Shellfish Immunol.* **2023**, *138*, 108853. [CrossRef]
72. Andjelkovic, M.; Buha Djordjevic, A.; Antonijevic, E.; Antonijevic, B.; Stanic, M.; Kotur-Stevuljevic, J.; Spasojevic-Kalimanovska, V.; Jovanovic, M.; Boricic, N.; Wallace, D.; et al. Toxic Effect of Acute Cadmium and Lead Exposure in Rat Blood, Liver, and Kidney. *Int. J. Environ. Res. Public Health* **2019**, *16*, 274. [CrossRef]
73. Cui, J.; Liu, Y.; Hao, Z.; Liu, Y.; Qiu, M.; Kang, L.; Teng, X.; Tang, Y. Cadmium Induced Time-Dependent Kidney Injury in Common Carp via Mitochondrial Pathway: Impaired Mitochondrial Energy Metabolism and Mitochondrion-Dependent Apoptosis. *Aquat. Toxicol.* **2023**, *261*, 106570. [CrossRef] [PubMed]
74. Raikwar, M.; Kumar, P.; Singh, M. Toxic Effect of Heavy Metals in Livestock Health. *Vet. World* **2008**, *1*, 28. [CrossRef]
75. Bąkowska, M.; Pilarczyk, B.; Tomza-Marciniak, A.; Pilarczyk, R.; Udała, J. Cadmium in Selected Organs of Game Animals from Areas with Different Degrees of Industrialisation and Its Intake by Human Consumers. *Animals* **2024**, *14*, 305. [CrossRef]
76. Acheampong, S. Heavy Metals' Poisoning in Farm Animals. In *Heavy Metals-Recent Advances*; IntechOpen: London, UK, 2023. [CrossRef]
77. Satarug, S. Cadmium Sources and Toxicity. *Toxics* **2019**, *7*, 25. [CrossRef]
78. Sethy, K.; Pati, S.; Jena, D.; Panda, S.; Pradhan, S.; Mishra, C. Heavy Metal Toxicity in Animals: A Review. *Pharma Innov. J.* **2020**, *9*, 134–137.
79. Dobson, S.; World Health Organization. *Cadmium, Environmental Aspects*; World Health Organization: Geneva, Switzerland, 1992; ISBN 9241571357.
80. Nordberg, M.; Nordberg, G.F. Metallothionein and Cadmium Toxicology—Historical Review and Commentary. *Biomolecules* **2022**, *12*, 360. [CrossRef]
81. Liu, Y.; Chen, Q.; Li, Y.; Bi, L.; Jin, L.; Peng, R. Toxic Effects of Cadmium on Fish. *Toxics* **2022**, *10*, 622. [CrossRef]
82. Mahjoub, M.; Fadlaoui, S.; El Maadoudi, M.; Smiri, Y. Mercury, Lead, and Cadmium in the Muscles of Five Fish Species from the Mechraâ-Hammadi Dam in Morocco and Health Risks for Their Consumers. *J. Toxicol.* **2021**, *2021*, 8865869. [CrossRef]
83. Djedjibegovic, J.; Marjanovic, A.; Tahirovic, D.; Caklovica, K.; Turalic, A.; Lugusic, A.; Omeragic, E.; Sober, M.; Caklovica, F. Heavy Metals in Commercial Fish and Seafood Products and Risk Assessment in Adult Population in Bosnia and Herzegovina. *Sci. Rep.* **2020**, *10*, 13238. [CrossRef]
84. Adamkovicova, M.; Toman, R.; Martiniakova, M.; Omelka, R.; Babosova, R.; Krajcovicova, V.; Grosskopf, B.; Massanyi, P. Sperm Motility and Morphology Changes in Rats Exposed to Cadmium and Diazinon. *Reprod. Biol. Endocrinol.* **2016**, *14*, 42. [CrossRef] [PubMed]
85. Drapal, J.; Steinhauser, L.; Stastny, K.; Faldyna, M. Cadmium Concentration in Cattle Tissues in the Czech Republic. *Vet. Med.* **2021**, *66*, 369–375. [CrossRef]
86. Tahir, I.; Alkheraije, K.A. A Review of Important Heavy Metals Toxicity with Special Emphasis on Nephrotoxicity and Its Management in Cattle. *Front. Vet. Sci.* **2023**, *10*, 1149720. [CrossRef] [PubMed]
87. Chirinos-Peinado, D.; Castro-Bedriñana, J.; Ríos-Ríos, E.; Mamani-Gamarra, G.; Quijada-Caro, E.; Huacho-Jurado, A.; Nuñez-Rojas, W. Lead and Cadmium Bioaccumulation in Fresh Cow's Milk in an Intermediate Area of the Central Andes of Peru and Risk to Human Health. *Toxics* **2022**, *10*, 317. [CrossRef] [PubMed]
88. Fay, M.J.; Alt, L.A.C.; Ryba, D.; Salamah, R.; Peach, R.; Papaeliou, A.; Zawadzka, S.; Weiss, A.; Patel, N.; Rahman, A.; et al. Cadmium Nephrotoxicity Is Associated with Altered MicroRNA Expression in the Rat Renal Cortex. *Toxics* **2018**, *6*, 16. [CrossRef] [PubMed]
89. Smereczański, N.M.; Brzóska, M.M. Current Levels of Environmental Exposure to Cadmium in Industrialized Countries as a Risk Factor for Kidney Damage in the General Population: A Comprehensive Review of Available Data. *Int. J. Mol. Sci.* **2023**, *24*, 8413. [CrossRef] [PubMed]
90. Charkiewicz, A.E.; Omeljaniuk, W.J.; Nowak, K.; Garley, M.; Nikliński, J. Cadmium Toxicity and Health Effects—A Brief Summary. *Molecules* **2023**, *28*, 6620. [CrossRef] [PubMed]
91. Skipper, A.; Sims, J.N.; Yedjou, C.G.; Tchounwou, P.B. Cadmium Chloride Induces DNA Damage and Apoptosis of Human Liver Carcinoma Cells via Oxidative Stress. *Int. J. Environ. Res. Public Health* **2016**, *13*, 88. [CrossRef]
92. Chavatte, L.; Juan, M.; Mounicou, S.; Leblanc Noblesse, E.; Pays, K.; Nizard, C.; Bulteau, A.-L. Elemental and Molecular Imaging of Human Full Thickness Skin after Exposure to Heavy Metals. *Metallomics* **2020**, *12*, 1555–1562. [CrossRef]
93. Sarkar, A.; Ravindran, G.; Krishnamurthy, V. A brief review on the effect of cadmium toxicity: From cellular to organ level. *Int. J. Biotechnol. Res.* **2013**, *3*, 17–36.
94. Egger, A.E.; Grabmann, G.; Gollmann-Tepeköylü, C.; Pechriggl, E.J.; Artner, C.; Türkcan, A.; Hartinger, C.G.; Fritsch, H.; Keppler, B.K.; Brenner, E.; et al. Chemical Imaging and Assessment of Cadmium Distribution in the Human Body. *Metallomics* **2019**, *11*, 2010–2019. [CrossRef]

95. Satchanska, G. Cadmium Toxicity on Human: A Review. *J. Balk. Ecol.* **2012**, *15*, 349–357. Available online: http://en.ecobalk.com/ (accessed on 21 August 2024).
96. Chavez, E.; He, Z.L.; Stoffella, P.J.; Mylavarapu, R.S.; Li, Y.C.; Moyano, B.; Baligar, V.C. Concentration of Cadmium in Cacao Beans and Its Relationship with Soil Cadmium in Southern Ecuador. *Sci. Total Environ.* **2015**, *533*, 205–214. [CrossRef] [PubMed]
97. Satarug, S.; Gobe, G.C.; A. Vesey, D.; Phelps, K.R. Cadmium and Lead Exposure, Nephrotoxicity, and Mortality. *Toxics* **2020**, *8*, 86. [CrossRef] [PubMed]
98. Mei, S.; Lin, K.; Williams, D.V.; Liu, Y.; Dai, H.; Cao, F. Cadmium Accumulation in Cereal Crops and Tobacco: A Review. *Agronomy* **2022**, *12*, 1952. [CrossRef]
99. Zhou, M.; Li, Z. Recent Advances in Minimizing Cadmium Accumulation in Wheat. *Toxics* **2022**, *10*, 187. [CrossRef] [PubMed]
100. Kim, K.; Melough, M.M.; Vance, T.M.; Noh, H.; Koo, S.I.; Chun, O.K. Dietary Cadmium Intake and Sources in the US. *Nutrients* **2019**, *11*, 2. [CrossRef]
101. Ai, H.; Wu, D.; Li, C.; Hou, M. Advances in Molecular Mechanisms Underlying Cadmium Uptake and Translocation in Rice. *Front. Plant Sci.* **2022**, *13*, 1003953. [CrossRef] [PubMed]
102. Nogawa, K.; Suwazono, Y.; Nishijo, M.; Sakurai, M.; Ishizaki, M.; Morikawa, Y.; Watanabe, Y.; Kido, T.; Nakagawa, H. Increase of Lifetime Cadmium Intake Dose-Dependently Increased All Cause of Mortality in Female Inhabitants of the Cadmium-Polluted Jinzu River Basin, Toyama, Japan. *Environ. Res.* **2018**, *164*, 379–384. [CrossRef]
103. Nishijo, M.; Nogawa, K.; Suwazono, Y.; Kido, T.; Sakurai, M.; Nakagawa, H. Lifetime Cadmium Exposure and Mortality for Renal Diseases in Residents of the Cadmium-Polluted Kakehashi River Basin in Japan. *Toxics* **2020**, *8*, 81. [CrossRef]
104. Uetani, M.; Kobayashi, E.; Suwazono, Y.; Kido, T.; Nogawa, K. Cadmium Exposure Aggravates Mortality More in Women than in Men. *Int. J. Environ. Health Res.* **2006**, *16*, 273–279. [CrossRef]
105. Vergara-Murillo, F.; Martinez-Yanez, K.; Fortich-Revollo, A.; Paternina-Caicedo, A.; Johnson-Restrepo, B. Biochemical and Hematological Markers in Workers with Chronical Exposure to Lead and Cadmium in Colombia. *Toxics* **2022**, *10*, 524. [CrossRef] [PubMed]
106. Conterato, G.M.M.; Bulcão, R.P.; Sobieski, R.; Moro, A.M.; Charão, M.F.; Monteiro, F.; Maria, F.; Moreira, A.; Roehrs, M.; Tonello, R.; et al. Blood Thioredoxin Reductase Activity, Oxidative Stress and Hematological Parameters in Painters and Battery Workers: Relationship with Lead and Cadmium Levels in Blood. *J. Appl. Toxicol.* **2013**, *33*, 142–150. [CrossRef] [PubMed]
107. Satarug, S.; Moore, M.R. Adverse Health Effects of Chronic Exposure to Low-Level Cadmium in Foodstuffs and Cigarette Smoke. *Environ. Health Perspect.* **2004**, *112*, 1099–1103. [CrossRef]
108. Barregard, L.; Fabricius-Lagging, E.; Lundh, T.; Mölne, J.; Wallin, M.; Olausson, M.; Modigh, C.; Sallsten, G. Cadmium, Mercury, and Lead in Kidney Cortex of Living Kidney Donors: Impact of Different Exposure Sources. *Environ. Res.* **2010**, *110*, 47–54. [CrossRef] [PubMed]
109. Gray, N.; Halstead, M.; Gonzalez-Jimenez, N.; Valentin-Blasini, L.; Watson, C.; Pappas, R.S. Analysis of Toxic Metals in Liquid from Electronic Cigarettes. *Int. J. Environ. Res. Public Health* **2019**, *16*, 4450. [CrossRef] [PubMed]
110. Liu, H.; Gao, H.; Long, M.; Fu, H.; Alvarez, P.J.J.; Li, Q.; Zheng, S.; Qu, X.; Zhu, D. Sunlight Promotes Fast Release of Hazardous Cadmium from Widely-Used Commercial Cadmium Pigment. *Environ. Sci. Technol.* **2017**, *51*, 6877–6886. [CrossRef]
111. Bridges, C.C.; Zalups, R.K. Molecular and Ionic Mimicry and the Transport of Toxic Metals. *Toxicol. Appl. Pharmacol.* **2005**, *204*, 274–308. [CrossRef] [PubMed]
112. Thévenod, F.; Fels, J.; Lee, W.-K.; Zarbock, R. Channels, Transporters and Receptors for Cadmium and Cadmium Complexes in Eukaryotic Cells: Myths and Facts. *Biometals* **2019**, *32*, 469–489. [CrossRef] [PubMed]
113. Tellez-Plaza, M.; Guallar, E.; Fabsitz, R.R.; Howard, B.V.; Umans, J.G.; Francesconi, K.A.; Goessler, W.; Devereux, R.B.; Navas-Acien, A. Cadmium Exposure and Incident Peripheral Arterial Disease. *Circ. Cardiovasc. Qual. Outcomes* **2013**, *6*, 626–633. [CrossRef]
114. Qu, F.; Zheng, W. Cadmium Exposure: Mechanisms and Pathways of Toxicity and Implications for Human Health. *Toxics* **2024**, *12*, 388. [CrossRef]
115. Tinkov, A.A.; Filippini, T.; Ajsuvakova, O.P.; Skalnaya, M.G.; Aaseth, J.; Bjørklund, G.; Gatiatulina, E.R.; Popova, E.V.; Nemereshina, O.N.; Huang, P.-T.; et al. Cadmium and Atherosclerosis: A Review of Toxicological Mechanisms and a Meta-Analysis of Epidemiologic Studies. *Environ. Res.* **2018**, *162*, 240–260. [CrossRef] [PubMed]
116. Ledda, C.; Cannizzaro, E.; Lovreglio, P.; Vitale, E.; Stufano, A.; Montana, A.; Li Volti, G.; Rapisarda, V. Exposure to Toxic Heavy Metals Can Influence Homocysteine Metabolism? *Antioxidants* **2020**, *9*, 30. [CrossRef] [PubMed]
117. Massányi, M.; Satarug, S.; Madeddu, R.; Stawarz, R.; Massányi, P. Evidence for Ovarian and Testicular Toxicities of Cadmium and Detoxification by Natural Substances. *Stresses* **2022**, *2*, 1–16. [CrossRef]
118. Saha, R.; Roychoudhury, S.; Varghese, A.; Nandi, P.; Kar, K.; Choudhury, A.P.; Kalita, J.C.; Lukac, N.; Massanyi, P.; Kolesarova, A. Cadmium Induced Deterioration of Sperm Quality: Protective Role of Coenzyme Q10 in Rats. In *Biotechnology and Biological Sciences*; CRC Press: Boca Raton, FL, USA, 2019; pp. 127–132. [CrossRef]
119. Massányi, P.; Massányi, M.; Madeddu, R.; Stawarz, R.; Lukáč, N. Effects of Cadmium, Lead, and Mercury on the Structure and Function of Reproductive Organs. *Toxics* **2020**, *8*, 94. [CrossRef] [PubMed]
120. Dettwiler, M.; Flynn, A.C.; Rigutto-Farebrother, J. Effects of Non-Essential "Toxic" Trace Elements on Pregnancy Outcomes: A Narrative Overview of Recent Literature Syntheses. *Int. J. Environ. Res. Public Health* **2023**, *20*, 5536. [CrossRef] [PubMed]

121. Kumar, S.; Sharma, A. Cadmium Toxicity: Effects on Human Reproduction and Fertility. *Rev. Environ. Health* **2019**, *34*, 327–338. [CrossRef]
122. Marini, H.R.; Micali, A.; Squadrito, G.; Puzzolo, D.; Freni, J.; Antonuccio, P.; Minutoli, L. Nutraceuticals: A New Challenge against Cadmium-Induced Testicular Injury. *Nutrients* **2022**, *14*, 663. [CrossRef]
123. Taylor, B.A.; Heiniger, H.J.; Meier, H. Genetic Analysis of Resistance to Cadmium-Induced Testicular Damage in Mice. *Exp. Biol. Med.* **1973**, *143*, 629–633. [CrossRef] [PubMed]
124. Xiong, R.; Wu, Q.; Trbojevich, R.; Muskhelishvili, L.; Davis, K.; Bryant, M.; Richter, P.; Cao, X. Disease-Related Responses Induced by Cadmium in an In Vitro Human Airway Tissue Model. *Toxicol. Lett.* **2019**, *303*, 16–27. [CrossRef]
125. Lampe, B.J.; Park, S.K.; Robins, T.; Mukherjee, B.; Litonjua, A.A.; Amarasiriwardena, C.; Weisskopf, M.; Sparrow, D.; Hu, H. Association between 24-h Urinary Cadmium and Pulmonary Function among Community-Exposed Men: The va Normative Aging Study. *Environ. Health Perspect.* **2008**, *116*, 1226–1230. [CrossRef]
126. Sasaki, T.; Horiguchi, H.; Matsukawa, T.; Kobayashi, M.; Omori, Y.; Oguma, E.; Komatsuda, A. A Suspected Case of "Itai-Itai Disease" in a Cadmium-Polluted Area in Akita Prefecture, Japan. *Environ. Health Prev. Med.* **2024**, *29*, 40. [CrossRef] [PubMed]
127. Satarug, S.; Ruangyuttikarn, W.; Nishijo, M.; Ruiz, P. Urinary Cadmium Threshold to Prevent Kidney Disease Development. *Toxics* **2018**, *6*, 26. [CrossRef] [PubMed]
128. Afridi, H.I.; Kazi, T.G.; Kazi, N.; Jamali, M.K.; Arain, M.B.; Jalbani, N.; Baig, J.A.; Sarfraz, R.A. Evaluation of Status of Toxic Metals in Biological Samples of Diabetes Mellitus Patients. *Diabetes Res. Clin. Pract.* **2008**, *80*, 280–288. [CrossRef] [PubMed]
129. Thomas, L.D.K.; Hodgson, S.; Nieuwenhuijsen, M.; Jarup, L. Early Kidney Damage in a Population Exposed to Cadmium and Other Heavy Metals. *Environ. Health Perspect.* **2009**, *117*, 181–184. [CrossRef]
130. Tobwala, S.; Wang, H.-J.; Carey, J.W.; Banks, W.A.; Ercal, N. Effects of Lead and Cadmium on Brain Endothelial Cell Survival, Monolayer Permeability, and Crucial Oxidative Stress Markers in an in Vitro Model of the Blood-Brain Barrier. *Toxics* **2014**, *2*, 258–275. [CrossRef]
131. Agnihotri, S.K.; Agrawal, U.; Ghosh, I. Brain Most Susceptible to Cadmium Induced Oxidative Stress in Mice. *J. Trace Elem. Med. Biol.* **2015**, *30*, 184–193. [CrossRef]
132. Leal, R.B.; Rieger, D.K.; Peres, T.V.; Lopes, M.W.; Gonçalves, C.A.S. Cadmium Neurotoxicity and Its Role in Brain Disorders. In *Metal Ion in Stroke*; Springer: Berlin/Heidelberg, Germany, 2012; pp. 751–766. [CrossRef]
133. Zheng, W. Toxicology of Choroid Plexus: Special Reference to Metal-Induced Neurotoxicities. *Microsc. Res. Tech.* **2001**, *52*, 89–103. [CrossRef]
134. Oggiano, R.; Pisano, A.; Sabalic, A.; Farace, C.; Fenu, G.; Lintas, S.; Forte, G.; Bocca, B.; Madeddu, R. An Overview on Amyotrophic Lateral Sclerosis and Cadmium. *Neurol. Sci.* **2020**, *42*, 531–537. [CrossRef] [PubMed]
135. Branca, J.J.V.; Fiorillo, C.; Carrino, D.; Paternostro, F.; Taddei, N.; Gulisano, M.; Pacini, A.; Becatti, M. Cadmium-Induced Oxidative Stress: Focus on the Central Nervous System. *Antioxidants* **2020**, *9*, 492. [CrossRef]
136. Fern, R.; Black, J.A.; Ransom, B.R.; Waxman, S.G. Cd(2+)-Induced Injury in CNS White Matter. *J. Neurophysiol.* **1996**, *76*, 3264–3273. [CrossRef] [PubMed]
137. Almazan, G.; Liu, H.-N.; Khorchid, A.; Sundararajan, S.; Martinez-Bermudez, A.K.; Chemtob, S. Exposure of Developing Oligodendrocytes to Cadmium Causes HSP72 Induction, Free Radical Generation, Reduction in Glutathione Levels, and Cell Death. *Free Radic. Biol. Med.* **2000**, *29*, 858–869. [CrossRef] [PubMed]
138. Arruebarrena, M.A.; Hawe, C.T.; Lee, Y.M.; Branco, R.C. Mechanisms of Cadmium Neurotoxicity. *Int. J. Mol. Sci.* **2023**, *24*, 16558. [CrossRef]
139. Viaene, M.K.; Roels, H.A.; Leenders, J.; De Groof, M.; Swerts, L.J.; Lison, D.; Masschelein, R. Cadmium: A Possible Etiological Factor in Peripheral Polyneuropathy. *Neurotoxicology* **1999**, *20*, 7–16. [PubMed]
140. Zhong, Q.; Zhou, W.; Lin, J.; Sun, W.; Qin, Y.; Li, X.; Xu, H. Independent and Combined Associations of Blood Manganese, Cadmium and Lead Exposures with the Systemic Immune-Inflammation Index in Adults. *Toxics* **2023**, *11*, 659. [CrossRef]
141. Mirkov, I.; Popov Aleksandrov, A.; Ninkov, M.; Tucovic, D.; Kulas, J.; Zeljkovic, M.; Popovic, D.; Kataranovski, M. Immunotoxicology of Cadmium: Cells of the Immune System as Targets and Effectors of Cadmium Toxicity. *Food Chem. Toxicol.* **2021**, *149*, 112026. [CrossRef] [PubMed]
142. Thomas, P.T.; Ratajczak, H.V.; Aranyi, C.; Gibbons, R.D.; Fenters, J.D. Evaluation of Host Resistance and Immune Function in Cadmium-Exposed Mice. *Toxicol. Appl. Pharmacol.* **1985**, *80*, 446–456. [CrossRef] [PubMed]
143. Dakeshita, S.; Kawai, T.; Uemura, H.; Hiyoshi, M.; Oguma, E.; Horiguchi, H.; Kayama, F.; Aoshima, K.; Shirahama, S.; Rokutan, K.; et al. Gene Expression Signatures in Peripheral Blood Cells from Japanese Women Exposed to Environmental Cadmium. *Toxicology* **2009**, *257*, 25–32. [CrossRef]
144. Bertin, G.; Averbeck, D. Cadmium: Cellular Effects, Modifications of Biomolecules, Modulation of DNA Repair and Genotoxic Consequences (a Review). *Biochimie* **2006**, *88*, 1549–1559. [CrossRef]
145. Teschke, R. Aluminum, Arsenic, Beryllium, Cadmium, Chromium, Cobalt, Copper, Iron, Lead, Mercury, Molybdenum, Nickel, Platinum, Thallium, Titanium, Vanadium, and Zinc: Molecular Aspects in Experimental Liver Injury. *Int. J. Mol. Sci.* **2022**, *23*, 12213. [CrossRef]
146. Luevano, J.; Damodaran, C. A Review of Molecular Events of Cadmium-Induced Carcinogenesis. *J. Environ. Pathol. Toxicol. Oncol.* **2014**, *33*, 183–194. [CrossRef]

147. Templeton, D.M.; Liu, Y. Multiple Roles of Cadmium in Cell Death and Survival. *Chem.-Biol. Interact.* **2010**, *188*, 267–275. [CrossRef] [PubMed]
148. Ortega, R.; Carmona, A. Neurotoxicity of Environmental Metal Toxicants: Special Issue. *Toxics* **2022**, *10*, 382. [CrossRef] [PubMed]
149. Đukić-Ćosić, D.; Baralić, K.; Javorac, D.; Djordjevic, A.B.; Bulat, Z. An Overview of Molecular Mechanisms in Cadmium Toxicity. *Curr. Opin. Toxicol.* **2020**, *19*, 56–62. [CrossRef]
150. Carmona, A.; Roudeau, S.; Ortega, R. Molecular Mechanisms of Environmental Metal Neurotoxicity: A Focus on the Interactions of Metals with Synapse Structure and Function. *Toxics* **2021**, *9*, 198. [CrossRef] [PubMed]
151. Hansen, K.B.; Yi, F.; Perszyk, R.E.; Furukawa, H.; Wollmuth, L.P.; Gibb, A.J.; Traynelis, S.F. Structure, Function, and Allosteric Modulation of NMDA Receptors. *J. Gen. Physiol.* **2018**, *150*, 1081–1105. [CrossRef]
152. Forcella, M.; Lau, P.; Oldani, M.; Melchioretto, P.; Bogni, A.; Gribaldo, L.; Fusi, P.; Urani, C. Neuronal Specific and Non-Specific Responses to Cadmium Possibly Involved in Neurodegeneration: A Toxicogenomics Study in a Human Neuronal Cell Model. *NeuroToxicology* **2020**, *76*, 162–173. [CrossRef]
153. Satchanska, G. Growing Environmental Bacterium Biofilms In PEO Cryogels for Environmental Biotechnology Application. In *Focus on Bacterial Biofilms*; IntechOpen: London, UK, 2022. [CrossRef]
154. Chwastowski, J.; Bradło, D.; Żukowski, W. Adsorption of Cadmium, Manganese and Lead Ions from Aqueous Solutions Using Spent Coffee Grounds and Biochar Produced by Its Pyrolysis in the Fluidized Bed Reactor. *Materials* **2020**, *13*, 2782. [CrossRef]
155. Satchanska, G.; Topalova, Y.; Dimkov, R.; Petrov, P.; Tsvetanov, C.; Selenska-Pobell, S.; Gorbovskaya, A.; Golovinsky, E. Phenol biodegradation by two xenobiotics-tolerant bacteria immobilized in polyethylene oxide cryogels. *Comptes Rendus Acad. Bulg. Sci.* **2009**, *62*, 957–964.
156. Alfadaly, R.A.; Elsayed, A.; Hassan, R.Y.A.; Noureldeen, A.; Darwish, H.; Gebreil, A.S. Microbial Sensing and Removal of Heavy Metals: Bioelectrochemical Detection and Removal of Chromium(VI) and Cadmium(II). *Molecules* **2021**, *26*, 2549. [CrossRef]
157. Vicas, S.I.; Laslo, V.; Timar, A.V.; Balta, C.; Herman, H.; Ciceu, A.; Gharbia, S.; Rosu, M.; Mladin, B.; Chiana, L.; et al. Nano Selenium—Enriched Probiotics as Functional Food Products against Cadmium Liver Toxicity. *Materials* **2021**, *14*, 2257. [CrossRef]
158. George, F.; Mahieux, S.; Daniel, C.; Titécat, M.; Beauval, N.; Houcke, I.; Neut, C.; Allorge, D.; Borges, F.; Jan, G.; et al. Assessment of Pb(II), Cd(II), and Al(III) Removal Capacity of Bacteria from Food and Gut Ecological Niches: Insights into Biodiversity to Limit Intestinal Biodisponibility of Toxic Metals. *Microorganisms* **2021**, *9*, 456. [CrossRef] [PubMed]
159. Zhai, Q.; Tian, F.; Zhao, J.; Zhang, H.; Narbad, A.; Chen, W. Oral Administration of Probiotics Inhibits Absorption of the Heavy Metal Cadmium by Protecting the Intestinal Barrier. *Appl. Environ. Microbiol.* **2016**, *82*, 4429–4440. [CrossRef] [PubMed]
160. Capriglione, F.; Maiuolo, J.; Celano, M.; Damante, G.; Russo, D.; Bulotta, S.; Maggisano, V. Quercetin Protects Human Thyroid Cells against Cadmium Toxicity. *Int. J. Mol. Sci.* **2021**, *22*, 6849. [CrossRef] [PubMed]
161. Cheng, H.; Yang, B.; Ke, T.; Li, S.; Yang, X.; Aschner, M.; Chen, P. Mechanisms of Metal-Induced Mitochondrial Dysfunction in Neurological Disorders. *Toxics* **2021**, *9*, 142. [CrossRef]
162. Smereczański, N.M.; Brzóska, M.M.; Rogalska, J.; Hutsch, T. The Protective Potential of *Aronia melanocarpa* L. Berry Extract against Cadmium-Induced Kidney Damage: A Study in an Animal Model of Human Environmental Exposure to This Toxic Element. *Int. J. Mol. Sci.* **2023**, *24*, 11647. [CrossRef] [PubMed]
163. Al-Baqami, N.M.; Hamza, R.Z. Protective Effect of Resveratrol against Hepatotoxicity of Cadmium in Male Rats: Antioxidant and Histopathological Approaches. *Coatings* **2021**, *11*, 594. [CrossRef]
164. Grosicki, A.; Małagocki, P.; Kycko, A.; Monkiewicz, J.; Korol, W. Magnesium Supplements Affect Selected Cadmium Toxic Actions and Uptake of Repeated Doses of Cadmium. *Bull. Vet. Inst. Pulawy* **2015**, *59*, 541–546. [CrossRef]
165. Kong, Z.; Liu, C.; Olatunji, O.J. Asperuloside Attenuates Cadmium-Induced Toxicity by Inhibiting Oxidative Stress, Inflammation, Fibrosis and Apoptosis in Rats. *Sci. Rep.* **2023**, *13*, 5698. [CrossRef]
166. Satchanska, G.; Selenska-Pobell, S.; Golovinsky, E. Bacterial diversity in heavy metal polluted soil sample KCM-B collected near Pb-Zn smelter KCM, Bulgaria. *Acta Microbiol. Bulg.* **2023**, *37*, 350–358. [CrossRef]
167. Wei, W.; Smith, N.; Wu, X.; Kim, H.; Seravalli, J.; Khalimonchuk, O.; Lee, J. YCF1-Mediated Cadmium Resistance in Yeast Is Dependent on Copper Metabolism and Antioxidant Enzymes. *Antioxid. Redox Signal.* **2014**, *21*, 1475–1489. [CrossRef]
168. Feng, T.; He, X.; Zhuo, R.; Qiao, G.; Han, X.; Qiu, W.; Chi, L.; Zhang, D.; Liu, M. Identification and Functional Characterization of ABCC Transporters for Cd Tolerance and Accumulation in Sedum Alfredii Hance. *Sci. Rep.* **2020**, *10*, 20928. [CrossRef] [PubMed]
169. Camacho-Chab, J.C.; Castañeda-Chávez, M.D.R.; Chan-Bacab, M.J.; Aguila-Ramírez, R.N.; Galaviz-Villa, I.; Bartolo-Pérez, P.; Lango-Reynoso, F.; Tabasco-Novelo, C.; Gaylarde, C.; Ortega-Morales, B.O. Biosorption of Cadmium by Non-Toxic Extracellular Polymeric Substances (EPS) Synthesized by Bacteria from Marine Intertidal Biofilms. *Int. J. Environ. Res. Public Health* **2018**, *15*, 314. [CrossRef]
170. Satchanska, G.; Topalova, Y.; Ivanov, I.; Golovinsky, E. Xenobiotic Biotransformation Potential of *Pseudomonas rhodesiae* KCM-R5 and *Bacillus subtilis* KCM-RG5, Tolerant to Heavy Metals and Phenol Derivatives. *Biotechnol. Biotechnol. Equip.* **2006**, *20*, 97–102. [CrossRef]

toxics

MDPI

Article

Gender Differences in the Severity of Cadmium Nephropathy

Supabhorn Yimthiang [1], David A. Vesey [2,3], Glenda C. Gobe [2,4,5], Phisit Pouyfung [1], Tanaporn Khamphaya [1] and Soisungwan Satarug [2,*]

1 Occupational Health and Safety, School of Public Health, Walailak University, Nakhon Si Thammarat 80160, Thailand; ksupapor@mail.wu.ac.th (S.Y.); phisit.po@mail.wu.ac.th (P.P.); tanaporn.kh@mail.wu.ac.th (T.K.)
2 The Centre for Kidney Disease Research, Translational Research Institute, Brisbane 4102, Australia; david.vesey@health.qld.gov.au (D.A.V.); g.gobe@uq.edu.au (G.C.G.)
3 Department of Kidney and Transplant Services, Princess Alexandra Hospital, Brisbane 4102, Australia
4 School of Biomedical Sciences, The University of Queensland, Brisbane 4072, Australia
5 NHMRC Centre of Research Excellence for CKD QLD, UQ Health Sciences, Royal Brisbane and Women's Hospital, Brisbane 4029, Australia
* Correspondence: sj.satarug@yahoo.com.au

Abstract: The excretion of β_2-microglobulin (β_2M) above 300 μg/g creatinine, termed tubulopathy, was regarded as the critical effect of chronic exposure to the metal pollutant cadmium (Cd). However, current evidence suggests that Cd may induce nephron atrophy, resulting in a reduction in the estimated glomerular filtration rate (eGFR) below 60 mL/min/1.73 m^2. Herein, these pathologies were investigated in relation to Cd exposure, smoking, diabetes, and hypertension. The data were collected from 448 residents of Cd-polluted and non-polluted regions of Thailand. The body burden of Cd, indicated by the mean Cd excretion (E_{Cd}), normalized to creatinine clearance (C_{cr}) as (E_{Cd}/C_{cr}) × 100 in women and men did not differ (3.21 vs. 3.12 μg/L filtrate). After adjustment of the confounding factors, the prevalence odds ratio (POR) for tubulopathy and a reduced eGFR were increased by 1.9-fold and 3.2-fold for every 10-fold rise in the Cd body burden. In women only, a dose–effect relationship was seen between β_2M excretion ($E_{\beta 2M}/C_{cr}$) and E_{Cd}/C_{cr} (F = 3.431, η^2 0.021). In men, $E_{\beta 2M}/C_{cr}$ was associated with diabetes (β = 0.279). In both genders, the eGFR was inversely associated with $E_{\beta 2M}/C_{cr}$. The respective covariate-adjusted mean eGFR values were 16.5 and 12.3 mL/min/1.73 m^2 lower in women and men who had severe tubulopathy (($E_{\beta 2M}/C_{cr}$) × 100 ≥ 1000 μg/L filtrate). These findings indicate that women were particularly susceptible to the nephrotoxicity of Cd, and that the increment of $E_{\beta 2M}/C_{cr}$ could be attributable mostly to Cd-induced impairment in the tubular reabsorption of the protein together with Cd-induced nephron loss, which is evident from an inverse relationship between $E_{\beta 2M}/C_{cr}$ and the eGFR.

Keywords: β_2-microglobulin; cadmium; diabetes; GFR; hypertension; smoking; tubular proteinuria

1. Introduction

Cadmium (Cd) is a toxic metal pollutant that preferentially accumulates in the proximal tubule of kidneys, where it causes tubular cell injury, cell death, nephron atrophy, and eventually, a reduction in the estimated glomerular filtration rate (eGFR) below 60 mL/min/1.73 m^2 [1–4]. As exposure to Cd in the diet is inevitable for most populations, it has become an environmental toxicant of significant worldwide public health concern. A total diet study undertaken in Japan between 2013 and 2018 reported that rice and its products, green vegetables, and cereals and seeds plus potatoes constituted 38%, 17%, and 11% of total dietary exposure, respectively [5].

To safeguard against excessive exposure to Cd in the human diet, guidelines, referred to as a tolerable intake level of Cd, were created by the Joint FAO/WHO Expert Committee on Food Additives and Contaminants (JECFA) [6]. Notably, the "tolerable" intake level was based on the risk assessment model that assumed tubular proteinuria, reflected by an increase in the excretion of the low-molecular-weight protein β2-microglubulin (β_2M, $E_{\beta 2M}$) above 300 μg/g creatinine, to be an early warning sign of the nephrotoxicity of Cd. Consequently, tubulopathy is the most frequently reported adverse effect of Cd exposure. Numerous studies, however, have cast considerable doubt on the utility of $E_{\beta 2M}$ for such purposes.

This study aims to examine whether the exposure to Cd adversely impacts kidney toxicity differently in men and women. To this end, tubular dysfunction and changes in the eGFR were quantified in residents of Cd-polluted and non-polluted regions of Thailand and analyzed in relation to Cd exposure levels. The confounding impact of smoking, diabetes, and hypertension were also evaluated. The exposure to Cd was assessed via the measurement of blood Cd concentration ($[Cd]_b$) and urinary Cd excretion (E_{Cd}). Tubular dysfunction was assessed via $E_{\beta 2M}$. The equations developed by the Chronic Kidney Disease Epidemiology Collaboration (CKD-EPI) were used to compute the estimated GFR (eGFR) [7].

2. Materials and Methods

2.1. Participants

To obtain a group with a wide range of environmental Cd exposure amenable to dose–effect relationship assessment, we assembled data from 334 women and 114 men who participated in the cross-sectional studies conducted in a high-exposure area of the Mae Sot district, Tak province [8], and a low-exposure locality in Pakpoon Municipality of Nakhon Si Thammarat Province [9]. Based on the data from a nationwide survey of Cd levels in soils and food crops [10], environmental exposure to Cd in Nakhon Si Thammarat was low.

The study protocol for the Mae Sot group was approved by the Institutional Ethical Committees of Chiang Mai University and the Mae Sot Hospital [8]. The study protocol for the Nakhon Si Thammarat group was approved by the Office of the Human Research Ethics Committee of Walailak University in Thailand [9].

All participants gave informed consent prior to participation. They had been living at their current addresses for at least 30 years. Exclusion criteria were pregnancy, breast-feeding, a history of metal work, and a hospital record or physician's diagnosis of an advanced chronic disease. Diabetes was defined as having fasting plasma glucose levels $\geq$ 126 mg/dL (https://www.cdc.gov/diabetes/basics/getting-tested.html (accessed on 25 June 2023)) or a physician's prescription of anti-diabetic medications. Hypertension was defined as systolic blood pressure $\geq$ 140 mmHg, diastolic blood pressure $\geq$ 90 mmHg [11], a physician's diagnosis, or prescription of anti-hypertensive medications.

2.2. Collection and Analysis of Blood and Urine Samples

Second morning urine samples were collected after an overnight fast, and whole blood samples were obtained within 3 h after the urine sampling. Aliquots of urine, whole blood, and plasma were stored at −20 or −80 °C prior to analysis. The assay for urine and plasma concentrations of creatinine ($[cr]_u$ and $[cr]_p$) was based on the Jaffe reaction. The assay of urinary β_2M concentration ($[\beta_2M]_u$) was based on the latex immunoagglutination method (LX test, Eiken 2MGII; Eiken and Shionogi Co., Tokyo, Japan) or the ELISA method (Sino Biological Inc., Wayne, PA, USA).

Urinary Cd concentrations ($[Cd]_u$) were determined using an atomic absorption spectrophotometer. Urine standard reference material No. 2670 (National Institute of Standards, Washington, DC, USA) or the reference urine metal control levels 1, 2, and 3 (Lyphocheck, Bio-Rad, Hercules, CA, USA) were used for quality control, analytical accuracy, and precision assurance. The limit of detection (LOD) of Cd quantitation was defined as 3 times the standard deviation of blank measurements. When $[Cd]_u$ was below its detection limit (0.1 μg/L), the Cd concentration assigned was the LOD divided by the square root of 2 [12].

2.3. Estimated Glomerular Filtration Rates (eGFRs)

The GFR is the product of the nephron number and mean single nephron GFR, and in theory, the GFR is indicative of nephron function [13–15]. In practice, the GFR is estimated from established chronic kidney disease epidemiology collaboration (CKD-EPI) equations and is reported as the eGFR [15].

Male eGFR = 141 × [plasma creatinine/0.9]^{Y} × 0.993^{age}, where Y = −0.411 if $[cr]_p \leq 0.9$ mg/dL, and Y = −1.209 if $[cr]_p > 0.9$ mg/dL. Female eGFR = 144 × [plasma creatinine/0.7]^{Y} × 0.993^{age}, where Y = −0.329 if $[cr]_p \leq 0.7$ mg/dL, and Y = −1.209 if $[cr]_p > 0.7$ mg/dL. CKD stages 1, 2, 3a, 3b, 4, and 5 corresponded to eGFRs of 90–119, 60–89, 45–59, 30−44, 15–29, and <15 mL/min/1.73 m^2, respectively.

2.4. Normalization of Excretion Rate

E_x was normalized to E_{cr} as $[x]_u/[cr]_u$, where x = Cd or β_2M; $[x]_u$ = urine concentration of x (mass/volume); and $[cr]_u$ = urine creatinine concentration (mg/dL). The ratio $[x]_u/[cr]_u$ was expressed in µg/g of creatinine.

E_x was normalized to C_{cr} as $E_x/C_{cr} = [x]_u[cr]_p/[cr]_u$, where x = Cd or β_2M; $[x]_u$ = urine concentration of x (mass/volume); $[cr]_p$ = plasma creatinine concentration (mg/dL); and $[cr]_u$ = urine creatinine concentration (mg/dL). E_x/C_{cr} was expressed as the excretion of x per volume of filtrate [7].

2.5. Statistical Analysis

Data were analyzed using IBM SPSS Statistics 21 (IBM Inc., New York, NY, USA). The Mann–Whitney U test was used to assess differences in mean values in women and men, and Pearson's chi-squared test was used to assess differences in percentages. The one-sample Kolmogorov–Smirnov test was used to identify departures of continuous variables from a normal distribution, and logarithmic transformation was applied to variables that showed rightward skewing before they were subjected to parametric statistical analysis.

The multivariable logistic regression analysis was used to determine the prevalence odds ratio (POR) for categorical outcomes. Reduced eGFR was assigned when eGFR ≤ 60 mL/min/1.73 m^2. For C_{cr}-normalized data, tubular dysfunction was defined as $(E_{\beta 2M}/C_{cr}) \times 100 \geq 300$ µg/L of filtrate. For E_{cr}-normalized data, tubular dysfunction was defined as $E_{\beta 2M}/E_{cr} \geq 300$ µg/g creatinine [6]. Univariate analysis of covariance via Bonferroni correction in multiple comparisons was used to obtain covariate-adjusted mean E_{Cd}/C_{cr} and mean $E_{\beta 2M}/C_{cr}$. For all tests, p-values ≤ 0.05 were considered to indicate statistical significance.

3. Results

3.1. Descriptive Characteristics of Participants

This cohort consisted of 334 women (mean age 51.5 years) and 114 men (mean age 49.9 years) (Table 1).

Table 1. Characteristics of study subjects.

Parameters	All Subjects, n = 448	Women, n = 334	Men, n = 114	p
Age, years	51.1 ± 8.6	51.5 ± 9.0	49.9 ± 7.2	0.344
BMI, kg/m^2	24.8 ± 4.0	25.2 ± 4.0	23.7 ± 3.6	<0.001
Smoking, %	31.3	18.6	68.4	<0.001
Hypertension, %	48.7	50.6	43.0	0.160
Diabetes, %	15.4	16.2	13.2	0.442
eGFR [a], mL/min/1.73 m^2	90 ± 18	89 ± 19	93 ± 16	0.145
Reduced eGFR [b], %	6.9	8.1	3.5	0.097
Plasma creatinine, mg/dL	0.82 ± 0.22	0.78 ± 0.21	0.95 ± 0.21	<0.001
Urine creatinine, mg/dL	114 ± 74	108 ± 74	132 ± 72	<0.001

Table 1. *Cont.*

Parameters	All Subjects, n = 448	Women, n = 334	Men, n = 114	p
Blood Cd, µg/L	2.75 ± 3.19	2.58 ± 3.10	3.25 ± 3.41	0.038
Urine Cd, µg/L	4.22 ± 5.67	4.36 ± 6.14	3.82 ± 4.01	0.875
Urine β_2M, µg/L	3122 ± 18,836	2596 ± 17,238	4665 ± 22,903	0.544
Normalized to C_{cr} (E_x/C_{cr}) [c]				
(E_{Cd}/C_{cr}) × 100, µg/L filtrate	3.19 ± 3.72	3.21 ± 3.79	3.12 ± 3.55	0.639
($E_{\beta 2M}/C_{cr}$) × 100, µg/L filtrate	3839 ± 30,422	3078 ± 26,986	6072 ± 38,837	0.212
($E_{\beta 2M}/C_{cr}$) × 100, µg/L filtrate, %				
<300	41.1	38.9	47.4	
300−999	34.8	35.9	31.6	
≥1000	24.1	25.1 *	21.1 ***	
Normalized to E_{cr} (E_x/E_{cr}) [d]				
E_{Cd}/E_{cr}, µg/g creatinine	4.02 ± 4.41	4.26 ± 4.62	3.30 ± 3.68	0.028
$E_{\beta 2M}/E_{cr}$, µg/g creatinine	3220 ± 21,847	3005 ± 22,812	3850 ± 18,815	0.017
$E_{\beta 2M}/E_{cr}$, µg/g creatinine, %				
<300	35.5	32.3	45.6	
300−999	34.6	34.7	34.2	
≥1000	29.7	32.9	20.2 **	

n, number of subjects; BMI, body mass index; eGFR, estimated glomerular filtration rate; β_2M, β_2-microglobulin; E_x, excretion of x; cr, creatinine; C_{cr}, creatinine clearance; Cd, cadmium; [a] eGFR was determined by established CKD-EPI equations [15]; [b] reduced eGFR corresponds to eGFR ≤ 60 mL/min/1.73 m^2; [c] $E_x/E_{cr} = [x]_u/[cr]_u$; [d] $E_x/C_{cr} = [x]_u[cr]_p/[cr]_u$, where x = Cd or β_2M [7]. Data for all continuous variables are arithmetic means ± standard deviation (SD). For all tests, $p \leq 0.05$ identifies statistical significance, determined by Pearson's chi-squared test for % differences and by the Mann–Whitney U test for mean differences between women and men. * p = 0.005; ** p = 0.004; *** p = 0.002.

Of the total of 334 women, 224 and 110 were from the high- and low-exposure regions, respectively. In comparison, of the total 114 men, 84 and 30 males were from the high- and low-exposure regions, respectively.

The respective overall percentages of smoking, hypertension, diabetes, and reduced eGFR were 31.3%, 48.7%, 15.4%, and 6.9%. More than half of the men (68.4%) smoked cigarettes, while only 18.6% of women did. The % of all other ill health conditions in men and women did not differ, nor did their mean age differ.

With the exception of BMI, the mean plasma creatinine, mean urine creatinine, and mean blood Cd were all lower in women than in men. The mean eGFR, mean urine Cd, and mean β_2M concentrations in women and men were not statistically different.

For the C_{cr}-normalized data, the mean E_{Cd}/C_{cr} and mean $E_{\beta 2M}/C_{cr}$ in women and men both did not differ statistically. However, there were statistically significant differences in the % of women and men across three $E_{\beta 2M}/C_{cr}$ groups.

For the E_{cr}-normalized data, the mean E_{Cd}/E_{cr} and mean $E_{\beta 2M}/E_{cr}$ were higher in women than in men. The % of men across three $E_{\beta 2M}/E_{cr}$ groups differed, but there was no difference in the % of women across the $E_{\beta 2M}/E_{cr}$ groups.

3.2. Cadmium Exposure Characterization

Figure 1 provides scatterplots relating two Cd exposure indicators, namely the blood Cd concentration and the excretion rate of Cd, represented as E_{Cd}/C_{cr}.

A strong positive association between log([Cd]$_b$ × 10^3) and [log[(E_{Cd}/C_{cr}) × 10^5] was evident in both women and men (Figure 1a). After the adjustment for the covariates and interactions, the Cd body burden ([log[(E_{Cd}/C_{cr}) × 10^5] explained a larger proportion of the variation in the blood Cd concentrations (log([Cd]$_b$ × 10^3) in men (η^2 = 0.407) than in women (η^2 = 0.105) (Figure 1b).

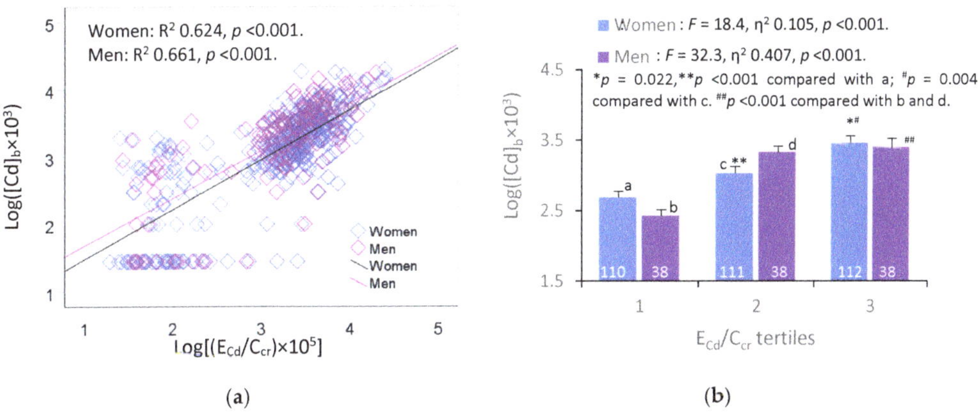

Figure 1. Blood cadmium and urinary cadmium excretion relationship. Scatterplots relating log([Cd]$_b$ × 10^3] to log[(E_{Cd}/C_{cr}) × 10^5] in women and men (**a**). Coefficients of determination (R^2) and *p*-values are provided for all scatterplots. Bar graph (**b**) depicts mean log([Cd]$_b$ × 10^3] values in women and men across E_{Cd}/C_{cr} tertiles. Letters a and c refer to groups of women whose E_{Cd}/Ccr values were in low and middle E_{Cd}/C_{cr} tertiles, respectively. Letters b and d refer to groups of men whose E_{Cd}/C_{cr} values were in low and middle E_{Cd}/C_{cr} tertiles, respectively. All means were obtained via univariate analysis with adjustment for covariates and interactions. For women, respective arithmetic means and standard deviations (SD) for (E_{Cd}/C_{cr}) × 100 tertiles 1, 2, and 3 are 0.37 (0.47), 2.34 (0.52), and 6.84 (4.46) µg/L of filtrate. For men, respective arithmetic means and standard deviations (SD) for (E_{Cd}/C_{cr}) × 100 tertiles quartiles 1, 2, and 3 are 0.36 (0.42), 2.14 (0.63), and 6.81 (3.78) µg/L of filtrate. For all tests, *p*-values ≤ 0.05 indicate statistically significant differences.

To further address the variables/factors that may influence the blood Cd levels, we conducted multiple regression and univariate analyses of variance that incorporated age, BMI, log[(E_{Cd}/C_{cr}) × 10^5], smoking, diabetes, and hypertension as independent variables. Table 2 provides the results of these analyses.

Table 2. Determinants of blood cadmium concentration in women versus men.

Independent Variables/Factors	Log([Cd]$_b$ × 10^3), µg/L					
	Women, *n* = 334			Men, *n* = 114		
	β	η^2	*p*	β	η^2	*p*
Age, years	−0.170	0.057	<0.001	−0.004	1 × 10^{-6}	0.946
BMI, kg/m^2	−0.019	0.001	0.586	−0.079	0.022	0.180
Log[(E_{Cd}/C_{cr}) × 10^5], µg/L filtrate	0.619	0.367	<0.001	0.581	0.420	<0.001
Smoking	0.123	0.028	0.001	0.184	0.055	0.002
Diabetes	−0.053	0.010	0.182	−0.246	0.095	<0.001
Hypertension	0.048	0.007	0.162	−0.061	0.004	0.287
DM × HTN	n/a	0.016	0.023	n/a	n/a	n/a
SMK × DM × HTN	n/a	n/a	n/a	n/a	0.042	0.036
Adjusted R^2	0.624	n/a	<0.001	0.661	n/a	<0.001

β, standardized regression coefficient; adjusted R^2, coefficient of determination; DM, diabetes; HTN, hypertension; SMK, smoking; n/a, not applicable. β indicates strength of association of log([Cd]$_b$ × 10^3) with independent variables (first column). Adjusted R^2 indicates a fractional variation of log([Cd]$_b$ × 10^3) explained by all independent variables. Eta square (η^2) indicates the fraction of the variability of each dependent variable explained by a corresponding independent variable. *p*-values ≤ 0.05 indicate a statistically significant contribution of variation of an independent variable to variation of a dependent variable.

In women, higher [Cd]$_b$ values were strongly associated with higher E_{Cd}/C_{cr} (β = 0.619), and were moderately associated with smoking (β = 0.123) and younger age (β = −0.170). E_{Cd}/C_{cr}, age, and smoking explained, respectively, 36.7%, 5.7%, and 2.8% of the variation of [Cd]$_b$ in women, while the interaction between diabetes and hypertension contributed to 1.6% of the $[Cd]_b$ variability. In men, higher $[Cd]_b$ values were strongly associated with higher E_{Cd}/C_{cr} (β = 0.581), and were moderately associated with smoking (β = 0.184) and not having diabetes (β = −0.246). E_{Cd}/C_{cr}, smoking, and diabetes accounted, respectively, for 42%%, 5.5%, and 9.5% of the variation of $[Cd]_b$ in men, while the interaction between smoking, diabetes, and hypertension contributed to 4.2% of the $[Cd]_b$ variability.

3.3. *Effects of Cadmium Exposure on β_2M Excretion*

We assessed the effects of Cd exposure on $E\beta_2M$ using multiple linear regression and univariate/covariance analyses, where the indicators of Cd exposure ($[Cd]_b$ and E_{Cd}) were incorporated as the independent variables together with age, BMI, smoking, diabetes, and hypertension (Table 3).

Table 3. Associations of β_2-mcirogloubulin excretion with cadmium exposure measurements.

Independent Variables/Factors	$\text{Log}[(E_{\beta 2M}/C_{cr}) \times 10^3]$, µg/L Filtrate					
	All Subjects		Women		Men	
	β	p	β	p	β	p
Age, years	0.137	0.013	0.131	0.041	0.128	0.238
BMI, kg/m^2	−0.089	0.065	−0.102	0.062	−0.043	0.664
$\text{Log}([Cd]_b \times 10^3)$, µg/L filtrate	−0.016	0.824	−0.083	0.328	0.217	0.180
$\text{Log}[(E_{Cd}/C_{cr}) \times 10^5]$, µg/L filtrate	0.283	<0.001	0.306	<0.001	0.175	0.247
Gender	0.052	0.318	−	−	−	−
Smoking	0.063	0.255	0.093	0.094	−0.065	0.526
Diabetes	0.323	<0.001	0.349	<0.001	0.279	0.017
Hypertension	0.015	0.745	−0.023	0.660	0.142	0.139
Adjusted R^2	0.105	<0.001	0.125	<0.001	0.059	0.060

β, standardized regression coefficient; adjusted R^2, coefficient of determination. β indicates strength of association of $\log[(E_{\beta 2M}/C_{cr}) \times 10^3]$ with independent variables (first column). Adjusted R^2 indicates a fractional variation of $\log[(E_{\beta 2M}/C_{cr}) \times 10^3]$ explained by all independent variables. For each test, p-values $\leq$ 0.05 indicate a statistically significant contribution of an independent variable to $\log[(E_{\beta 2M}/C_{cr}) \times 10^3]$ variability.

In all subjects, $E_{\beta 2M}/C_{cr}$ was associated with age (β = 0.137), E_{Cd}/C_{cr} (β = 0.283), and diabetes (β = 0.323). In women, the associations of $E_{\beta 2M}/C_{cr}$ with these three independent variables were evident. In men, $E_{\beta 2M}/C_{cr}$ only showed a significant association with diabetes (β = 0.279).

We next examined the association between the E_{Cd}/C_{cr} and $E_{\beta 2M}/C_{cr}$ with the scatterplots and the covariate-adjusted mean $E_{\beta 2M}$ in subjects grouped by E_{Cd}/C_{cr} tertiles (Figure 2).

The relationship between $E_{\beta 2M}/C_{cr}$ and E_{Cd}/C_{cr} was weak and statistically insignificant in all subjects (Figure 1a), as well as in women and men (Figure 2c). However, with the adjustment for the covariates that included age and BMI, diabetes, hypertension, and smoking (Figure 2b), a significant contribution of the Cd body burden to the variability of $E_{\beta 2M}$ became evident when all subjects were included in an analysis (F = 4.473, η^2 0.012, p = 0.021). The $E_{\beta 2M}$ in the subjects of the high E_{Cd}/C_{cr} tertile was higher compared with those of the middle and low E_{Cd}/C_{cr} tertiles (Figure 1b). In the subgroup analysis (Figure 1d), a dose–effect relationship of E_{Cd} and $E_{\beta 2M}$ was seen in women only (F = 3.431, η^2 0.021, p = 0.034).

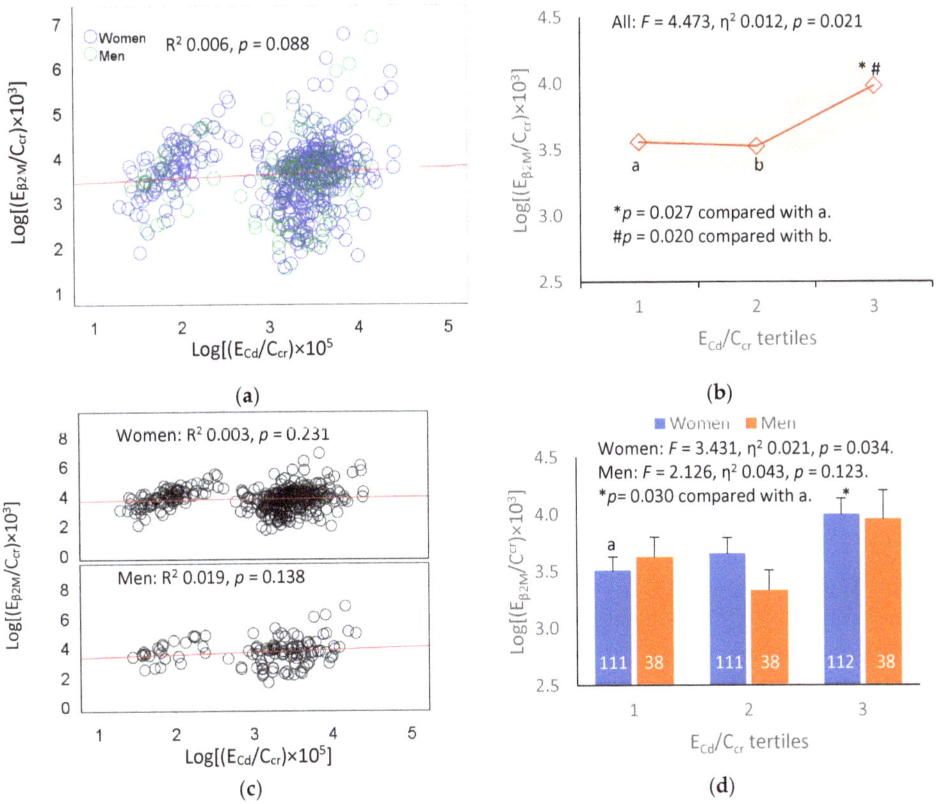

Figure 2. Dose–effect relationship of β_2-microgloubulin and cadmium excretion. Scatterplots relate $\log[(E_{\beta M}/C_{cr}) \times 10^3]$ to $\log[(E_{Cd}/C_{cr}) \times 10^5]$ in all subjects (**a**), and in women and men (**c**). Coefficients of determination (R^2) and p-values are provided for all scatterplots. The color-coded area graph (**b**) depicts means of $\log[(E_{\beta 2M}/C_{cr}) \times 10^3]$ across E_{Cd}/C_{cr} tertiles. Shaded areas indicate variability of means. Bar graph (**d**) depicts mean $\log[(E_{\beta 2M}/C_{cr}) \times 10^3]$ in women and men in each E_{Cd}/C_{cr} tertile. The respective numbers of subjects in E_{Cd}/C_{cr} tertiles 1, 2, and 3 are 149, 149, and 150. In Figure 2d, a letter a refers to a group of women whose ECd/C_{cr} values were in low E_{Cd}/C_{cr} tertile. All means were obtained via univariate analysis with adjustment for covariates and interaction. For women, respective arithmetic means and standard deviations (SD) for $(E_{Cd}/C_{cr}) \times 100$ tertiles 1, 2, and 3 are 0.37 (0.47), 2.34 (0.52), and 6.84 (4.46) µg/L of filtrate. For men, respective arithmetic means and standard deviations (SD) for $(E_{Cd}/C_{cr}) \times 100$ tertiles 1, 2, and 3 are 0.36 (0.42), 2.14 (0.63), and 6.81 (3.78) µg/L of filtrate. For all tests, p-values ≤ 0.05 indicate statistically significant differences.

3.4. *Effects of Cadmium Exposure on the Prevalence Odds of Tubulopathy*

Table 4 provides the results of the logistic regression analysis of abnormal $E_{\beta 2M}$ that incorporated age, BMI, $\log[(E_{Cd}/C_{cr}) \times 10^5]$, gender, smoking, diabetes, and hypertension as independent variables.

Among seven independent variables, the prevalence odds ratios (POR) for $(E_{\beta 2M}/C_{cr}) \times 100 \geq 300$–999 and $\geq$1000 µg/L filtrate were increased with age, $\log[(E_{Cd}/C_{cr}) \times 10^5]$, and diabetes. All of the other four independent variables did not show a significant association with abnormal β_2M excretion. For every 10-fold rise in E_{Cd}/C_{cr}, the POR for $(E_{\beta 2M}/C_{cr}) \times 100$ of $\geq$300 and $\geq$1000 µg/L were increased by 1.94-fold and 3.34-fold, respectively.

Table 4. Prevalence odds for excessive excretion of β_2M in relation to cadmium excretion and other variables.

Independent Variables/Factors	Number of Subjects	$(E_{\beta 2M}/C_{cr}) \times 100 \geq 300$ µg/L		$(E_{\beta 2M}/C_{cr}) \times 100 \geq 1000$ µg/L	
		POR (95% CI)	*p*	POR (95% CI)	*p*
Age, years	448	1.036 (1.007, 1.067)	0.016	1.062 (1.027, 1.098)	<0.001
BMI, kg/m^2	448	0.971 (0.919, 1.025)	0.284	0.958 (0.896, 1.023)	0.203
Log[$(E_{Cd}/C_{cr}) \times 10^5$], µg/L filtrate	448	1.940 (1.344, 2.802)	<0.001	3.343 (2.036, 5.488)	<0.001
Gender (F/M)	334/114	1.406 (0.835, 2.367)	0.200	1.299 (0.687, 2.458)	0.421
Smoking	140	1.067 (0.645, 1.765)	0.801	1.388 (0.763, 2.522)	0.282
Diabetes	69	5.294 (2.526, 11.09)	<0.001	11.52 (5.004, 26.50)	<0.001
Hypertension	218	1.066 (0.714, 1.592)	0.753	1.535 (0.942, 2.501)	0.085

POR, prevalence odds ratio, CI, confidence interval. The units of $(E_{\beta 2M}/C_{cr}) \times 100$ and log[$(E_{Cd}/C_{cr}) \times 10^5$] are µg/L filtrate; data were generated from logistic regression analyses, relating POR for excessive β_2M excretion to seven independent variables (first column). For all tests, *p*-values ≤ 0.05 indicate a statistically significant association of POR with a given independent variable.

3.5. Effects of Cadmium Exposure on eGFR

Similarly, we assessed the effects of Cd exposure on the estimated glomerular filtration rate (eGFR) via multiple linear regression and logistic regression analyses, where $[Cd]_b$ and E_{Cd} were incorporated as the independent variables together with age, BMI, smoking, diabetes, and hypertension (Table 5).

Table 5. Associations of eGFR with cadmium exposure measurements and other variables.

Independent Variables/Factors	eGFR, mL/min/1.73 m^2					
	All Subjects		Women		Men	
	β	*p*	β	*p*	β	*p*
Age, years	−0.517	<0.001	−0.511	<0.001	−0.506	<0.001
BMI, kg/m^2	−0.064	0.136	−0.048	0.327	−0.149	0.095
Log($[Cd]_b \times 10^3$), µg/L filtrate	0.053	0.420	0.102	0.182	−0.153	0.291
Log[$(E_{Cd}/C_{cr}) \times 10^5$], µg/L filtrate	−0.148	0.026	−0.185	0.018	0.011	0.933
Gender	−0.001	0.977	−	−	−	−
Smoking	0.025	0.610	0.018	0.717	0.071	0.438
Diabetes	−0.109	0.023	−0.128	0.021	−0.055	0.593
Hypertension	−0.079	0.055	−0.050	0.295	−0.212	0.014
Adjusted R^2	0.279	<0.001	0.281	<0.001	0.249	<0.001

eGFR, estimated glomerular filtration rate; β, standardized regression coefficient; adjusted R^2, coefficient of determination. β indicates strength of association of eGFR with independent variables (first column). Adjusted R^2 indicates a fractional variation of eGFR explained by all independent variables. For each test, *p*-values ≤ 0.05 indicate a statistically significant contribution of an independent variable to eGFR variability.

In all subjects, the eGFR was inversely associated with age (β = −0.517), E_{Cd} (β = −0.148), and diabetes (β = −0.109). In the subgroup analysis, inverse associations of the eGFR with these three independent variables (age, E_{Cd}, and diabetes) were seen only in women. In men, the eGFR was not associated with E_{Cd}, but this parameter showed inverse associations with age (β = −0.506) and hypertension (β = −0.212).

In the logistic regression of a reduced eGFR (eGFR $\leq$ 60 mL/min/1.73 m^2), age, BMI, log[$(E_{Cd}/C_{cr}) \times 10^5$], gender, smoking, diabetes, and hypertension were incorporated as independent variables (Table 6).

The POR values for a reduced eGFR were increased with age, log[$(E_{Cd}/C_{cr}) \times 10^5$], and diabetes. For every 10-fold rise in E_{Cd}/C_{cr}, the POR for a reduced eGFR was increased by 3.2-fold. There was a 4.2-fold increase in the POR for a reduced eGFR among those with diabetes.

Table 6. Prevalence odds for a reduced eGFR in relation to cadmium excretion and other variables.

Independent Variables/ Factors	Reduced eGFR [a]				
	β Coefficients (SE)	POR	95% CI		p
			Lower	Upper	
Age, years	0.136 (0.027)	1.146	1.086	1.209	<0.001
BMI, kg/m^2	0.020 (0.051)	1.021	0.923	1.128	0.688
Log[(E_{Cd}/C_{cr}) × 10^5], µg/L filtrate	1.154 (0.358)	3.172	1.572	6.402	0.001
Gender	−0.542 (0.649)	0.582	0.163	2.075	0.404
Smoking	−0.228 (0.583)	0.796	0.254	2.493	0.695
Diabetes	1.439 (0.524)	4.217	1.510	11.78	0.006
Hypertension	0.115 (0.427)	1.122	0.486	2.591	0.787

[a] Reduced eGFR is defined as the estimated glomerular filtration rate ≤ 60 mL/min/1.73 m^2; β, regression coefficient; POR, prevalence odds ratio; SE, standard error of mean; CI, confidence interval. Data were generated from logistic regression, relating POR for a reduced eGFR to seven independent variables (first column). For each test, p-values ≤ 0.05 indicate a statistically significant contribution of individual independent variables to the POR for a reduced eGFR.

3.6. Inverse Relationship of β_2M Excretion and eGFR

Figure 3 provides scatterplots relating the eGFR to $E_{\beta2M}$ among the study subjects together with the covariate-adjusted means of the eGFR in women and men.

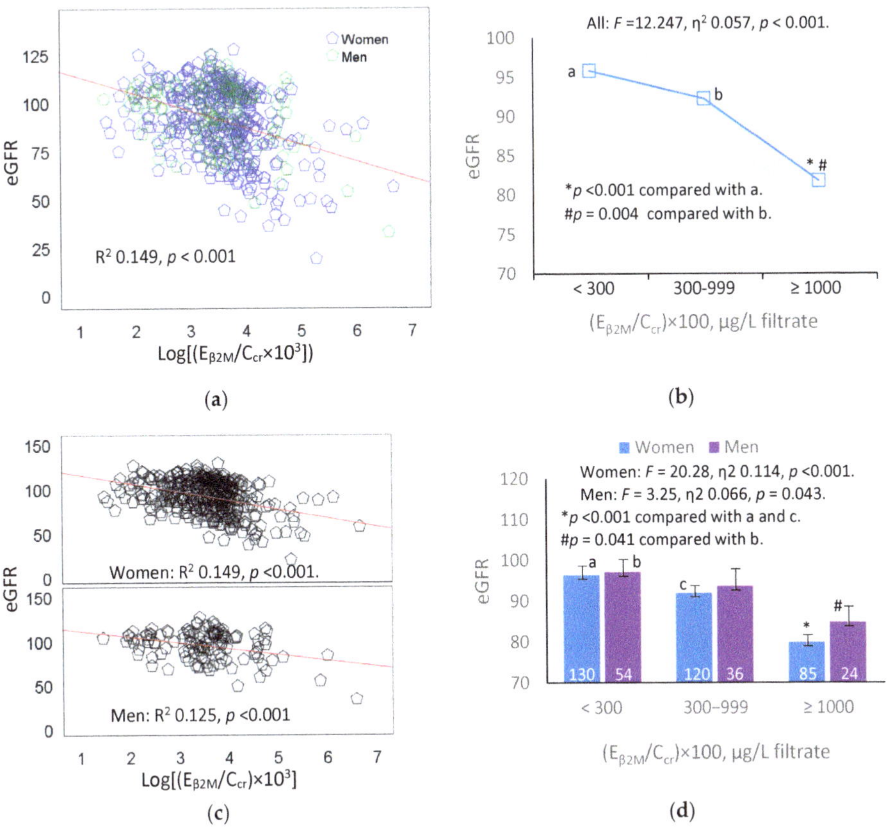

Figure 3. An inverse relationship of eGFR with β_2-microgloubulin excretion. Scatterplots relate eGFR to log[($E_{\beta M}/C_{cr}$) × 10^3] in all subjects (**a**), and in women and men (**c**). Coefficients of determination

(R^2) and p-values are provided for all scatterplots. The color-coded area graph (**b**) depicts means of eGFR across three $E_{\beta 2M}/C_{cr}$ ranges. Shaded areas indicate variability of means. Bar graph (**d**) depicts means of eGFR in women and men in each $E_{\beta 2M}/C_{cr}$ range. The respective numbers of subjects in $(E_{\beta 2M}/C_{cr}) \times 100 < 300$, 300−999 and ≥1000 μg/L of filtrate are 184, 156, and 109. In Figure 3d, letters a and c refer to groups of women whose $(E_{\beta 2M}/C_{cr}) \times 100 < 300$ and 300–999 μg/L filtrate, respectively. The letter b refers to a group of men whose $(E_{\beta 2M}/C_{cr}) \times 100 < 300$ μg/L filtrate, respectively. All means were obtained via univariate analysis with adjustment for covariates and interaction. For all tests, p-values ≤ 0.05 indicate statistically significant differences.

A statistically significant inverse relationship between the eGFR and $E_{\beta 2M}$ was seen in all subjects (Figure 3a), as well as in women and men (Figure 3c). In all subjects (Figure 3b), the eGFR explained 5.7% of the variation in $E_{\beta 2M}/C_{cr}$ ($F = 12.247$, $p < 0.001$).

For simplicity, the degree of tubulopathy, assessed via $E_{\beta 2M}$, was graded into three levels, where levels 1, 2, and 3 of tubulopathy corresponded to $(E_{\beta 2M}/C_{cr}) \times 100 < 300$, 300–999, and ≥1000 μg/L filtrate, respectively.

The covariate-adjusted mean eGFR was 14.0 and 10.5 mL/min/1.73 m^2 lower in subjects with level 3 tubulopathy, compared to those with tubular dysfunction levels 2 and 1, respectively (Figure 3b).

In the subgroup analysis (Figure 3d), the η^2 values indicated a nearly twice larger effect size of the eGFR on $E_{\beta 2M}$ in women ($\eta^2 = 0.114$), compared to men ($\eta^2 = 0.066$). In women, those with level 3 tubulopathy had a covariate-adjusted mean eGFR 16.5 and 12.0 mL/min/1.73 m^2 lower compared to those with levels 1 and 2 tubulopathy, respectively. In men, those with level 3 tubulopathy had a covariate-adjusted mean eGFR 12.3 mL/min/1.73 m^2 lower compared to those with level 1 tubulopathy. The adjusted mean eGFR values in men with tubulopathy levels 3 and 2 did not differ statistically.

In another logistic regression, a relative contribution of Cd exposure and tubular dysfunction levels to the prevalence of a reduced eGFR was determined. E_{Cd} was entered as a continuous variable, while $E_{\beta 2M}$ was categorized into levels 1, 2, and 3, as previously stated. Table 7 provides the results of such analysis.

The POR values for a reduced eGFR rose with age (POR = 1.14), E_{Cd}/C_{cr} (POR = 2.25), tubulopathy level 2 (POR = 8.31), and tubulopathy level 3 (POR = 33.7). All other independent variables, such as diabetes and hypertension, did not show a significant association with the POR for a reduced eGFR.

Table 7. Prevalence odds of a reduced eGFR in relation to cadmium and β_2M excretion rates normalized to C_{cr}.

Independent Variables/ Factors	Reduced eGFR [a]				
	Number of Subjects	POR	95% CI		p
			Lower	Upper	
Age, years	448	1.140	1.072	1.211	<0.001
BMI, kg/m^2	448	1.066	0.950	1.198	0.278
Log[$(E_{Cd}/C_{cr}) \times 10^5$], μg/L filtrate	448	2.251	1.043	4.858	0.039
Gender (F/M)	334/114	0.674	0.176	2.583	0.565
Smoking	140	0.885	0.270	2.899	0.840
Diabetes	69	1.216	0.394	3.753	0.734
Hypertension	218	1.199	0.487	2.957	0.693
$(E_{\beta 2M}/C_{cr}) \times 100$, μg/L filtrate					
<300	184	Referent			
300–999	156	8.310	2.655	26.01	<0.001
≥1000	108	33.731	4.193	271.3	0.001

[a] Reduced eGFR is defined as the estimated glomerular filtration rate ≤ 60 mL/min/1.73 m^2; POR, prevalence odds ratio; CI, confidence interval. Data were generated from multivariable logistic regression analyses, relating the POR for a reduced eGFR to eight independent variables (first column). p-values < 0.05 indicate a statistically significant increase in the POR for a reduced eGFR.

An equivalent logistic regression was conducted using E_{cr}-normalized E_{Cd} and $E_{\beta 2M}$ data (Table 8).

Table 8. Prevalence odds of a reduced eGFR in relation to cadmium and β_2M excretion rates normalized to Ecr.

Independent Variables/ Factors	Reduced eGFR [a]				
	Number of Subjects	POR	95% CI		*p*
			Lower	Upper	
Age, years	448	1.101	1.045	1.160	<0.001
BMI, kg/m^2	448	1.032	0.929	1.147	0.555
$Log[(E_{Cd}/E_{cr}) \times 10^3]$, µg/g creatinine	448	1.278	0.630	2.593	0.497
Gender (F/M)	334/114	0.671	0.181	2.484	0.550
Smoking	140	0.762	0.240	2.418	0.644
Diabetes	69	1.614	0.581	4.484	0.359
Hypertension	218	1.041	0.449	2.415	0.926
$(E_{\beta 2M}/C_{cr}) \times 100$, µg/g creatinine					
<300	160				
300–999	155	3.204	1.226	8.375	0.018
≥1000	133	19.042	2.387	151.907	0.005

[a] Reduced eGFR is defined as the estimated glomerular filtration rate ≤ 60 mL/min/1.73 m^2; POR, prevalence odds ratio; CI, confidence interval. Data were generated from multivariable logistic regression analyses, relating the POR for a reduced eGFR to eight independent variables (first column). *p*-values < 0.05 indicate a statistically significant increase in the POR for a reduced eGFR.

The POR values for a reduced eGFR rose with age (POR = 1.10), the severity of tubular dysfunction, $E_{\beta 2M}/E_{cr}$ 300–999 µg/g creatinine (POR = 3.20), and $E_{\beta 2M}/E_{cr} \geq 100$ µg/g creatinine (POR = 19.0). Associations of POR for a reduced eGFR with E_{Cd}/E_{cr} and all other variables were statistically insignificant.

4. Discussion

This study used a cross-sectional analysis of kidney dysfunction, tubular proteinuria, and eGFR decline to determine the differential impact of Cd exposure in men and women. Whereas many previous studies focused primarily on Cd-induced tubulopathy in women, we investigated these health outcomes in both men and women along with confounding risk factors, smoking, diabetes, and hypertension. The excretion rate of Cd and β_2M (E_{Cd} and $E_{\beta 2M}$) were normalized to the surrogate measure of the GFR, creatinine clearance (C_{cr}). This C_{cr}-normalization of E_{Cd} and $E_{\beta 2M}$ as E_{Cd}/C_{cr} and $E_{\beta 2M}/C_{cr}$ depicts the excretion rates per functional nephron; thereby, it corrects for differences in the number of functioning nephrons among the study subjects [7]. This C_{cr}-normalized excretion rate also corrects for urine dilution, but it is unaffected by creatinine excretion (E_{cr}). Thus, E_{Cd}/C_{cr} and $E_{\beta 2M}/C_{cr}$ provide an accurate quantification of the kidney burden of Cd and its toxicity to kidney tubular cells.

We selected subjects from two population-based studies, undertaken in an area with endemic Cd contamination in the Mae Sot district, Tak province [8], and in a control, non-contaminated area in the Nakhon-Si-Thammarat province of Thailand [9,10]. The Cd content of the paddy soil samples from the Mae Sot district exceeded the standard of 0.15 mg/kg, and the rice samples collected from households contained four times the amount of the permissible Cd level of 0.1 mg/kg [16].

4.1. *Exposure Levels of Cadmium in Women Versus Men*

Men and women in this cohort carry the same body burden of Cd, which is evident from a nearly identical mean $(E_{Cd}/C_{cr}) \times 100$ values of 3.12 vs. 3.21 µg/L filtrate. The sources of Cd could be differentiated through an analysis of blood–urine Cd relationships.

$[Cd]_b$ and E_{Cd}/C_{cr} correlated strongly with each other in women (R^2 = 0.624) and men (R^2 = 0.661) (Figure 1a), and the covariate-adjusted means $[Cd]_b$ showed a stepwise increase through the E_{Cd}/C_{cr} tertiles in both genders. Notably, E_{Cd}/C_{cr} explained a larger fraction of the variation in $[Cd]_b$ in men than it did in women (η^2 0.407 vs. 0.105) (Figure 1b). The variability in $[Cd]_b$ was associated mostly with E_{Cd}/C_{cr} in both genders, while smoking explained a larger fraction of the $[Cd]_b$ variability among men than among women (5.5% vs.

2.8%). This result was expected, given the high % of smokers in the male group (68.4% vs. 18.6%) and the higher mean $[Cd]_b$ in men than in women (3.25 vs. 4.36 μg/L). In men only, the $[Cd]_b$ variation was associated with diabetes.

4.2. The Toxic Manifestation of Cadmium Exposure in Women Versus Men

An independent health survey reported that the prevalence of chronic kidney disease (CKD), defined as the eGFR $\leq$ 60 mL/min/1.73 m^2, among Mae Sot residents was 16.1%, while the prevalence of tubulopathy, referred to as tubular proteinuria, was 36.1% [17]. This reported tubular proteinuria was based on the cut-off value of $E_{\beta 2M}/E_{cr}$ at 300 μg/g creatinine [6], which equates to $E_{\beta 2M}/C_{cr}$ of 2–3 μg/ L filtrate or $(E_{\beta 2M}/C_{cr}) \times 100$ of 200–300 μg/ L filtrate. Notably, the cut-off value for $E_{\beta 2M}/E_{cr}$ at 300 μg/g creatinine was used as the critical effect of exposure to Cd in the human diet [6].

In this cohort, tubular proteinuria affected more than half of women (61.1%) and men (52.6%). One of four women had severe tubular impairment $[(E_{\beta 2M}/C_{cr}) \times 100 \geq 1000$ μg/L filtrate], whereas one of five men had this abnormality.

In women, $E_{\beta 2M}/C_{cr}$ showed a moderate positive association with age (β = 0.131), and an equally strong association with E_{Cd}/C_{cr} (β = 0.306) and diabetes (β = 0.349) (Table 3). In men, $E_{\beta 2M}/C_{cr}$ did not show a significant association with age or E_{Cd}/C_{cr}, but this tubular defect was associated with diabetes only (β = 0.279). In the covariance analysis, the contribution of E_{Cd}/C_{cr} to the variability of $E_{\beta 2M}/C_{cr}$ in women was demonstrable together with a dose–effect relationship after the adjustment of the covariates and interactions (Figure 2d). In contrast, the contribution of E_{Cd}/C_{cr} to the variation of $E_{\beta 2M}/C_{cr}$ in men was statistically insignificant (Figure 2d). An association of the marker of tubular dysfunction ($E_{\beta M}$) and diabetes seen in both men and women is in line with the current knowledge that diabetes adversely affects both glomerular (GFR) and tubular function, termed diabetic tubulopathy [18,19].

The overall mean eGFR was 90 mL/min/1.73 m^2, and the overall prevalence of eGFR $\leq$ 60 mL/min/1.73 m^2 in this cohort was 6.9% (Table 1). The difference in the % of the reduced eGFR in women and men (8.1% vs. 3.5%) did not reach a statistical significance level (p = 0.097), nor did the difference in the mean eGFR in women and men (p = 0.145). The weaker effect of Cd exposure on the eGFR in men, compared to women, remains to be confirmed with a sufficiently large sample group of men. However, the regression analysis also indicated gender differences in susceptibility to the nephrotoxicity of Cd (Table 5). In women, the eGFR was inversely associated with age (β = −0.511), E_{Cd}/C_{cr} (β = −0.185), and diabetes (β = −0.128). In comparison, the eGFR in men was not associated with E_{Cd}/C_{cr}, while showing an inverse association with age (β = −0.506) and hypertension (β = −0.212). Adverse effects of hypertension and diabetes on the eGFR have been noted in a cross-sectional study of the general U.S. population, where a Cd-induced GFR reduction was more pronounced in those who had diabetes and/or hypertension [20].

We speculate that gender differences in the levels of some protective factors, notably body status of nutritionally essential metals such as iron and zinc, may contribute to the increased susceptibility to Cd nephrotoxicity that was seen in women. Similarly, environmental Cd exposure has been linked to a reduction in the eGFR among participants in various cycles of the U.S. National Health and Nutrition Examination Survey (NHANES) undertaken over 18 years (1999 to 2016) [20–22]. Lin et al. (2014) reported that the risk for a reduced eGFR was higher in those with lower serum zinc (OR 3.38) compared to those with similar Cd exposure levels and serum zinc > 74 μg/dL (OR 2.04) [22].

4.3. Increment of β_2M Excretion as GFR Falls

The protein β_2M with the molecular weight of 11,800 Da is filtered freely by the glomeruli and is reabsorbed almost completely by the kidney's tubular epithelial cells [13]. Thus, the defective tubular re-absorption of β_2M will result in an enhanced excretion rate of β_2M [23–26]. The loss of nephrons also raises the excretion of β_2M for the following reasons [23,26]. When the reabsorption rate of β_2M per nephron remains constant, its excretion will vary directly with its production. If the production and reabsorption per nephron remain constant as nephrons are lost, the excretion of β_2M will rise [27].

It can thus be expected that the excretion of β_2M will increase when the GFR falls for any causes. Indeed, $E_{\beta 2M}/C_{cr}$ was inversely associated with the eGFR in both women and men (Figure 3c), although the causes of their eGFR decreases seemed to be different. In a quantitative analysis (Figure 3d), the η^2 value describing the effect size of $E_{\beta 2M}/C_{cr}$ on the eGFR variability was 1.7-fold larger in women than in men (0.114 vs. 0.066). In both women and men, the eGFR was the lowest in those who had $(E_{\beta 2M}/C_{cr}) \times 100 \geq 1000$ μg/L filtrate, indicative of severely impaired tubular function.

Notably, the POR for a reduced eGFR was increased by 8.3-fold and 33.7-fold in those with $(E_{\beta 2M}/C_{cr}) \times 100$ of 300–999 and ≥1000 μg/L filtrate, respectively (Table 7), compared to those with $E_{\beta 2M}/C_{cr}) \times 100 < 300$ μg/L filtrate. A substantial loss of nephron function was a likely cause of the massive increases in $E_{\beta 2M}/C_{cr}$ that was seen in those who had an eGFR below 60 mL/min/1.73 m^2.

4.4. The Pitfall of Adjusting Excretion Rate to Ecr and Implication for Health Risk Estimation

The C_{cr}-normalized data indicate that women and men shared the same burden of Cd (Table 1). The data also indicate that the % of women and men across the three categories of tubulopathy were all statistically different, thereby linking Cd exposure to the severity of tubulopathy in both genders. The logistic regression data (Table 7) show that the likelihood of having a reduced eGFR was increased by 8.3-fold and 33.7-fold in those who had $(E_{\beta 2M}/C_{cr}) \times 100$ of 300–999 and ≥1000 μg/L filtrate compared to those with $(E_{\beta 2M}/C_{cr}) \times 100 < 300$ μg/L filtrate.

The E_{cr}-normalized data indicated that the mean E_{Cd}/E_{cr} in women was statistically higher than that of men (4.26 vs. 3.30 μg/g creatinine). They also indicated that the difference in % of women across the three tubulopathy categories was minuscule, and that the % distribution of men across the tubulopathy categories was statistically significant. These data suggest an association of Cd exposure and the severity of tubulopathy in men only. The logistic regression data (Table 8) show that the likelihood of having a reduced eGFR was increased by 3.2-fold and 19-fold in those who had $E_{\beta 2M}/E_{cr}$ of 300–999 and ≥1000 μg/g creatinine compared to those with $E_{\beta 2M}/E_{cr} < E_{\beta 2M}/E_{cr}$.

Previously, $E_{\beta 2M}/E_{cr}$ of 100–299, 300–999, and ≥1000 μg/g creatinine were found to be associated with 4.7-fold, 6.2-fold, and 10.5-fold increases in the prevalence odds of a reduced eGFR [28,29]. Similarly, a rise in $E_{\beta 2M}/E_{cr}$ to levels not higher than 100 μg/g creatinine was associated with an increased risk of hypertension in the general Japanese population [30], while the prospective cohort data showed that $E_{\beta 2M}/E_{cr}$ was associated with a 79% increase in the likelihood of having a large decline in the eGFR (10 mL/min/1.73 m^2) over a five-year period [31]. Thus, a cut-off value for $E_{\beta 2M}/E_{cr}$ above 300 μg/g creatinine does not reflect an early warning sign of the nephrotoxicity of Cd. The utility of this $E_{\beta 2M}/E_{cr}$ value as a toxicity criterion to derive a toxicity threshold level for Cd is inappropriate.

In summary, adjusting E_{Cd} and $E_{\beta 2M}$ to E_{cr} produces an erroneous interpretation of the effect of Cd exposure on the eGFR, while underestimating the severity of Cd-induced tubulopathy, especially among women. These data call into question the utility of $E_{\beta 2M}/E_{cr}$ of 300 μg/g creatinine to represent the critical effect of exposure to Cd in the human diet. New health guidance values need to be established for this toxic metal, and new public measures are needed to minimize the Cd contamination of food chains.

4.5. Strength and Limitation

In this cohort, the levels of environmental exposure among the participants were assessed by measuring the blood Cd and urinary Cd excretion rates. Strong correlations between these two parameters were seen in both men and women (Figure 1). Both the tubular and glomerular function were examined concurrently together with confounding factors, smoking, hypertension, and type 2 diabetes. These are the strengths of our study.

The small number of males from high- (n = 84) and low-exposure (n = 30) locations is a limitation. This precludes an analysis of both genders separately from both locations that may help to rule out any other environmental effects on adverse kidney outcomes in women. In addition, the heterogeneity in the hormonal status, notably estrogen, in female participants who are menopausal and post-menopausal is a limitation [32].

5. Conclusions

The excretion of β_2M above 300 μg/g creatinine (≈2–3 μg/L filtrate) and a reduction in the glomerular function, indicated by an eGFR below 60 mL/min/1.73 m^2, are the manifestations of severe kidney toxicities due to chronic exposure to Cd that are more prevalent and more severe in women than men of the same body burden.

Author Contributions: Conceptualization, S.S.; methodology, S.S. and S.Y.; formal analysis, S.S.; investigation, S.Y., P.P. and T.K.; resources, G.C.G. and D.A.V.; writing—original draft preparation, S.S.; writing—review and editing, G.C.G. and D.A.V.; project administration, S.S. and S.Y. All authors have read and agreed to the published version of the manuscript.

Funding: This research received no external funding.

Institutional Review Board Statement: This is not applicable for this study, which used archived data [8,9].

Informed Consent Statement: Informed consent was obtained from all participants in the study prior to their participation.

Data Availability Statement: All data are contained within this article.

Acknowledgments: This work was supported with resources from the Centre for Kidney Disease Research, Translational Research Institute, and the Department of Kidney and Transplant Services, Princess Alexandra Hospital.

Conflicts of Interest: The authors declare no conflict of interest.

References

1. Satarug, S.; Baker, J.R.; Reilly, P.E.; Moore, M.R.; Williams, D.J. Cadmium levels in the lung, liver, kidney cortex, and urine samples from Australians without occupational exposure to metals. *Arch. Environ. Health* **2002**, *57*, 69–77. [CrossRef] [PubMed]
2. Barregard, L.; Fabricius-Lagging, E.; Lundh, T.; Mölne, J.; Wallin, M.; Olausson, M.; Modigh, C.; Sallsten, G. Cadmium, mercury, and lead in kidney cortex of living kidney donors: Impact of different exposure sources. *Environ. Res.* **2010**, *110*, 47–54. [CrossRef] [PubMed]
3. Barregard, L.; Sallsten, G.; Lundh, T.; Mölne, J. Low-level exposure to lead, cadmium and mercury, and histopathological findings in kidney biopsies. *Environ. Res.* **2022**, *211*, 113119. [CrossRef] [PubMed]
4. Satarug, S.; Vesey, D.A.; Gobe, G.C.; Phelps, K.R. Estimation of health risks associated with dietary cadmium exposure. *Arch. Toxicol.* **2023**, *97*, 329–358. [CrossRef]
5. Watanabe, T.; Kataoka, Y.; Hayashi, K.; Matsuda, R.; Uneyama, C. Dietary exposure of the Japanese general population to elements: Total diet study 2013–2018. *Food Saf.* **2022**, *10*, 83–101. [CrossRef]
6. JECFA. In *Summary and Conclusions, Proceedings of the Joint FAO/WHO Expert Committee on Food Additives and Contaminants, Seventy-Third Meeting, Geneva, Switzerland, 8–17 June 2010*; JECFA/73/SC; Food and Agriculture Organization of the United Nations/World Health Organization: Geneva, Switzerland, 2011; Available online: https://apps.who.int/iris/handle/10665/44521 (accessed on 25 June 2023).
7. Phelps, K.R.; Gosmanova, E.O. A generic method for analysis of plasma concentrations. *Clin. Nephrol.* **2020**, *94*, 43–49. [CrossRef]
8. Satarug, S.; Swaddiwudhipong, W.; Ruangyuttikarn, W.; Nishijo, M.; Ruiz, P. Modeling cadmium exposures in low- and high-exposure areas in Thailand. *Environ. Health Perspect.* **2013**, *121*, 531–536. [CrossRef]
9. Yimthiang, S.; Pouyfung, P.; Khamphaya, T.; Kuraeiad, S.; Wongrith, P.; Vesey, D.A.; Gobe, G.C.; Satarug, S. Effects of environmental exposure to cadmium and lead on the risks of diabetes and kidney dysfunction. *Int. J. Environ. Res. Public Health* **2022**, *19*, 2259. [CrossRef]
10. Zarcinas, B.A.; Pongsakul, P.; McLaughlin, M.J.; Cozens, G. Heavy metals in soils and crops in Southeast Asia. 2. Thailand. *Environ. Geochem. Health* **2004**, *26*, 359–371. [CrossRef]
11. Bloch, M.J.; Basile, J.N. Review of recent literature in hypertension: Updated clinical practice guidelines for chronic kidney disease now include albuminuria in the classification system. *J. Clin. Hypertens* **2013**, *15*, 865–867. [CrossRef]
12. Hornung, R.W.; Reed, L.D. Estimation of average concentration in the presence of nondetectable values. *Appl. Occup. Environ. Hyg.* **1990**, *5*, 46–51. [CrossRef]

13. Murton, M.; Goff-Leggett, D.; Bobrowska, A.; Garcia Sanchez, J.J.; James, G.; Wittbrodt, E.; Nolan, S.; Sörstadius, E.; Pe-coits-Filho, R.; Tuttle, K. Burden of chronic kidney disease by KDIGO categories of glomerular filtration rate and albuminuria: A Systematic review. *Adv. Ther.* **2021**, *38*, 180–200. [CrossRef] [PubMed]
14. Soveri, I.; Berg, U.B.; Björk, J.; Elinder, C.G.; Grubb, A.; Mejare, I.; Sterner, G.; Bäck, S.E.; SBU GFR Review Group. Measuring GFR: A systematic review. *Am. J. Kidney Dis.* **2014**, *64*, 411–424. [CrossRef] [PubMed]
15. Levey, A.S.; Becker, C.; Inker, L.A. Glomerular filtration rate and albuminuria for detection and staging of acute and chronic kidney disease in adults: A systematic review. *JAMA* **2015**, *313*, 837–846. [CrossRef]
16. Suwatvitayakorn, P.; Ko, M.S.; Kim, K.W.; Chanpiwat, P. Human health risk assessment of cadmium exposure through rice consumption in cadmium-contaminated areas of the Mae Tao sub-district, Tak, Thailand. *Environ. Geochem. Health* **2020**, *42*, 2331–2344. [CrossRef]
17. Swaddiwudhipong, W.; Nguntra, P.; Kaewnate, Y.; Mahasakpan, P.; Limpatanachote, P.; Aunjai, T.; Jeekeeree, W.; Punta, B.; Funkhiew, T.; Phopueng, I. Human health effects from cadmium exposure: Comparison between persons living in cadmium-contaminated and non-contaminated areas in northwestern Thailand. *Southeast Asian J. Trop. Med. Publ. Health* **2015**, *46*, 133–142.
18. Zhou, X.; Xu, C.; Dong, J.; Liao, L. Role of renal tubular programed cell death in diabetic kidney disease. *Diabetes Metab. Res. Rev.* **2023**, *39*, e3596. [CrossRef]
19. Yao, L.; Liang, X.; Qiao, Y.; Chen, B.; Wang, P.; Liu, Z. Mitochondrial dysfunction in diabetic tubulopathy. *Metabolism* **2022**, *131*, 155195. [CrossRef]
20. Madrigal, J.M.; Ricardo, A.C.; Persky, V.; Turyk, M. Associations between blood cadmium concentration and kidney function in the U.S. population: Impact of sex, diabetes and hypertension. *Environ. Res.* **2018**, *169*, 180–188. [CrossRef]
21. Navas-Acien, A.; Tellez-Plaza, M.; Guallar, E.; Muntner, P.; Silbergeld, E.; Jaar, B.; Weaver, V. Blood cadmium and lead and chronic kidney disease in US adults: A joint analysis. *Am. J. Epidemiol.* **2009**, *170*, 1156–1164. [CrossRef]
22. Ferraro, P.M.; Costanzi, S.; Naticchia, A.; Sturniolo, A.; Gambaro, G. Low level exposure to cadmium increases the risk of chronic kidney disease: Analysis of the NHANES 1999–2006. *BMC Public Health* **2010**, *10*, 304. [CrossRef] [PubMed]
23. Lin, Y.S.; Ho, W.C.; Caffrey, J.L.; Sonawane, B. Low serum zinc is associated with elevated risk of cadmium nephrotoxicity. *Environ. Res.* **2014**, *134*, 33–38. [CrossRef] [PubMed]
24. Argyropoulos, C.P.; Chen, S.S.; Ng, Y.H.; Roumelioti, M.E.; Shaffi, K.; Singh, P.P.; Tzamaloukas, A.H. Rediscovering beta-2 microglobulin as a biomarker across the spectrum of kidney diseases. *Front. Med.* **2017**, *4*, 73. [CrossRef] [PubMed]
25. Portman, R.J.; Kissane, J.M.; Robson, A.M. Use of B2-microglobulin to diagnose tubulo-interstitial renal lesions in children. *Kidney Int.* **1986**, *30*, 91–98. [CrossRef]
26. Gauthier, C.; Nguyen-Simonnet, H.; Vincent, C.; Revillard, J.-P.; Pellet, M.V. Renal tubular absorption of beta 2 microglobulin. *Kidney Int.* **1984**, *26*, 170–175. [CrossRef]
27. Peterson, P.A.; Evrin, P.-E.; Berggard, I. Differentiation of glomerular, tubular, and normal proteinuria: Determination of urinary excretion of B2-microglobulin, albumin, and total protein. *J. Clin. Investig.* **1969**, *48*, 1189–1198. [CrossRef]
28. Satarug, S.; Vesey, D.A.; Gobe, G.C.; Đorđević, A.B. The validity of benchmark dose limit analysis for estimating permissible accumulation of cadmium. *Int. J. Environ. Res. Public Health* **2022**, *19*, 15697. [CrossRef]
29. Satarug, S.; Vesey, D.A.; Nishijo, M.; Ruangyuttikarn, W.; Gobe, G.C. The inverse association of glomerular function and urinary β2-MG excretion and its implications for cadmium health risk assessment. *Environ. Res.* **2019**, *173*, 40–47. [CrossRef]
30. Mashima, Y.; Konta, T.; Kudo, K.; Takasaki, S.; Ichikawa, K.; Suzuki, K.; Shibata, Y.; Watanabe, T.; Kato, T.; Kawata, S.; et al. Increases in urinary albumin and beta2-microglobulin are independently associated with blood pressure in the Japanese general population: The Takahata Study. *Hypertens. Res.* **2011**, *34*, 831–835. [CrossRef]
31. Kudo, K.; Konta, T.; Mashima, Y.; Ichikawa, K.; Takasaki, S.; Ikeda, A.; Hoshikawa, M.; Suzuki, K.; Shibata, Y.; Watanabe, T.; et al. The association between renal tubular damage and rapid renal deterioration in the Japanese population: The Takahata study. *Clin. Exp. Nephrol.* **2011**, *15*, 235–241. [CrossRef]
32. Vahter, M.; Berglund, M.; Akesson, A. Toxic metals and the menopause. *J. Br. Menopause Soc.* **2004**, *10*, 60–64. [CrossRef] [PubMed]

toxics

MDPI

Article

Long-Term Exposure to Cadmium Causes Hepatic Iron Deficiency through the Suppression of Iron-Transport-Related Gene Expression in the Proximal Duodenum

Maki Tokumoto [1], Jin-Yong Lee [1], Yasuyuki Fujiwara [2] and Masahiko Satoh [1,*]

[1] Laboratory of Pharmaceutical Health Sciences, School of Pharmacy, Aichi Gakuin University, Nagoya 464-8650, Japan; maki@dpc.agu.ac.jp (M.T.); leejy@dpc.agu.ac.jp (J.-Y.L.)
[2] Department of Environmental Health, School of Pharmacy, Tokyo University of Pharmacy and Life Sciences, Hachioji 192-0392, Japan; yasuyuki@toyaku.ac.jp
* Correspondence: masahiko@dpc.agu.ac.jp; Tel.: +81-52-757-6790

Abstract: Cadmium (Cd) is an environmental pollutant that damages various tissues. Cd may cause a depletion of iron stores and subsequently an iron deficiency state in the liver. However, the molecular mechanism of decreased iron accumulation in the liver induced by long-term exposure to Cd is unknown. In this study, we investigated the hepatic accumulation of iron and the proximal duodenal expression of the genes involved in iron transport using mice chronically exposed to Cd. Five-week-old female C57BL/6J mice were fed a diet containing 300 ppm Cd for 12, 15, 19 and 21 months. The iron concentration in the liver was markedly decreased by Cd. Among iron-transport-related genes in the proximal duodenum, the gene expression of *HCP1* and *Cybrd1* was significantly decreased by Cd. *HCP1* is an influx transporter of heme iron. *Cybrd1* is a reductase that allows non-heme iron to enter cells. The expression of iron-transport-related genes on the duodenal basolateral membrane side was hardly altered by Cd. These results suggest that long-term exposure to Cd suppresses the expression of *HCP1* and *Cybrd1* in the proximal duodenum, resulting in reduced iron absorption and iron accumulation in the liver.

Keywords: cadmium; iron deficiency; iron storage; iron-transport-related factor; proximal duodenum

1. Introduction

Cadmium (Cd) is a toxic heavy metal and well-known causative agent of Itai-itai disease, a pollution disease identified in Japan in the 1950s. Cd causes damages in the liver, kidney and bone via oral exposure, and respiratory damage through inhalation exposure [1–3]. The main target of chronic exposure to Cd is renal proximal tubules, leading to the development of osteomalacia in Itai-itai disease. Cd is widely present in the environment and accumulates easily in rice and seafood. Therefore, humans are exposed to Cd through their diet, which lasts a lifetime. Additionally, the absorbed Cd is retained in the body for a long time, because Cd has a very long biological half-life of 15–30 years [1]. While Cd causes various chronic toxicities, including renal toxicity, it also causes iron deficiency and iron deficiency anemia. In humans, there is a high correlation between blood Cd concentration and iron deficiency or iron deficiency anemia [4–7]. Previous studies have also demonstrated that oral exposure to Cd decreases the iron concentration in the liver of mice (10 mg Cd/kg body weight for 14 days [8], 100 and 200 ppm for 85 days [9]), rats (0.6 mg Cd per kg of feed for 30 days [10], 2.5, 5, 10, 25, 50 and 100 ppm Cd for 12 weeks [11]), lambs (5, 15, 30 and 60 ppm of Cd for 191 days) [12] and chickens (100 ppm Cd for 4 or 9 weeks) [13].

The disruption of iron homeostasis due to long-term exposure to Cd affects human health because iron is an essential metal. According to WHO (2023) [14], anemia was estimated to affect 40% of children aged 6–59 months, 37% of pregnant women and 30% of women aged 15–49 years globally in 2019. Anemia is also the most prevalent in low- and middle-income countries, especially among those living in situations of poverty and social exclusion. It is estimated that nearly a quarter of the world's population (1.8 billion people) suffered from anemia in 2019. The effects of anemia include a reduction in cognitive and motor development in children, a reduction in fertility and an increase in morbidity and mortality. In addition, anemia can exacerbate the symptoms of other diseases. Iron is absorbed from the duodenum via certain transporters. Orally ingested iron is mainly absorbed from duodenum epithelial cells, and iron homeostasis is regulated by multiple iron-transport-related factors. Two types of iron are in foods: heme and non-heme iron. Heme iron is a complex composed of divalent iron [Fe(II)] and porphyrin, and is abundant in animal sources of food. Non-heme iron is inorganic iron and is found in plant-based foods as trivalent iron [Fe(III)]. Heme iron is more readily absorbed in the duodenum than non-heme iron. On the luminal side of the duodenum, a heme iron transporter, heme carrier protein 1 (HCP1) [15] and a non-heme iron transporter called divalent metal transporter 1 (*Dmt1*) [16] are expressed. *HCP1* is encoded by *Slc46a1* (solute carrier family 46 member 1). *Dmt1* is encoded by *Slc11a2* [solute carrier family 11 (proton-coupled divalent metal ion transporters), member 2]. Heme iron absorbed through *HCP1* is degraded to Fe(II), biliverdin and carbon monoxide by heme oxygenase-1 (HO-1) in enterocytes. Non-heme iron is Fe(III) and is not directly taken up into enterocytes. Fe(III) is reduced to Fe(II) by cytochrome b reductase 1 (Cybrd1) [17], and Fe(II) is taken up into enterocytes via *Dmt1*. Absorbed Fe(II) is released to the vascular side via solute carrier family 40 member 1, also known as efflux transporter ferroportin (Fpn1) [18]. Fe(II) is oxidized to Fe(III) by ceruloplasmin (Cp) and hephaestin (Heph) [19]. The majority of Cp is secreted extracellularly into the blood, while Cp linked to glycosylphosphatidylinositol (GPI) binds to the plasma membrane. In contrast, *Heph* is highly expressed in the basolateral membrane of the small intestine. *Fpn1* expression is post-translationally regulated by hepcidin antimicrobial peptide (Hamp) produced in the liver. When blood iron is in excess, *Hamp* is produced and *Fpn1* is degraded, thereby regulating the iron delivered into the blood [20,21]. Blood iron is transported to various tissues by binding to transferrin.

Our recent study demonstrated that single-dose oral administration of Cd (50 mg Cd/kg body weight) to female mice decreased the serum iron concentration to 54% of the control group and inhibited the expression of iron-transport-related genes in the duodenum 24 h after administration [22]. *Dmt1*, *Fpn1* and *Cybrd1* mRNA levels were markedly higher in the proximal duodenum (2 cm from just below the stomach) than in other parts of the intestine [distal duodenum (2 cm from just below the proximal duodenum), jejunum, ileum, colon, and rectum]. mRNA levels of *Dmt1*, *Fpn1* and *Cybrd1* were 2.5, 4 and 7 times those of the distal duodenum, respectively [22]. Moreover, the expression of iron-transport-related genes was also inhibited in human colon cancer Caco-2 cells, which are a cell model of small intestinal enterocytes, exposed to Cd [22].

The molecular mechanism of decreased iron accumulation in the liver due to long-term exposure to Cd is unknown. Therefore, in this study, we investigated the effect of long-term exposure to Cd not only on the iron concentration in the liver but also on the expression of iron-transport-related genes in the proximal duodenum and liver of mice.

2. Materials and Methods

2.1. Animals

Mouse feed containing 300 ppm Cd was prepared by Oriental Yeast (Tokyo, Japan) with a certified diet (MF, Oriental Yeast) and $CdCl_2$ (Fujifilm Wako Pure Chemical., Osaka, Japan). Four-week-old female C57BL/6J mice were purchased from CLEA Japan (Tokyo, Japan) and randomly assigned to control (n = 24) or Cd-exposed groups (n = 21) (approval protocol code: 14-024). These mice were maintained in the laboratory animal facility of the School of Pharmacy, Aichi Gakuin University, at 23 ± 1 °C, with 55 ± 15% relative humidity. This laboratory animal facility is checked twice a year for contamination by pathogenic microorganisms, and no pathogenic microorganisms were detected during this study period. Mice were acclimated with MF and tap water *ad libitum* for 1 week. All maintenance of mice and experiments with mice were performed in accordance with the guidelines established by the Animal Care and Use Committee of the School of Pharmacy, Aichi Gakuin University.

2.2. Animal Treatments

Five-week-old mice were fed MF containing 300 ppm Cd and tap water *ad libitum* for 12, 15, 19 and 21 months. The number of mice in each group was as follows: (1) 12-month control (n = 5), (2) 12-month Cd (n = 5), (3) 15-month control (n = 7), (4) 15-month Cd (n = 5), (5) 19-month control (n = 6), (6) 19-month Cd (n = 5), (7) 21-month control (n = 6), (8) 21-month Cd (n = 6). Mice were sacrificed under anesthesia and blood (approximately 0.5–1.0 mL per mouse) was collected in a capiject (Terumo, Tokyo, Japan), a container for the collection of a small amount of blood. Serum was separated from the blood in the capiject via centrifugation. The liver and duodenum were removed from mice. The portion just below the stomach (2 cm) was defined as the proximal duodenum. Tissues were snap-frozen and stored at −80 °C until analysis.

2.3. Hepatotoxicity

As indicators of hepatotoxicity, aspartate aminotransferase (AST) and alanine transaminase (ALT) activities in the serum were examined using spectrophotometry, SPOTCHEM EZ SP-4460 (ARKRAY, Kyoto, Japan).

2.4. Real-Time RT-PCR Analysis

Initially, 5% (w/v) homogenates of the proximal duodenum or liver with Lysis Buffer of QuickGene RNA tissue kit (Kurabo, Osaka, Japan) were produced. Subsequently, total RNA was extracted from the 5% (w/v) homogenate using the QuickGene RNA tissue kit and QuickGene (Kurabo) in accordance with the manufacturer's protocols. The quantity and purity of RNA were measured using a NanoDrop device (Thermo Fisher Scientific, Waltham, MA, USA). RNA was subjected to a PrimeScript RT reagent kit (Takara Bio, Shiga, Japan) to generate cDNA. Real-time PCR was carried out using SYBR Premix Ex Taq II (Takara Bio) on a Thermal Cycler Dice Real Time System (Takara Bio) in accordance with the manufacturer's instructions. The sequences of specific primers for mouse genes are listed in Table 1.

Table 1. Sequences of mouse primers (5′–3′).

Gene Names	Forward	Reverse	Product Size (bp)
Slc46a1 (*HCP1*)	ATCTACCCGGCCACTCTGAA	GGAACTCCGGGTGTCGATTA	121
Hmox1	ACCTTCCCGAACATCGACAG	GGAAGGCGGTCTTAGCCTCT	121
Slc11a2 (*Dmt1*)	GGCTTTCTTATGAGCATTGCCTA	GGAGCACCCAGAGCAGCTTA	97
Cybrd1	GCAGCGGGCTCGAGTTTA	TTCCAGGTCCATGGCAGTCT	103
Slc40a1 (*Fpn1*)	CTGTCGGCCAGATTATGACA	GAGCAGGGGTCTTCTGGTAA	126
Heph	TTGTCTCATGAAGAACATTTACAGCAC	CATATGGCAATCAAAGCAGAAGA	161
Hamp	GGCAGACATTGCGATACCAA	TGGCTCTAGGCTATGTTTTGCA	128
Actin, beta (*Actb*)	CCTAAGGCCAACCGTGAAAA	AGGCATACAGGGACAGCACA	100

2.5. Cd and Fe Concentrations

For metal analysis, serum (20 μL) and liver homogenates (100 μL of 5% (w/v) homogenates) were digested by wet ashing procedure using nitric acid for metal analysis (60%) (KANTO CHEMICAL, Tokyo, Japan) and hydrogen peroxide (30%) for atomic absorption spectrometry (KANTO CHEMICAL). Serum and liver homogenates added with

2 mL nitric acid were heated using a dry block heater with a Teflon (polytetrafluoroethylene) ball on top of a glass test tube at 80 °C for 30 min, 100 °C for 1 h, 110 °C for 1 h, 120 °C for 2 h and then 140 °C until volatilized without the Teflon ball. Next, 0.5 mL nitric acid and 2 mL hydrogen peroxide were added, followed by heating with a Teflon ball at 80 °C for 1 h, 90 °C for 1 h, 100 °C for 1 h, 110 °C for 1 h, 120 °C for 1 h and then 140 °C until volatilized with 2.5 mL mixture of nitric acid and hydrogen peroxide without the Teflon ball. After sample digestion, Cd and Fe concentrations were measured using the graphite furnace atomic absorption instrument, 280Z AA (Agilent Technologies, Santa Clara, CA, USA). The injection volume was 5 μL. The heating program for Cd was as follows: 85 °C for 5 s, 95 °C for 40 s, 120 °C for 10 s, 250 °C for 6 s with 0.3 L/min argon gas flow, 250 °C for 2 s, 1800 °C for 2.8 s without argon gas and 1800 °C for 2 s with 0.3 L/min argon gas flow. The heating program for Fe was as follows: 85 °C for 5 s, 95 °C for 40 s, 120 °C for 10 s, 700 °C for 6 s with 0.3 L/min argon gas flow, 700 °C for 2 s, 2300 °C for 2.8 s without argon gas and 2300 °C for 2 s with 0.3 L/min argon gas flow. The measurement wavelengths for Cd and Fe were 228.8 nm and 248.3 nm, respectively.

2.6. Serum UIBC and TIBC Quantification

The unsaturated iron-binding capacity (UIBC) in the serum was measured using a Microassay UIBC quantification kit (Bathophenanthroline method) (Metallogenics, Chiba, Japan) in accordance with the manufacturer's instructions. The total iron-binding capacity (TIBC) in the serum was calculated via the summation of the serum iron concentration and serum UIBC.

2.7. Statistical Analysis

All values are expressed as the mean ± standard deviation (S.D.). Statistical significance was assessed using analysis of variance and Bonferroni's multiple *t*-test. *p*-values of less than 0.05 were statistically significant.

3. Results

3.1. Effect of Long-Term Exposure to Cd on Accumulation of Iron and Cd in the Liver

As shown in Figure 1a, the total iron concentration in the liver of the mice exposed to Cd for 12 months was significantly decreased compared to that in the control. Additionally, the iron concentration in the liver was reduced by Cd exposure until the 21st month. The Cd concentration in the liver of the mice exposed to Cd for a long time was >300 ppm and slightly increased in an exposure-time-dependent manner (Figure 1b).

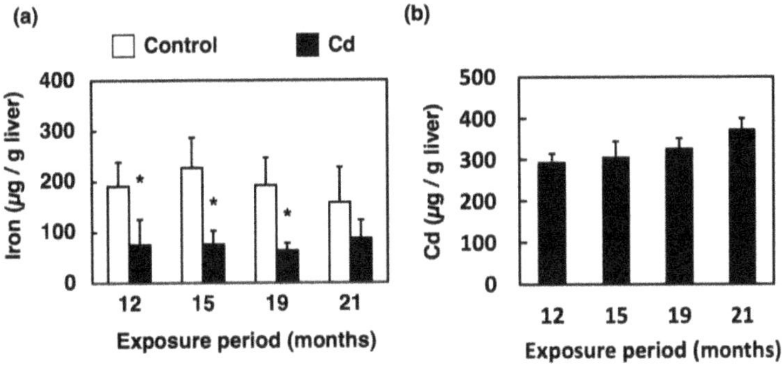

Figure 1. Concentration of iron and Cd in the liver of mice exposed to Cd for a long time. Concentrations of iron (**a**) and Cd (**b**) in the liver were determined after 12, 15, 19 and 21 months of exposure to Cd. Values are means ± S.D. (n = 4–6). * Significantly different from the corresponding control, $p < 0.05$. Cd in the liver of the control group was not detected.

3.2. Changes in Body Weight and Hepatotoxicity in Mice Exposed to Cd for a Long Time

The body weights of Cd-exposed mice were significantly lower than those of the control group (Figure 2a). AST activity was not altered by long-term exposure to Cd (Figure 2b). ALT activity was increased slightly by exposure to Cd for 12 and 15 months (Figure 2c).

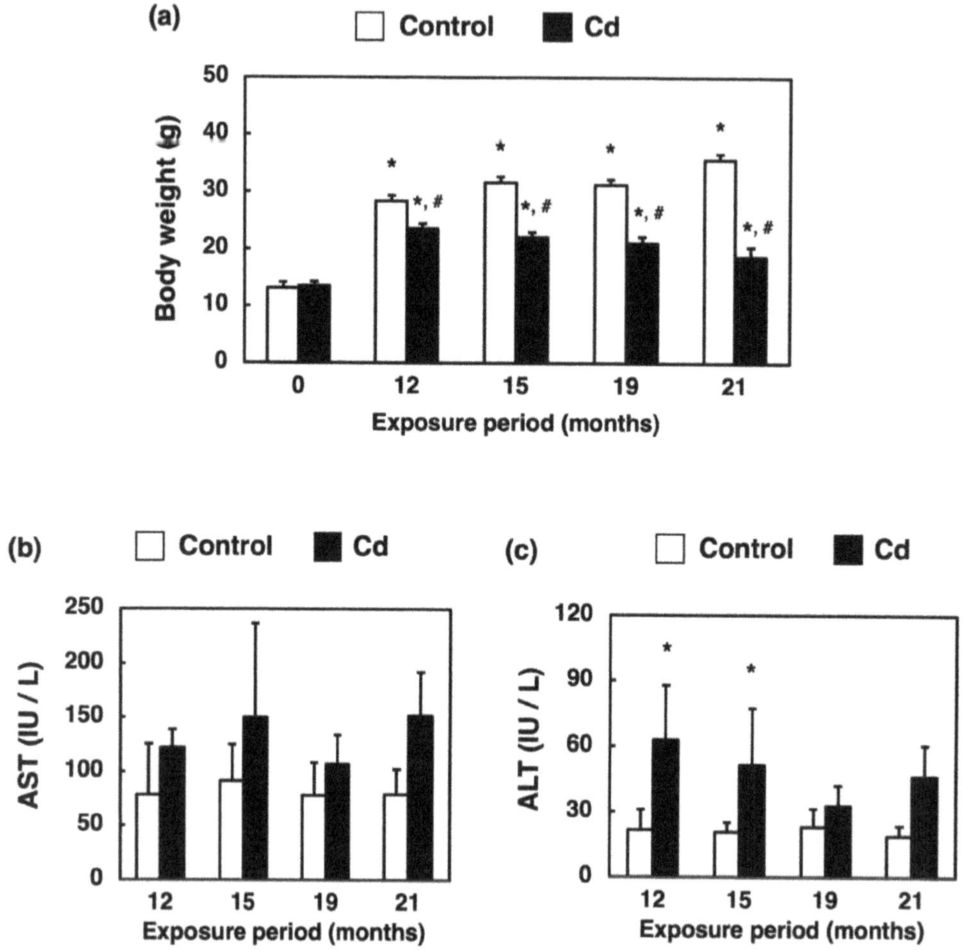

Figure 2. Changes in body weight and activities of AST and ALT in the serum of mice exposed to Cd for a long time. Changes in body weight (**a**) of mice after 12, 15, 19 and 21 months of exposure to Cd. Values are means ± S.D. (n = 5–7). * Significantly different from the corresponding 0-month-exposed group, $p < 0.05$. # Significantly different from corresponding control, $p < 0.05$. Activities of AST (**b**) and ALT (**c**) in the serum were determined after 12, 15, 19 and 21 months of exposure to Cd. Values are means ± S.D. (n = 4–7). * Significantly different from the corresponding control, $p < 0.05$.

3.3. Effect of Long-Term Exposure to Cd on Iron Concentration, UIBC and TIBC in the Serum

Iron in serum is bound to transferrin and transported to the whole body. In a healthy individual, only about one third of the total transferrin in the serum is bound to iron. The amount of iron that can bind to the remaining two thirds of transferrin is called UIBC. The sum of serum iron (iron bound to transferrin) and UIBC, i.e., the amount of iron that can bind to the total transferrin in the serum, is called TIBC [23]. The iron concentration in the serum was not changed by long-term exposure to Cd (Figure 3a). UIBC was significantly increased by long-term exposure to Cd except for the 15-month exposure (Figure 3b). TIBC was significantly increased by 19 and 21 months of exposure to Cd (Figure 3c).

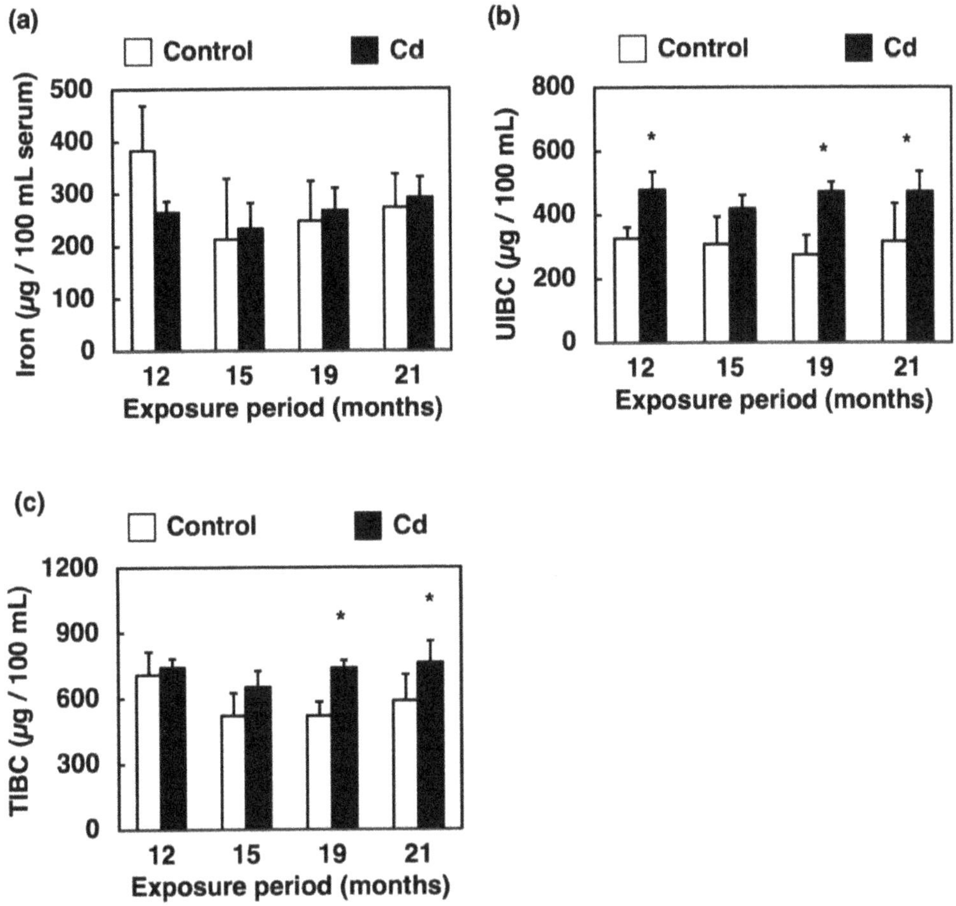

Figure 3. Iron concentration, UIBC and TIBC in the serum of mice exposed to Cd for a long time. Iron concentration (**a**), UIBC (**b**) and TIBC (**c**) in the serum were determined after 12, 15, 19 and 21 months of exposure to Cd. Values are means ± S.D. (n = 4–7). * Significantly different from the corresponding control, $p < 0.05$.

3.4. Effect of Long-Term Exposure to Cd on the Expression of Genes Involved in Absorption of Heme Iron in the Proximal Duodenum

To elucidate the cause of loss of stored iron in the liver induced by Cd, we measured the expression of heme iron influx transporters in the proximal duodenum. The mRNA level of *HCP1*, which encodes a heme iron influx transporter in the Cd-exposed group, was approximately half of that in the control, which was significantly decreased at 12 and 19 months of exposure to Cd (Figure 4a). The *Hmox1* is the gene encoding HO-1. The *Hmox1* mRNA level was drastically increased by Cd (Figure 4b).

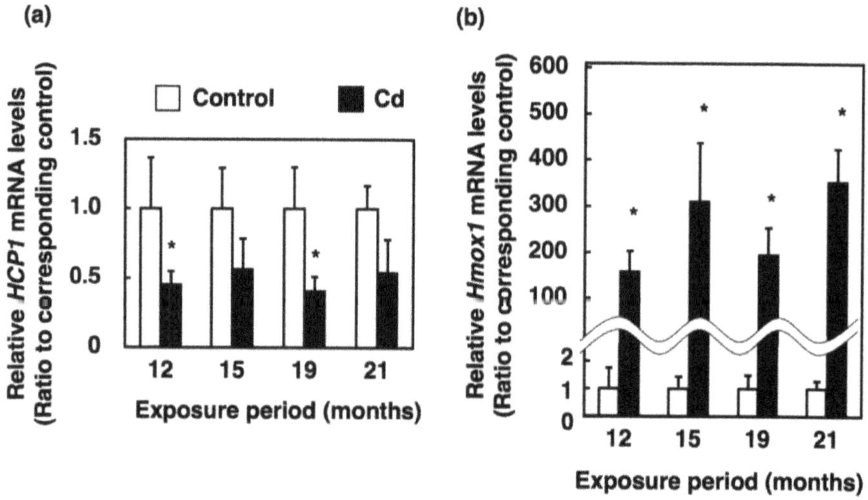

Figure 4. Expression of genes involved in heme iron absorption in the proximal duodenum of mice exposed to Cd for a long time. mRNA levels of *HCP1* (**a**) and *Hmox1* (**b**) in the proximal duodenum were determined after 12, 15, 19 and 21 months of exposure to Cd. mRNA levels were normalized to *Actb* mRNA expression. Values are means ± S.D. (n = 4–7). * Significantly different from the corresponding control, $p < 0.05$.

3.5. Effect of Long-Term Exposure to Cd on the Expression of Genes Involved in Absorption of Non-Heme Iron in the Proximal Duodenum

To elucidate the cause of loss of stored iron in the liver induced by Cd, we measured the expression of non-heme iron influx transporters in the proximal duodenum. The *Cybrd1* mRNA level was dramatically suppressed by long-term exposure to Cd (Figure 5a). However, the mRNA level of *Dmt1*, which encodes a non-heme iron influx transporter, was not affected by Cd (Figure 5b).

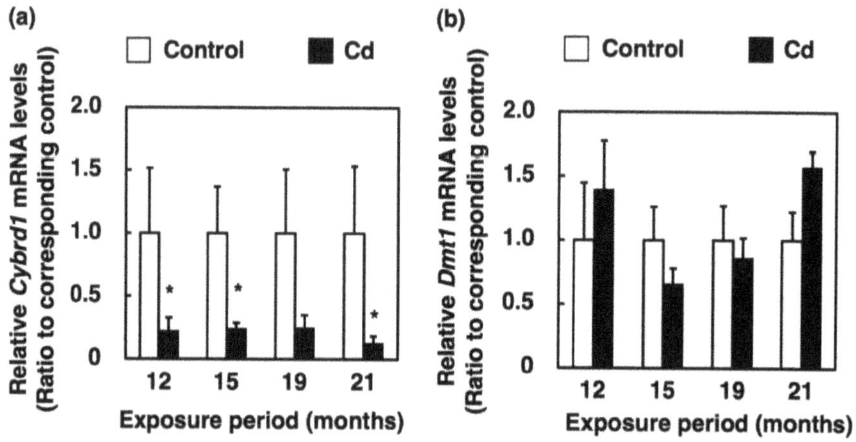

Figure 5. Expression of genes involved in non-heme iron absorption in the proximal duodenum of mice exposed to Cd for a long time. mRNA levels of *Cybrd1* (**a**) and *Dmt1* (**b**) in the proximal duodenum were determined after 12, 15, 19 and 21 months of exposure to Cd. mRNA levels were normalized to *Actb* mRNA expression. Values are means ± S.D. (n = 4–6). * Significantly different from the corresponding control, $p < 0.05$.

3.6. Effect of Long-Term Exposure to Cd on the Expression of Genes Involved in the Efflux of Iron from the Proximal Duodenum Enterocytes into Blood Vessels

Fpn1 and *Heph* mRNA levels were generally not altered by long-term exposure to Cd, but they were significantly reduced at some time points (Figure 6a,b).

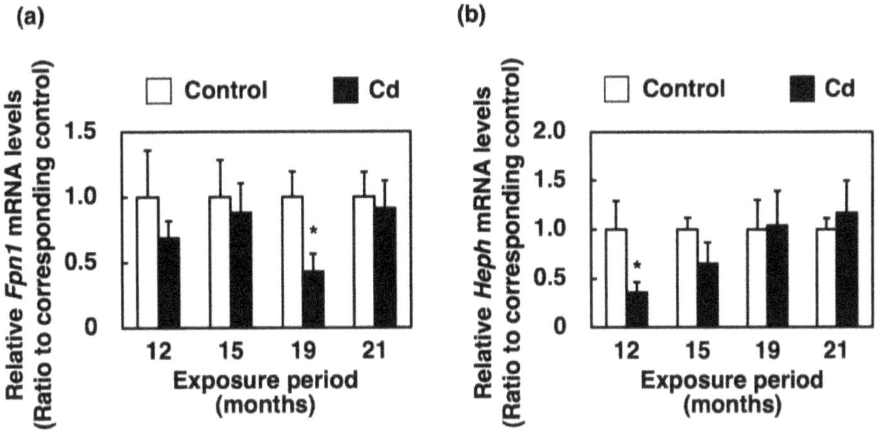

Figure 6. Expression of genes involved in iron efflux from the proximal duodenum enterocytes into blood vessels in mice exposed to Cd for a long time. mRNA levels of *Fpn1* (**a**) and *Heph* (**b**) in the proximal duodenum were determined after 12, 15, 19 and 21 months of exposure to Cd. mRNA levels were normalized to *Actb* mRNA expression. Values are means ± S.D. (n = 3–7). * Significantly different from the corresponding control, $p < 0.05$.

3.7. Effect of Long-Term Exposure to Cd on Gene Expression of Hamp and Fpn1 in the Liver

The mRNA level of Hamp, which restricts *Fpn1* expression, in the liver of mice was significantly decreased in all of the Cd exposure period groups (Figure 7a). The *Fpn1* mRNA level in the liver of mice exposed to Cd for a long time was not changed (Figure 7b).

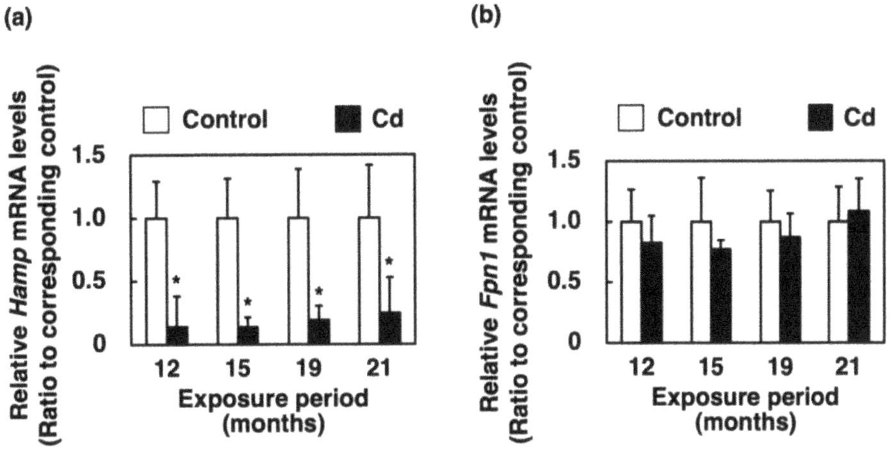

Figure 7. Gene expression of *Hamp* and *Fpn1* in the liver of mice exposed to Cd for a long time. mRNA levels of *Hamp* (**a**) and *Fpn1* (**b**) in the liver were determined after 12, 15 19 and 21 months of exposure to Cd. The mRNA level was normalized to *Actb* mRNA expression. Values are means ± S.D. (n = 3–7). * Significantly different from the corresponding control, $p < 0.05$.

4. Discussion

As shown in Figure 1b, hepatic Cd concentrations increased slightly, but only in a period-dependent manner. Thijssen et al. [24] reported that in mice exposed to 100 mg Cd/L drinking water for 23 weeks, renal Cd concentrations increased in a period-independent manner while hepatic Cd concentration peaked at 16 weeks of Cd exposure and began to decrease at 23 weeks. Although Cd accumulation in the liver increases in the early phase of exposure, Cd concentration in the liver eventually becomes stabilized due to the transportation of Cd as Cd-metallothionein (CdMT) from the liver to the kidney. Therefore, the present study suggests that hepatic Cd is maintained at the same concentration during long-term Cd exposure. In this study, significant increases in ALT were observed in the Cd 12- and 15-month exposure groups, but the mean values were about 60 IU/L (Figure 2c). These were not high enough to determine the appearance of hepatotoxicity. In addition, no variation in AST values due to Cd exposure was observed (Figure 2b). From these results, it can be concluded that no noteworthy hepatotoxicity appeared throughout the Cd exposure period. Therefore, long-term exposure to Cd showed little apparent hepatotoxicity.

Our previous study showed that the serum iron concentration and TIBC of mice were significantly decreased at 24 h after a single administration of Cd [22]. In the present study, on the other hand, the serum iron levels in mice exposed to Cd for a long time remained at the control levels (Figure 3a), which did not induce an iron deficiency. However, iron accumulation in the liver was markedly decreased (Figure 1a) and UIBC was significantly elevated after 12 months of exposure to Cd (Figure 3b). Moreover, Figure 3c indicates a significant increase in the TIBC of mice exposed to Cd for 19 and 21 months. As shown in Figure 3a, there was no "apparent" change in serum iron levels by the long-term exposure to Cd in this experiment. However, the iron concentration in the liver was significantly decreased (Figure 1a) and the UIBC and TIBC levels were significantly increased (Figure 3b,c) via long-term exposure to Cd. These findings imply that the biological response to ameliorate the iron deficiency state may occur by increasing the amount of transferrin that can bind to iron in the serum. Therefore, these results suggested that the mice exposed to Cd were severely iron-deficient. Excess iron absorbed is stored in the liver as ferritin and is released into the blood through *Fpn1* expressed in the liver when iron deficiency occurs [25]. *Fpn1* is highly expressed in duodenal enterocytes, liver Kupffer cells, splenic red pulp macrophages, periportal hepatocytes and the placental syncytiotrophoblast [25]. Subsequently, the Fe(II) excreted from the liver is oxidized to Fe(III) mainly by Cp, and then binds to transferrin (Figure 8). Therefore, long-term Cd exposure may consume iron in the liver while maintaining iron concentrations in the serum.

In the present study, the expression of *HCP1* and *Cybrd1* in the proximal duodenum where iron from food is absorbed, and the expression of *Hamp* in the liver, were significantly suppressed in female C57BL/6J mice orally administered with 300 ppm Cd for more than 1 year (Figures 4a, 5a and 7a). The gene expression of HCP1, a heme iron transporter, in the Cd exposed group was approximately half of the control group and significantly decreased after 12 and 19 months of exposure (Figure 4a). On the other hand, the gene expression of *Dmt1*, which is a non-heme iron transporter, did not change after exposure to Cd, compared to the control (Figure 5b). However, *Cybrd1* gene expression was significantly reduced throughout the periods of exposure to Cd (Figure 5a). *Cybrd1* is responsible for the reduction of Fe(III) to Fe(II) and is required for non-heme iron to be taken up into cells through *Dmt1*. These results suggest, therefore, that the uptake of both heme- and non-heme iron into enterocytes was diminished by Cd. It is well known that Cd readily induces HO-1 expression [26]. In the present study, long-term exposure to Cd also markedly elevated *Hmox1* expression in the enterocytes (Figure 4b). Since the expression level of *HCP1* was reduced by Cd, heme iron uptake into enterocytes was expected to be reduced, but the uptake of heme iron is expected to be rapidly degraded to Fe(II) by HO-1, whose expression was increased in the enterocytes. Iron taken up into enterocytes is released into blood via Fpn1. *Heph* and Cp are enzymes that oxidize Fe(II) released into the vascular side to Fe(III). It is known that *Heph* is a homolog of Cp with 50% homology at the amino acid level [19]. As shown in the review by Helman et al. [27], Cp particularly works in the liver and certain cell types in the brain and pancreas [28–30], whereas the major oxidative enzyme in the small intestine is *Heph* [31]. Therefore, we investigated the effect of long-term exposure to Cd on *Heph* in the duodenum of mice. Cd did not induce a change in the expression of *Fpn1* and *Heph* genes, which are involved in the release of iron from the duodenum into blood vessels (Figure 6a,b). Furthermore, gene expression of the peptide hormone Hamp, which degrades Fpn1, in the liver remained significantly reduced throughout the Cd exposure period (Figure 7a). Thus, it is likely that the reduced expression of *HCP1* and *Cybrd1* in the Cd-exposed group reduced the amount of iron taken up by intestinal cells, but that the small amount of iron taken up was released into the bloodstream as normally as in the control group. Moreover, the mRNA level of *Fpn1* in the liver was not altered by Cd exposure (Figure 7b). *Hamp* produced in the liver is also involved in the degradation of *Fpn1* in the liver. It is reasonable that the *Fpn1* mRNA level in the liver was not altered by Cd exposure since *Hamp* production was significantly suppressed by Cd exposure (Figure 7a). However, the iron concentration in the liver of Cd-exposed mice was about half of the control. Therefore, it was indicated that Cd-exposed mice may not have been able to accumulate as much iron in the liver as the control group because iron absorption was inhibited, and iron was released from the liver into the blood.

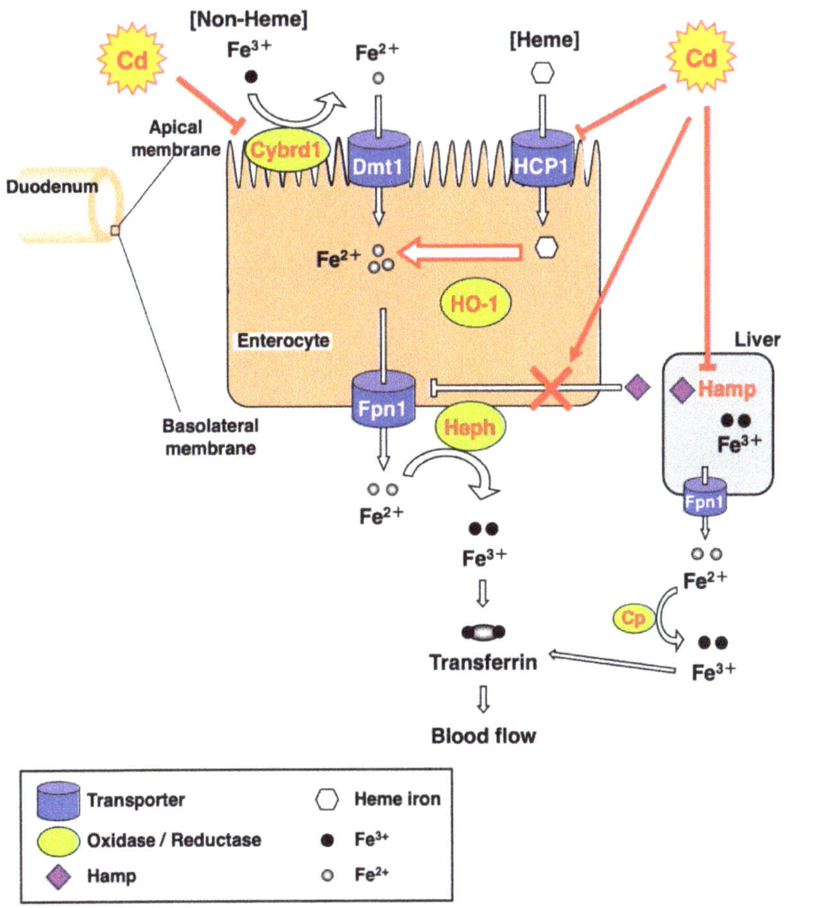

Figure 8. Proposed scheme which rationalizes the long-term exposure of mice to Cd on the metabolism of Fe.

Ohta and Ohba [32] reported that the *Dmt1* mRNA level in the duodenum of rats was significantly increased by oral exposure to Cd for 5 weeks. Similarly, orally ingested Cd increased the *Dmt1* mRNA level in the duodenum of pregnant rats [33]. However, in our previous study, it was reported that the expression of *HCP1*, *Dmt1*, *Cybrd1*, *Fpn1* and *Heph* in the duodenum of mice was significantly reduced at 24 h after a single oral administration of Cd [22]. Moreover, the gene expression of *Hamp* in the liver of mice in that study was significantly reduced at 24 h after exposure to Cd. Short-term Cd exposure sensitively reduced the expression of almost all iron-transport-related factors. However, long-term Cd exposure resulted in an expression comparable to control levels except for *HCP1* and *Cybrd1*, due to homeostasis that maintains normal in vivo levels of iron. Thus, it is suggested that *HCP1* and *Cybrd1* may be key factors in Cd-induced iron deficiency.

In conclusion, the present study found that long-term exposure to Cd decreases iron accumulation in the liver, and suppresses the expression of *HCP1* and *Cybrd1*, which are associated with iron absorption into duodenal epithelial cells, indicating that iron absorption may be inhibited (Figure 8). Moreover, these results suggest that iron stored in the liver is released into serum, leading to a potential iron deficiency and suggesting the need for aggressive iron supplementation to counteract the chronic toxicity of Cd in mammals. Anemia is a health problem of global concern. Our findings suggest that long-term exposure to Cd may exacerbate iron deficiency. Therefore, it is proposed that populations at high risk for anemia have more of a reduced Cd intake from their diet and luxury items such as cigarettes, and take iron supplementation in order to prevent anemia and various health problems caused by anemia.

Author Contributions: Conceptualization, Y.F. and M.S.; methodology, M.T., J.-Y.L., Y.F. and M.S.; investigation, M.T.; data curation, M.T., J.-Y.L., Y.F. and M.S.; writing—original draft preparation, M.T.; writing—review and editing, M.T., J.-Y.L., Y.F. and M.S.; visualization, M.T.; supervision, M.S.; project administration, M.T., J.-Y.L. and M.S. All authors have read and agreed to the published version of the manuscript.

Funding: This study was partly supported by a research grant from the Institute of Pharmaceutical Life Sciences, Aichi Gakuin University.

Institutional Review Board Statement: The animal study protocol was approved by the Animal Care and Use Committee of the School of Pharmacy, Aichi Gakuin University (protocol code, 14-024; date of approval, 1 April 2014).

Informed Consent Statement: Not applicable.

Data Availability Statement: All data is contained within this manuscript.

Conflicts of Interest: The authors declare no conflict of interest.

References

1. Jarup, L.; Berglund, M.; Elinder, C.G.; Nordberg, G.; Vahter, M. Health effects of cadmium exposure—A review of the literature and a risk estimate. *Scand. J. Work Environ. Health* **1998**, *24* (Suppl. S1), 1–51. [PubMed]
2. Satoh, M.; Koyama, H.; Kaji, T.; Kito, H.; Tohyama, C. Perspectives on cadmium toxicity research. *Tohoku J. Exp. Med.* **2002**, *196*, 23–32. [CrossRef]
3. Zalups, R.K.; Ahmad, S. Molecular handling of cadmium in transporting epithelia. *Toxicol. Appl. Pharmacol.* **2003**, *186*, 163–188. [CrossRef] [PubMed]
4. Lee, B.K.; Kim, Y. Iron deficiency is associated with increased levels of blood cadmium in the Korean general population: Analysis of 2008–2009 Korean National Health and Nutrition Examination Survey data. *Environ. Res.* **2012**, *112*, 155–163. [CrossRef] [PubMed]
5. Silver, M.K.; Lozoff, B.; Meeker, J.D. Blood cadmium is elevated in iron deficient U.S. children: A cross-sectional study. *Environ. Health* **2013**, *12*, 117. [CrossRef]
6. Suh, Y.J.; Lee, J.E.; Lee, D.H.; Yi, H.G.; Lee, M.H.; Kim, C.S.; Nah, J.W.; Kim, S.K. Prevalence and Relationships of Iron Deficiency Anemia with Blood Cadmium and Vitamin D Levels in Korean Women. *J Korean Med. Sci.* **2016**, *31*, 25–32. [CrossRef]
7. Turgut, S.; Polat, A.; Inan, M.; Turgut, G.; Emmungil, G.; Bican, M.; Karakus, T.Y.; Genc, O. Interaction between anemia and blood levels of iron, zinc, copper, cadmium and lead in children. *Indian J. Pediatr.* **2007**, *74*, 827–830. [CrossRef]

8. Djukić-Cosić, D.; Curcić Jovanović, M.; Plamenac Bulat, Z.; Ninković, M.; Malicević, Z.; Matović, V. Relation between lipid peroxidation and iron concentration in mouse liver after acute and subacute cadmium intoxication. *J. Trace Elem. Med. Biol.* **2008**, *22*, 66–72. [CrossRef]
9. Sugawara, N. Effect of dietary cadmium on serum and liver lipids in mice in relation to interaction of essential metals, zinc, copper and iron. *J. Toxicol. Sci.* **1984**, *9*, 29–36. [CrossRef]
10. Timofeeva, S.N.; Kadikov, I.R.; Korchemkin, A.A.; Vafin, I.F. The effects of zinc and magnesium ingestion on essential trace-element accumulation in the organs of white rats upon cadmium exposure. *BIO Web Conf.* **2020**, *27*, 00110. [CrossRef]
11. Whanger, P.D. Cadmium effects in rats on tissue iron, selenium, and blood pressure; blood and hair cadmium in some oregon residents. *Environ. Health Perspect.* **1979**, *28*, 115–121. [CrossRef]
12. Doyle, J.J.; Pfander, W.H. Interactions of cadmium with copper, iron, zinc, and manganese in ovine tissues. *J. Nutr.* **1975**, *105*, 599–606. [CrossRef] [PubMed]
13. Rama, R.; Planas, J. Dietary cadmium effect on iron metabolism in chickens. *Biol. Trace Elem. Res.* **1981**, *3*, 169–183. [CrossRef] [PubMed]
14. World Health Organization. Accelerating Anaemia Reduction: A Comprehensive Framework for Action. 2023. Available online: https://apps.who.int/iris/handle/10665/367661 (accessed on 23 June 2023).
15. Shayeghi, M.; Latunde-Dada, G.O.; Oakhill, J.S.; Laftah, A.H.; Takeuchi, K.; Halliday, N.; Khan, Y.; Warley, A.; McCann, F.E.; Hider, R.C.; et al. Identification of an intestinal heme transporter. *Cell* **2005**, *122*, 789–801. [CrossRef]
16. Gunshin, H.; Mackenzie, B.; Berger, U.V.; Gunshin, Y.; Romero, M.F.; Boron, W.F.; Nussberger, S.; Gollan, J.L.; Hediger, M.A. Cloning and characterization of a mammalian proton-coupled metal-ion transporter. *Nature* **1997**, *388*, 482–488. [CrossRef]
17. McKie, A.T.; Barrow, D.; Latunde-Dada, G.O.; Rolfs, A.; Sager, G.; Mudaly, E.; Mudaly, M.; Richardson, C.; Barlow, D.; Bomford, A.; et al. An iron-regulated ferric reductase associated with the absorption of dietary iron. *Science* **2001**, *291*, 1755–1759. [CrossRef]
18. Donovan, A.; Brownlie, A.; Zhou, Y.; Shepard, J.; Pratt, S.J.; Moynihan, J.; Paw, B.H.; Drejer, A.; Barut, B.; Zapata, A.; et al. Positional cloning of zebrafish ferroportin1 identifies a conserved vertebrate iron exporter. *Nature* **2000**, *403*, 776–781. [CrossRef]
19. Vulpe, C.D.; Kuo, Y.M.; Murphy, T.L.; Cowley, L.; Askwith, C.; Libina, N.; Gitschier, J.; Anderson, G.J. Hephaestin, a ceruloplasmin homologue implicated in intestinal iron transport, is defective in the sla mouse. *Nat. Genet.* **1999**, *21*, 195–199. [CrossRef] [PubMed]
20. Nemeth, E.; Ganz, T. Regulation of iron metabolism by hepcidin. *Annu. Rev. Nutr.* **2006**, *26*, 323–342. [CrossRef]
21. Nemeth, E.; Tuttle, M.S.; Powelson, J.; Vaughn, M.B.; Donovan, A.; Ward, D.M.; Ganz, T.; Kaplan, J. Hepcidin regulates cellular iron efflux by binding to ferroportin and inducing its internalization. *Science* **2004**, *306*, 2090–2093. [CrossRef]
22. Fujiwara, Y.; Lee, J.Y.; Banno, H.; Imai, S.; Tokumoto, M.; Hasegawa, T.; Seko, Y.; Nagase, H.; Satoh, M. Cadmium induces iron deficiency anemia through the suppression of iron transport in the duodenum. *Toxicol. Lett.* **2020**, *332*, 130–139. [CrossRef]
23. Auerbach, M.; Adamson, J.W. How we diagnose and treat iron deficiency anemia. *Am. J. Hematol.* **2016**, *91*, 31–38. [CrossRef]
24. Thijssen, S.; Maringwa, J.; Faes, C.; Lambrichts, I.; Van Kerkhove, E. Chronic exposure of mice to environmentally relevant, low doses of cadmium leads to early renal damage, not predicted by blood or urine cadmium levels. *Toxicology* **2007**, *229*, 145–156. [CrossRef]
25. Drakesmith, H.; Nemeth, E.; Ganz, T. Ironing out Ferroportin. *Cell Metab.* **2015**, *22*, 777–787. [CrossRef] [PubMed]
26. Taketani, S.; Kohno, H.; Yoshinaga, T.; Tokunaga, R. The human 32-kDa stress protein induced by exposure to arsenite and cadmium ions is heme oxygenase. *FEBS Lett.* **1989**, *245*, 173–176. [CrossRef]
27. Helman, S.L.; Zhou, J.; Fuqua, B.K.; Lu, Y.; Collins, J.F.; Chen, H.; Vulpe, C.D.; Anderson, G.J.; Frazer, D.M. The biology of mammalian multi-copper ferroxidases. *Biometals* **2023**, *36*, 263–281. [CrossRef] [PubMed]
28. Hellman, N.E.; Gitlin, J.D. Ceruloplasmin metabolism and function. *Annu. Rev. Nutr.* **2002**, *22*, 439–458. [CrossRef] [PubMed]
29. Jeong, S.Y.; David, S. Glycosylphosphatidylinositol-anchored ceruloplasmin is required for iron efflux from cells in the central nervous system. *J. Biol. Chem.* **2003**, *278*, 27144–27148. [CrossRef] [PubMed]
30. Linder, M.C. Ceruloplasmin and other copper binding components of blood plasma and their functions: An update. *Metallomics* **2016**, *8*, 887–905. [CrossRef]
31. Fuqua, B.K.; Lu, Y.; Darshan, D.; Frazer, D.M.; Wilkins, S.J.; Wolkow, N.; Bell, A.G.; Hsu, J.; Yu, C.C.; Chen, H.; et al. The multicopper ferroxidase hephaestin enhances intestinal iron absorption in mice. *PLoS ONE* **2014**, *9*, e98792. [CrossRef]
32. Ohta, H.; Ohba, K. Involvement of metal transporters in the intestinal uptake of cadmium. *J. Toxicol. Sci.* **2020**, *45*, 539–548. [CrossRef] [PubMed]
33. Leazer, T.M.; Liu, Y.; Klaassen, C.D. Cadmium absorption and its relationship to divalent metal transporter-1 in the pregnant rat. *Toxicol. Appl. Pharmacol.* **2002**, *185*, 18–24. [CrossRef] [PubMed]

toxics

MDPI

Article

Toxic Metal and Essential Element Concentrations in the Blood and Tissues of Pancreatic Ductal Adenocarcinoma Patients

Giovanni Forte [1,†], Andrea Pisano [2,†], Beatrice Bocca [1,*], Grazia Fenu [2], Cristiano Farace [2,3], Federica Etzi [2], Teresa Perra [4], Angela Sabalic [2], Alberto Porcu [4] and Roberto Madeddu [2,3]

1 Department of Environment and Health, Italian National Institute of Health, 00161 Rome, Italy; giovanni.forte@iss.it
2 Department of Biomedical Science—Histology, University of Sassari, 07100 Sassari, Italy; pisanoandrea87@gmail.com (A.P.); grazia.fenu94@gmail.com (G.F.); cfarace@uniss.it (C.F.); f.etzi@studenti.uniss.it (F.E.); angelasabalic08@gmail.com (A.S.); rmadeddu@uniss.it (R.M.)
3 National Institute of Biostructures and Biosystems, Interuniversity Consortium INBB, 00136 Rome, Italy
4 Department of Medicine, Surgery and Pharmacy, Unit of General Surgery, University of Sassari, 07100 Sassari, Italy; teresaperra1992@gmail.com (T.P.); alberto@uniss.it (A.P.)
* Correspondence: beatrice.bocca@iss.it
† These authors contributed equally.

Abstract: Background: Pancreatic ductal adenocarcinoma (PDAC) is a highly aggressive lethal neoplasm, and it has an average 5-year survival rate of less than 10%. Although the factors that influence PDAC development remain unclear, exposure to toxic metals or the imbalance in essential elements may have a role in PDAC-associated metabolic pathways. Methods: This study determined the concentrations of Cd, Cr, Cu, Fe, Mn, Ni, Pb, Se and Zn in whole blood, cancer and non-cancer tissues of patients affected by PDAC, and compared them with levels in healthy controls using inductively coupled plasma mass spectrometry. Results: Results of the whole blood showed significantly higher levels of Cr, Cu and Cu/Zn ratio in PDAC patients compared to the controls. In addition, the concentrations of Cu, Se, Fe and Zn significantly increased in cancer tissue compared to the healthy counterparts. Conclusions: This study revealed evidence of altered metal levels in the blood and pancreatic tissues of PDAC patients with respect to healthy controls. These changes may contribute to multiple mechanisms involved in metal-induced carcinogenesis, including oxidative stress, DNA damage, genetic alteration, decreased antioxidant barriers and inflammatory responses. Thus, the analysis of metals can be used in the diagnosis and monitoring of PDAC neoplasms.

Keywords: toxic metals; essential elements; pancreatic ductal adenocarcinoma; whole blood; pancreatic tissue

1. Introduction

The pancreas produces some important hormones including insulin and glucagon able to maintain healthy blood sugar levels. Moreover, the pancreas secretes various enzymes, such as trypsin, lipase and amylase that contribute to the proper digestion and absorption of fats, proteins and carbohydrates. About 70% of pancreatic cancers develop in the head of the organ, and most of them originate in ducts that carry digestive enzymes. Having an insufficient amount of pancreatic enzymes is very common among people with pancreatic cancer. Pancreatic ductal adenocarcinoma (PDAC) is the most common histologic type of pancreatic cancer and is one of the most aggressive malignant neoplasms with poor outcomes, and subjects most at risk are those in the age range between 50 and 80 years [1]. According to the WHO, in 2019, deaths in Italy due to PADC amounted to ca. 13,000, the age standardized death rate was 7.9%, the death rate for 100,000 subjects was 21.6% and the percentage of cause-specific deaths out of total deaths was 2% [2]. The onset and increase in PADC numbers have a multifactorial origin. In 10% of cases, there is a genetic basis with mutations of different genes, while in remaining cases, cigarette smoke, alcohol, incorrect diet, diabetes, obesity and environmental pollutants have been associated with an increased risk of PDAC neoplasms [1].

Among environmental pollutants, exposure to toxic metals such as Cd, Cr, Ni and Pb may have an impact in PDAC neoplasms. In particular, a study explored the effects of Cd on three different lines of evidence: a case control study on patients with pancreatic cancer; an animal study employing Cd-treated Wistar rats; and an in vitro study using Cd-exposed pancreatic cells such as hTERT-HPNE (human pancreatic Nestin-expressing cells) and AsPC-1 (human pancreatic tumor cell line). The results indicated Cd accumulation in human and rat pancreatic tissues, disturbances in the intrinsic pathway of apoptotic activity and oxidative stress elevation in pancreatic cells [3]. Moreover, Cd was a competitor for pancreatic Zn due to its similar physico-chemical properties. The same authors also reported that Cd acted in a mitogenic manner to pancreatic cells and was able to cause their transdifferentiation and increase both the synthesis of pancreatic DNA and oncogene activation [4].

Regarding Cr and Ni, the literature supports the concept that these metals may interfere with key steps in repairing DNA damage and they may also stimulate cell proliferation via the activation of early response genes or interference with genes down-regulating cell growth and senescence [5,6]. A study reported significantly higher levels of Ni in pancreatic cancerous tissues with respect to the controls [7].

Many microRNAs (miRNAs) have an essential role in the regulation of oncogenes or tumor suppressor genes in cell signaling pathways, and the dysregulation of miRNAs is a hallmark of cancer. In vitro studies found high expression levels of two miRNAs, miR-221 and miR-155, in pancreatic tumor cell lines exposed to increasing Ni concentrations, while the expression level of miR-126 was significantly decreased [7].

Authors investigated the relationship between toenail concentrations of trace elements and occupational history in pancreatic cancer patients. [8]. They observed that patients exposed to aromatic hydrocarbon solvents presented high levels of metals such as Cd, Mn, Pb and Fe, whilst patients exposed to pesticides showed increased Cd and Mn values, and patients exposed to formaldehyde and chlorinated hydrocarbon solvents reported higher levels of Pb [8].

Moreover, despite metals such as Cu, Fe, Mn, Se and Zn being essential for normal biological functioning [9,10], the alteration in their levels may increase cellular oxidative stress, posing the basis for the development of diseases. Regarding Cu and Se, both micronutrients are involved in many biochemical processes, including cellular respiration, cellular utilization of oxygen, DNA and RNA production, maintenance of cell membrane integrity and sequestration of free radicals. Copper is important for functions involved in cell proliferation or angiogenesis, which are central to tumorigenesis and cancer development [11]. For Se, a higher intake of Se might reduce the risk of pancreatic cancer [12]. Iron is an essential element for life due to its role in synthesizing oxygen transport proteins such as hemoglobin, myoglobin and other Fe-containing proteins. It is widely acknowledged that an excess accumulation of Fe can be harmful by increasing the production of reactive oxygen species (ROS), which ultimately cause DNA damage and genetic alterations, leading to the basis of pancreatic cancer growth [13]. Other authors reported on the role of Mn in carcinogenesis via the generation of oxygen radicals, suppression of apoptosis and binding competition between chromatin and metal ions in molecules [14]. Zinc is involved in a multitude of processes within the pancreas, including glucagon secretion, digestive enzyme activity and insulin packaging, secretion and signaling. A dysregulation of Zn metabolism within the pancreas impaired a multitude of key processes, including glycemic control, pancreatic cancer and chronic pancreatitis [15].

Literature data suggested that multiple Zn-regulated transporters (Zip) play a role in the intracellular concentrations of Zn during the development and progression of pancreatic cancer. There is evidence of the overexpression of Zrt-Irt-like protein 4 (Zip4) in pancreatic cancers allowing for Zn accumulation [16], whilst other studies reported a downregulation of Zip8 and Zip3 transporters in adenocarcinoma tissue sections accompanied by the loss of intracellular Zn during well-differentiated and progressing pancreatic malignancy [17].

Building upon the aforementioned background, increased levels of toxic metals in the body and variations in the concentrations of essential elements might be associated with the development of pancreatic cancer. To this end, the present study determined Cd, Cr, Cu, Fe, Mn, Ni, Pb, Se and Zn in the whole blood and cancer tissues of patients affected by PDAC with respect to the levels of these elements in healthy controls. The final objective was to investigate the use of metals for the diagnosis and monitoring of PDAC neoplasms.

2. Materials and Methods

2.1. Characteristics of the Subjects

Forty-six subjects affected by PDAC (mean age, 67.2 ± 9.4 years; 21 males and 25 females) and 20 healthy controls (HC; mean age, 60.5 ± 8.5 years; 11 females and 9 males) were enrolled in this study, and their details are reported in Table 1. Whole blood samples were collected in EDTA tubes for trace metals (BD Vacutainer, Franklin Lakes, NJ, USA) from all 46 PDAC patients, including 32 operated PDAC patients (OPPs) and 14 non-operated PDAC patients (NOPPs) that corresponded to cancer stage IV, and 20 HCs. Furthermore, PDAC tissues and adjacent non-cancer tissue samples were collected from the 32 OPPs. PDAC patients were also subdivided according to the severity of the disease as per the American Joint Committee on Cancer (AJCC) pancreatic cancer stages [18]. In particular:

- Stage 0 (carcinoma in situ): cancer cells are only present in the top layers of pancreatic duct cells and have not invaded deeper tissues;
- Stage Ia: cancer (max. 2 cm) is confined to the pancreas and has not spread beyond;
- Stage Ib: cancer (larger than 2 cm) is confined to the pancreas and has not spread beyond;
- Stage IIb: cancer has extended beyond the pancreas (from 1 to 3 regional lymph nodes) but has not invaded major blood vessels;
- Stage III: cancer has spread to nearby major blood vessels but has not spread to distant organs;
- Stage IV: cancer has spread to distant organs, such as the liver, lungs or other organs.

Table 1. Descriptive characteristics of PDAC patients and controls.

	PDAC	Controls
Subject (no.)		
Total population	46	20
Operated PDAC patients (OPPs)	32	
Non-operated PDAC patients (NOPPs)	14	
Mean age (years)	67.2 ± 9.4	60.5 ± 8.5
Sex (no.)		
Females	25	11
Males	21	9
Cancer stages (no.)		
Stage 0	5	
Stage Ia	6	
Stage Ib	5	
Stage IIb	10	
Stage III	6	
Stage IV	14	
Metastasis (no.)		
Yes	14	
No	32	

The numbers of patients within each stage are reported in Table 1.

All subjects were natives of Sardinia and had the same ethnic origin. Age and gender-matched controls were blood donors with no history of cancer, and resided in the same geographical area of the patients. The study protocol was approved by the Institutional Ethical Committee of the University of Cagliari (protocol number PG/2021/8575) and the Health Directorate of Sassari (protocol number 464), and informed written consent was obtained from each subject. A questionnaire was administered to all individuals (cases and controls) and relevant personal data (e.g., gender, age, smoke, alcohol, metallic prosthesis, occupational exposure to metals) were collected. All subjects were non-smokers and non-drinkers, none of the subjects were occupationally exposed to metals and none of them had metallic prosthesis in the body. The study was conducted according to the declaration of Helsinki.

2.2. Sample Preparation

One mL of whole blood from each subject was collected into a 15 mL polystyrene tube (Corning, Glendale, AZ, USA), and was added to 2 mL of ultrapure HNO_3 (VWR, Leuven, Belgium) and digested on a heat block (ModBlock, CPI International, Santa Rosa, CA, USA) at 80 °C until complete dissolution. The digests were further diluted with ultrapure deionized water (Micro Pure UV, Thermo Scientific Barnstead, Langenselbold, Germany).

Approximately 0.150 g of cancer and non-cancer tissues were dried at 105 °C overnight (the water content was ca. 50%). Further, samples were weighed in 15 mL polystyrene tubes, added with 2 mL of ultrapure HNO_3 (VWR), digested on a heat block at 80 °C until complete dissolution (ModBlock), and then diluted with ultrapure, deionized water for analysis.

2.3. Quantification of Metals

Thermo Scientific iCAP Qc inductively coupled plasma mass spectrometry (ICP-MS) (Bremen, Germany) was used to quantify the following metals: ^{114}Cd, ^{52}Cr, ^{63}Cu, ^{56}Fe, ^{55}Mn, ^{60}Ni, ^{82}Se, ^{208}Pb and ^{64}Zn. In order to efficiently reduce polyatomic interferences in analytical masses, metals were quantified using the He-pressurized QCell in the kinetic energy discrimination (KED) mode. Procedural blanks to assess the possible exogenous metal contamination from plastics and reagents were analyzed. The addition calibration method was used to quantify the elements, and ^{103}Rh at 1 ng/mL in the analytical solutions was used as an internal standard to account for possible instrumental drifts. The limits of detection (LoD) was calculated as three times the standard deviation of replicated measurements of the pooled digested samples (blood or tissue). Analytical details and ICP-MS settings are reported in Supplementary Materials and Table S1.

In whole blood analysis, certified reference material (CRM) ClinChek® Level I Whole Blood Control (Recipe, Munich, Germany) was analyzed to determine the recovery method and intra-day precision (Table S2 in Supplementary Materials). Recovery was in the range of 90–110% and intra-day precision between 3.1 and 7.5% for all elements. With reference to tissue analysis, CRM pig kidney ERM–BB186 and muscle tissue ERM-CE278k (IRMM, Geel, Belgium) were analyzed (Tables S3 and S4, respectively, in Supplementary Materials). For all elements, the recovery was between 94 and 117%, and intra-day precision ranged between 4.7 and 7.5%.

2.4. Statistics

Since data were not normally distributed, the results were expressed as the median (50th percentile) and 5th–95th percentiles. Differences between the metal levels in PDAC patients and controls, and metal–variables associations (sex, age, and stage of disease) were tested using non-parametric tests (U Mann–Whitney, Kruskal–Wallis, Spearman's ρ, Wilcoxson for coupled tissue samples) and p-values of <0.05 were considered statistically significant. Moreover, receiver operating characteristic (ROC) curves were used to establish a concentration threshold value in blood and cancer tissue able to distinguish the PDAC patients and healthy controls. The quality of ROC results was expressed in terms of the area under the ROC curve (AUC), specificity (proportion of correctly identified healthy controls, also referred to as the false positive rate) and sensitivity (proportion of correctly identified PDAC patients, also called the true positive rate). IBM SPSS Statistics 28 was used as the statistical package.

3. Results

Table 2 reports on the concentrations of toxic metals and essential elements in the whole blood of PDAC patients, OPPs, NOPPs and the HC group. The results of the blood showed significantly higher levels of Cr and Cu (both $p < 0.01$) in PDAC patients when compared to the HCs. Similarly, significantly higher concentrations of Cr and Cu were observed in OPPs (Cr, $p < 0.03$; Cu, $p < 0.01$) and NOPPs (Cr and Cu, both $p < 0.03$) with respect to the HCs.

Table 2. Concentration of toxic metals and essential elements (5th–50th–95th percentiles) in the whole blood of PDAC patients and controls.

	PDAC Patients (No. 46)	OPPs (No. 32)	NOPPs (No. 14)	HCs (No. 20)	Statistical Test (Patients vs. HCs)
Cd (ng/mL)	0.21–0.58–1.52	0.27–0.58–1.51	0.20–0.54–1.47	0.11–0.50–1.57	ns
Cr (ng/mL)	0.26–0.70–1.63 [a]	0.30–0.74–1.60 [b]	0.26–0.64–1.59 [b]	0.17–0.38–1.33	[a] $p < 0.01$; [b] $p < 0.05$
Cu (ng/mL)	853–1251–1682 [a]	914–1274–1775 [a]	851–1148–1670 [b]	779–982–1312	[a] $p < 0.01$; [b] $p < 0.05$
Fe (µg/mL)	234–400–608	236–398–588	240–407–610	324–426–689	ns
Mn (ng/mL)	3.79–7.24–15.8	3.78–7.24–14.6	4.13–7.43–15.8	5.63–9.81–16.4	ns
Ni (ng/mL)	0.44–0.83–2.36	0.44–0.79–2.54	0.47–0.84–1.77	0.41–0.77–2.02	ns
Pb (ng/mL)	8.48–29.6–94.3	8.38–23.4–53.4	9.43–32.2–112	6.76–16.8–63.5	ns
Se (ng/mL)	90.6–129–203	101–149–233	90.4–119–176	108–151–187	ns
Zn (µg/mL)	4.04–6.86–10.0	4.71–7.99–10.6	3.93–6.41–9.42	5.31–7.02–10.2	ns

OPPs = Operated PDAC patients; NOPPs = Non-operated PDAC patients; HCs = Healthy controls; ns: Not significant, a = patients vs. HCs significant at p value less than 0.01, b = patients vs. HCs significant at p value less than 0.05.

When considering gender, the levels of blood Pb were increased in males compared to females in all groups, i.e., PDAC patients (median; 43.4 ng/mL vs. 17.2 ng/mL; $p < 0.01$), OPPs (median; 44.7 ng/mL vs. 17.7 ng/mL; $p < 0.01$) and NOPPs (median; 40.1 ng/mL vs. 16.5 ng/mL; $p < 0.03$).

In PDAC females, the concentrations of blood Cu were higher than in HC females (median; 1269 ng/mL vs. 982 ng/mL; $p < 0.03$). In PDAC males, levels of blood Cu (median; 1202 ng/mL vs. 858 ng/mL; $p < 0.05$) and Pb (median; 43.3 ng/mL vs. 17.9 ng/mL; $p < 0.01$) increased respect to HC males. On the contrary, in PDAC males, the levels of blood Fe (median; 357 µg/mL vs. 485 µg/mL; $p < 0.05$), Mn (median; 6.78 ng/mL vs. 10.3 ng/mL; $p < 0.05$) and Se (median; 130 ng/mL vs. 151 ng/mL; $p < 0.03$) were lower than in HC males.

In OPP females, blood Cu increased compared with HC females (median; 1309 ng/mL vs. 982 ng/mL; $p = 0.01$). In OPP males, the contents of blood Cu (median; 1192 ng/mL vs. 858 ng/mL; $p < 0.03$) and Pb (median; 44.7 ng/mL vs. 17.9 ng/mL; $p < 0.03$) were higher with respect to HC males.

In NOPP males, the levels of blood Cr (median; 0.70 ng/mL vs. 0.35 ng/mL; $p < 0.05$), Cu (median; 1212 ng/mL vs. 858 ng/mL; $p < 0.05$) and Pb (median; 40.1 ng/mL vs. 17.9 ng/mL; $p < 0.05$) were higher, whilst blood Se was lower than the levels detected in HC males (median; 120 ng/mL vs. 151 ng/mL; $p < 0.01$).

Moreover, the whole blood levels of metals in PDAC patients with stage IV cancer were comparable with the levels in PDAC patients belonging to the other cancer stages.

Table 3 reports on the concentrations of toxic metals and essential elements determined in cancer and non-cancer tissues. Data showed that the levels of Cu ($p < 0.01$), Se ($p < 0.01$),

Fe ($p < 0.05$) and Zn ($p < 0.05$) significantly increased in cancer tissue (ca. 2 to 4 times higher) than in the healthy tissue of the same patient. Moreover, some differences were observed when patients were stratified by gender. In males, the levels of Cu (median; 5045 ng/g vs. 690 ng/g; $p < 0.01$), Se (median; 732 ng/g vs. 96.6 ng/g; $p < 0.01$) and Zn (median; 49.8 ng/g vs. 7.68 ng/g; $p < 0.01$) resulted ca. 6 times higher in cancer tissues with respect to healthy tissues. In females, the concentrations of Se (median; 576 ng/g vs. 269 ng/g; $p < 0.03$) were ca. 2 times higher in cancer tissues with respect to healthy tissues.

Table 3. Concentration of toxic metals and essential elements (5th–50th–95th percentiles) in cancer and non-cancer tissue collected from the same PDAC patients.

	Cancer Tissue (No. 32)	Non-Cancer Tissue (No. 32)	Statistical Test (Cancer vs. Non-Cancer)
Cd (ng/g)	84.2–1593–5080	43.8–346–6883	ns
Cr (ng/g)	7.90–31.3–590	12.2–32.6–309	ns
Cu (ng/g)	2232–4664–6307	485–1330–5557	$p < 0.01$
Fe (µg/g)	60.4–172–552	47.9–94.8–319	$p < 0.05$
Mn (ng/g)	214–1098–3532	25.3–296–3891	ns
Ni (ng/g)	18.0–89.7–1919	14.3–56.5–1806	ns
Pb (ng/g)	6.33–23.4–130	4.44–12.1–73.4	ns
Se (ng/g)	327–598–1350	40.2–223–738	$p < 0.01$
Zn (µg/g)	27.3–49.7–107	3.78–24.1–116	$p < 0.05$

ns: Not significant.

As observed in the whole blood, there were no differences in the metal content in tissues depending on the PDAC stage.

Furthermore, ROC analysis was used as a diagnostic test to determine the metal concentrations able to discriminate PDAC patients from HC. In blood, the concentrations of 900 ng/mL of Cu (AUC, 0.769; sensitivity, 87%; specificity, 60%) and 0.35 ng/mL of Cr (AUC, 0.706; sensitivity, 84%; specificity, 60%) accurately divided patients with PDAC neoplasms from the HCs. In cancer tissues, the concentrations of 131 ng/g of Se (AUC, 0.816; sensitivity, 95.7%; specificity, 62%), 998 ng/g of Cu (AUC, 0762; sensitivity, 96%; specificity, 61.9%), 13.9 µg/g of Zn (AUC, 0.700; sensitivity, 96%; specificity, 61.9%) and 88.8 µg/g of Fe (AUC, 0.683; sensitivity, 91.3%; specificity, 57.1%) accurately discriminated PDAC tissues of patients with respect to healthy tissues in the same patients.

4. Discussion

PDAC is a highly aggressive lethal neoplasm considering the challenges of early diagnosis and due to poor response to treatments. Among all pancreatic cancers, PDAC represents more than 90% of all cases, and notwithstanding the scientific progress on this disease, PDAC has an average 5-year survival rate of less than 10% [19]. PDAC is considered to have a multifactorial origin, and among the possible risk factors, exposure to toxic metals or the imbalance of essential elements may alter normal metal homeostasis in the body, posing the way to activate molecular mechanisms involved in metal-induced malignant transformation or metal effects on tumor behavior [20].

Analysis of blood samples provides information on metal exposure at the long and medium terms. Metals are found in red blood cells (e.g., Cd, Mn and Pb), plasma components (Se), or occur unbound in blood [21]. Blood metal composition may reflect changes due to a diseased condition such as in PDAC, or it may reflect contributions of factors ranging from genetics to lifestyles. On the other hand, the analysis of healthy tissue adjacent to cancer tissues from the same individual served as an internal control, thus reducing individual-specific (i.e., age, gender, environment and genetics) and anatomical site-specific effects.

In this study, whole blood Cr was found to be significantly increased in PDAC patients (median, 0.70 ng/mL), OPPs (median, 0.70 ng/mL) and NOPPs (median, 0.64 ng/mL) with respect to the HCs (P50, 0.38 ng/mL). Cr(VI) compounds are known carcinogens, and are included in Group 1 (carcinogenic to humans) by the International Agency for Research on Cancer (IARC) classification [22]. Nevertheless, Cr ions may activate Fenton-like reactions, generate hydroxy radicals that induce oxidative stress, activate transcription factors such as nuclear factor kappa B (NF-κB), activate protein-1 (AP-1), tumor protein P53 (p53) and hypoxia-inducible factor-1 (HIF-1), regulate the cell cycle and induce apoptosis [23,24]. In previous literature, a study reported a high level of Cr in the serum of subjects with acute pancreatitis [25]. In addition, significantly increased levels of Cr in pancreatic juice were quantified in three population groups, and the levels of Cr followed a decreasing order: pancreatic cancer > pancreatitis > controls [20]. On the other hand, other authors did not find associations between self-reported regular exposure to Cr and the risk of developing PDAC [26], and between the levels of Cr in the toenails of patients with pancreatic cancer [27]. Similarly, a study reported comparable toenail levels of Cr between the controls and PDAC patients with a mutated and wild-type Kirsten rat sarcoma (KRAS) viral oncogene homolog [28]. In addition, similar findings, with no variations in the urinary Cr content of PDAC patients and controls, were also reported [29].

Regarding Cu, it was found to be significantly increased in the whole blood of PDAC patients (median, 1251 ng/mL), OPPs (median, 1274 ng/mL) and NOPPs (median, 1148 ng/mL) with respect to the HCs (median, 982 ng/mL), and this result was confirmed in PDAC, OPP and NOPP males and females with respect to HC males and females. Significantly higher Cu levels were also observed in pancreatic cancer tissue (median, 4664 ng/g) with respect to the adjacent healthy tissue (median, 1330 ng/g), and the same significant finding was observed in PDAC males, but not in PDAC females. Copper is an integral part of Cu-based metalloenzymes (as cytochrome oxidase, NADH dehydrogenase-2, Cu/Zn superoxide dismutase) and has antioxidant capacity, lowering oxidative stress as well as stabilizing cellular functions. Therefore, when Cu homeostasis is not regulated, Cu accumulation may create the risk of cellular injuries [25]. Moreover, in cancer and inflammation, plasma Cu and ceruloplasmin concentrations were raised, and the rates of synthesis and secretion of ceruloplasmin by the liver were enhanced. The elevated ceruloplasmin concentrations in both conditions would provide additional Cu uptake by cells in normal tissues and cancer cells [30]. In addition, Cu-dependent amino oxidases may mediate angiogenesis that plays a critical role in the growth of cancer [30]. In the literature, two different studies observed elevated levels of Cu in the serum of pancreatic patients with respect to the controls [14,30]. In addition, a higher serum Cu content in patients with acute pancreatitis and higher urinary Cu levels in PDAC patients were observed [26,29]. Furthermore, Cu in pancreatic juice was 2 times higher with respect to the Cu levels detected in acute pancreatitis, but comparable to Cu contents in the controls [20]. Opposite findings were found by other authors that reported no differences in Cu concentrations in the toenails of PDAC patients and the controls [27,28].

Although Fe is essential for the function of many cellular enzymes, an excess of Fe may result in the production of ROS that can lead to peroxidation and apoptosis by attacking protein, lipids, nucleic acids and carbohydrates, processes closely related to tumorigenesis [9,31]. In this study, Fe levels in whole blood were decreased in male PDAC patients with respect to HC males. Contrarily, Fe concentrations in PDAC tissue (median, 173 µg/g) were elevated compared to healthy tissue (median, 94.8 µg/g). A growing body of evidence suggests that an Fe imbalance can be a key factor in pancreatic cancer. In general, the dependence of cancer cells on Fe uptake is fully recognized because this element is responsible for cell respiration, oxygen transport, oxygen metabolism, energy metabolism, DNA synthesis, etc. In particular, the incremented Fe level in cancer cells was linked to tumor-associated angiogenesis, which is the development of new blood vessels in response to proteins secreted by tumor cells [32,33]. Other studies suggested that the Fe overload in pancreatic islets impaired insulin secretion and β-cell function, and accelerated pancreatic β-cell death [34]. In addition, exposure to an excess of Fe may induce epithelial–mesenchymal transition (EMT) in pancreatic cancer cells, loss of p53 and suppression of p53 transcriptional activity [13]. Higher serum Fe was found to be associated with pancreatic cancer, although not consistently [35–37]. Authors reported higher levels of serum ferritin in pancreatic cancer, and Fe homeostasis genes and ferroptosis genes significantly altered with pancreatic tumor grades [9]. Epidemiological studies have suggested positive associations between Fe and red meat intake, and the risk of PDAC [38].

Although the IARC did not consider Pb to show a clear carcinogenic activity (Group 2A, probably carcinogenic to humans), this metal is extensively distributed in the environment and may cause numerous acute and chronic circulatory, neurological, hematological, gastrointestinal, reproductive and immunological pathologies [39]. Lead can inhibit delta-aminolevulinic acid dehydratase (δ-ALAD), disrupting heme synthesis in bone marrow erythroblasts. In addition, it can induce oxidative stress interfering with antioxidant activities by inhibiting functional SH groups in several enzymes such as superoxide dismutase (SOD), catalase (CAT), glutathione peroxidase (GPx), glucose-6-phosphate dehydrogenase (G6PD) and δ-ALAD [40]. In this study, Pb levels were comparable between diseased patients and the controls upon both whole blood and tissue analyses. On the other hand, differences in whole blood Pb levels were found as a function of gender. Indeed, blood Pb was significantly (ca. 2.5 times) higher in PDAC males with respect to PDAC females. The reason for this result is that males have more erythrocytes (ca. 10%) than females and there are different long-term Pb kinetics in men and women. In fact, Pb is stored in bone for 10–30 years, which hold approximately 90% of the body burden and, during the life span, it is released back into the blood (as an "endogenous source") via bone remodeling [41,42]. There is no conclusive evidence for the Fe involvement in PDAC neoplasms. Some authors found unchanged levels of Pb between PDAC patients and the controls in different matrices, such as urine [29], pancreatic juice [20] and blood [43]. On the other hand, significantly elevated levels of Pb in the toenails of pancreatic cancer patients [27] and significantly decreased serum levels of Pb in pancreatic cancer patients [14] were also observed. Another study showed that higher levels of Pb may be a risk factor for both KRAS-mutated and wild-type cases of pancreatic cancer [28].

Regarding Se, it is a micronutrient required for the proper functioning of the body and takes part in several anti-carcinogenic mechanisms, including inactivating oxygen free radicals and initiating major DNA repair pathways [12]. Furthermore, selenoproteins and Se metabolites reduced pancreatic carcinogenesis in an animal model [44]. Considering this study, no differences were noted in Se whole blood levels among the patients and controls, but some relevant results were observed according to gender. In particular, Se levels in the whole blood of PDAC males and NOPP males were significantly decreased with respect to Se content in the whole blood of HC males. On the other hand, Se levels in cancer tissues (median, 598 ng/g) were significantly higher than in the corresponding healthy tissues (median, 223 ng/g). Higher Se tissue concentrations were found in both females and males with respect to the adjacent healthy tissues. Regarding Se and PDAC, there are some controversial findings. Two studies found higher levels of Se in patients with pancreatic cancer [20,30]. Contrastingly, other investigations reported that high concentrations of Se or a high intake of Se might reduce the risk of pancreatic cancer [12,27]. Contrarily, other authors did not find associations between Se and pancreatic cancer risk [44].

There is a lot of information involving Zn dysregulation in the development of pancreatic cancer. For example, the overexpression of Zip4 protein was reported in 94% of clinical PDAC tissue compared with adjacent normal tissue, and malignant cells displayed significantly higher Zip4 expression compared with normal pancreatic cells [16,45]. Further, whereas Zip4 expression may be increased in some cancerous tissues, Zip3 and Zip8 were reported as down-regulated in some adenocarcinomas [17]. Decreased serum Zn levels were associated with chronic pancreatitis, and this disease was linked with lower levels of Cu/Zn-SOD and metallothionein (MT), responsible for the increase in ROS and elevation in oxidative stress [15]. In the present study, the levels of Zn in the whole blood were unchanged between PDAC patients and the controls. In PDAC, tissue concentrations of Zn were significantly elevated with respect to the levels in healthy tissues (median, 43.7 μg/g vs. 24.1 μg/g). This result was confirmed when the cancer and non-cancer tissue of males were compared. The present findings were consistent with those reported in a previous study, where the Zn content in tumor tissues was higher than that in adjacent non-cancer tissues, also indicating that the high concentration of Zn was beneficial for the occurrence and development of pancreatic cancer [46]. In addition, an increased level of Zn was found in the urine of PDAC patients [29]. These authors postulated that oxidative stress caused by cancer development and progression oxidizes the sulfhydryl groups in cysteine, decreasing the Zn-binding capacity with a consequent higher amount of free Zn that is rapidly eliminated through urine. Conflictingly, other authors reported decreased levels of Zn in subjects with acute pancreatitis and pancreatic cancer, or in rats with induced pancreatitis [14,25,43]. In addition, it was observed that the Zn concentration in pancreatic fluid was comparable in patients with chronic pancreatitis and pancreatic cancer compared to a normal pancreas [47].

The use of trace element ratios instead of single trace element content was previously reported in studies, for example, in lung cancer [48], open-angle glaucoma patients [49] and idiopathic pulmonary fibrosis [50], in order to reflect the dyshomeostasis of metals associated with disorders. Thus, in this study, the ratio between metals was investigated as a possible indicator of PDAC disease. In particular, the whole blood Cu/Zn ratio in PDAC (0.185) patients was found to be ca. 1.3-fold higher than in the controls (0.143), resulting in a statistically different level at $p < 0.01$. The disruption of the whole blood Cu/Zn ratio may be responsible for the increased oxidative stress and decreased blood antioxidant capacity in PDAC patients. This ratio was found to be increased in cases of acute pancreatitis in patients [25].

5. Conclusions

Environmental exposure to chemicals, genetics and lifestyle are among the possible risk factors of PDAC development and progression. Among these factors, exposure to toxic metals or the imbalance in essential elements may have an impact on PDAC neoplasm. The results of the present study suggest that whole blood (Cr and Cu) and tissue (Cu, Fe, Se, and Zn) levels of toxic metals and essential elements in PDAC patients were different from those observed in the healthy controls. These metals may have a role in PDAC-associated metabolic pathways, including the increase in ROS production, boosting lipid peroxidation, depleting of antioxidant barriers, enhancing the expression of inflammation genes, modifying DNA bases, etc. Moreover, the higher Cu/Zn ratio may indicate an increased level of oxidative stress and decreased antioxidant properties in PDAC neoplasms. Although studies with a high number of PDAC patients are required to support the present findings and better understand the underlying pathophysiological processes, the levels of metals in blood and cancer tissue can be utilized as biomarkers of PDAC diagnosis and monitoring.

Supplementary Materials: The following supporting information can be downloaded at: https://www.mdpi.com/article/10.3390/toxics12010032/s1. "Methodology for concentration of toxic metals and essential elements in blood and tissues of patients with Pancreatic Ductal Adenocarcinoma"; "Table S1: ICP-MS instrumental characteristics and settings"; "Table S2: Results of the CRM ClinChek® Whole Blood Control, Level I (Recipe, Munich, Germany) analysis"; "Table S3: Results of the CRM pig kidney ERM-BB186 (IRMM, Geel, Belgium) analysis"; "Table S4: Results of the CRM mussel tissue ERM-CE278k (IRMM, Geel Belgium) analysis".

Author Contributions: Conceptualization, R.M., C.F. and A.P. (Andrea Pisano); Methodology, A.P. (Alberto Porcu), T.P., G.F. (Grazia Fenu), F.E., A.S. and C.F.; Formal Analysis, B.B., G.F. (Giovanni Forte), G.F. (Grazia Fenu) and F.E.; Data Curation, B.B. and G.F. (Giovanni Forte); Writing—Original Draft Preparation, B.B., G.F. (Giovanni Forte), A.P. (Andrea Pisano) and R.M.; Writing—Review and Editing, R.M. and C.F. All authors have read and agreed to the published version of the manuscript.

Funding: This study was supported by the grant FAR2020-Madeddu, University of Sassari, Italy.

Institutional Review Board Statement: All subjects gave their informed consent for inclusion before they participated in the study. The study was conducted in accordance with the Declaration of Helsinki, and the protocol was approved by the Ethics Committee of Cagliari (protocol number PG/2021/8575) and the Health Directorate of Sassari (protocol number 464).

Informed Consent Statement: Informed consent was obtained from all subjects involved in the study.

Data Availability Statement: Data are unavailable due to privacy or ethical restrictions.

Conflicts of Interest: The authors declare no conflicts of interest.

References

1. AIRC, Associazione Italiana per la Ricerca sul Cancro. Available online: https://www.airc.it/cancro/informazioni-tumori/guida-ai-tumori/tumore-del-pancreas (accessed on 4 September 2023).
2. WHO, World Health Organization. Available online: https://platform.who.int/mortality/themes/theme-details/topics/indicator-groups/indicator-group-details/MDB/pancreas-cancer (accessed on 4 September 2023).
3. Djordjevic, V.R.; Wallace, D.R.; Schweitzer, A.; Boricic, N.; Knezevic, D.; Matic, S.; Grubor, N.; Kerkez, M.; Radenkovic, D.; Bulat, Z.; et al. Environmental cadmium exposure and pancreatic cancer: Evidence from case control, animal and in vitro studies. *Environ. Int.* **2019**, *128*, 353–361. [CrossRef] [PubMed]
4. Schwartz, G.G.; Reis, I.M. Is cadmium a cause of human pancreatic cancer? *Cancer Epidemiol. Biomark. Prev.* **2000**, *9*, 139–145.
5. Zhao, L.; Islam, R.; Wang, Y.; Zhang, X.; Liu, L.-Z. Epigenetic regulation in chromium-, nickel- and cadmium-induced carcinogenesis. *Cancers* **2022**, *14*, 5768. [CrossRef] [PubMed]
6. Hartwig, A.; Schwerdtle, T. Interactions by carcinogenic metal compounds with DNA repair processes: Toxicological implications. *Toxicol. Lett.* **2002**, *127*, 47–54. [CrossRef] [PubMed]
7. Mortoglou, M.; Manić, L.; Buha Djordjevic, A.; Bulat, Z.; Djordjevic, V.; Manis, K.; Valle, E.; York, L.; Wallace, D.; Uysal-Onganer, P. Nickel's role in pancreatic ductal adenocarcinoma: Potential involvement of microRNAs. *Toxics* **2022**, *10*, 148. [CrossRef] [PubMed]
8. Camargo, J.; Pumarega, J.A.; Alguacil, J.; Sanz-Gallén, P.; Gasull, M.; Delclos, G.L.; Amaral, A.F.S.; Porta, P. Toenail concentrations of trace elements and occupational history in pancreatic cancer. *Environ. Int.* **2019**, *127*, 216–225. [CrossRef] [PubMed]
9. Park, J.M.; Mau, C.-Z.; Chen, Y.-C.; Su, Y.-H.; Chen, H.-A.; Huang, S.-Y.; Chang, J.-S.; Chiu, C.-F. A case–control study in Taiwanese cohort and meta-analysis of serum ferritin in pancreatic cancer. *Sci. Rep.* **2021**, *11*, 21242. [CrossRef]
10. Bocca, B.; Madeddu, R.; Asara, Y.; Tolu, P.; Marchal, J.A.; Forte, G. Assessment of reference ranges for blood Cu, Mn, Se and Zn in a selected Italian population. *J. Trace Elem. Med. Biol.* **2011**, *25*, 19–26. [CrossRef]
11. Lelièvre, P.; Sancey, L.; Coll, J.-L.; Deniaud, A.; Busser, B. The multifaceted roles of copper in cancer: A trace metal element with dysregulated metabolism, but also a target or a bullet for therapy. *Cancers* **2020**, *12*, 3594. [CrossRef]
12. Wang, L.; Wang, J.; Liu, X.; Liu, Q.; Zhang, G.; Liang, L. Association between selenium intake and the risk of pancreatic cancer: A meta-analysis of observational studies. *Biosci. Rep.* **2016**, *36*, e00395. [CrossRef]
13. Bhutia, Y.D.; Ogura, J.; Grippo, P.J.; Torres, C.; Sato, T.; Wachtel, M.; Ramachandran, S.; Babu, E.; Sivaprakasam, S.; Rajasekaran, D.; et al. Chronic exposure to excess iron promotes EMT and cancer via p53 loss in pancreatic cancer. *Asian J. Pharm. Sci.* **2020**, *15*, 237–251. [CrossRef]
14. Karapinar, H.S.; Türkdoğan, M.K.; Kiliçel, F. Serum trace element levels of liver cirrhosis and pancreatic cancer patients. *Arch. Community Med. Public Health* **2022**, *8*, 055–061.
15. Kelleher, S.L.; McCormick, N.H.; Velasquez, V.; Lopez, V. Zinc in specialized secretory tissues: Roles in the pancreas, prostate, and mammary gland. *Adv. Nutr.* **2011**, *2*, 101–111. [CrossRef]
16. Li, M.; Zhang, Y.; Liu, Z.; Bharadwaj, U.; Wang, H.; Wang, X.; Zhang, S.; Liuzzi, J.P.; Chang, S.M.; Cousins, R.J.; et al. Aberrant expression of zinc transporter ZIP4 (SLC39A4) significantly contributes to human pancreatic cancer pathogenesis and progression. *Proc. Natl. Acad. Sci. USA* **2007**, *104*, 18636–18641. [CrossRef]
17. Richter, P.; Faroon, O.; Pappas, R.S. Cadmium and cadmium/zinc ratios and tobacco-related morbidities. *Int. J. Environ. Res. Public Health* **2017**, *14*, 1154. [CrossRef]
18. AJCC. *Cancer Staging Manual*, 8th ed.; Amin, M.B., Edge, S.B., Eds.; Springer International Publishing AG: New York, NY, USA, 2018; p. 1032.
19. Miller, K.D.; Siegel, R.L.; Lin, C.C.; Mariotto, A.B.; Kramer, J.L.; Rowland, J.H.; Stein, K.D.; Alteri, R.; Jemal, A. Cancer treatment and survivorship statistics, 2016. *CA Cancer J. Clin.* **2016**, *66*, 271–289. [CrossRef]
20. Carrigan, P.E.; Hentz, J.G.; Gordon, G.; Morgan, J.L.; Raimondo, M.; Anbar, A.D.; Miller, L.J. Distinctive heavy metal composition of pancreatic juice in patients with pancreatic carcinoma. *Cancer Epidemiol. Biomark. Prev.* **2007**, *16*, 2656–2663. [CrossRef]
21. Bocca, B.; Forte, G.; Oggiano, R.; Clemente, S.; Asara, Y.; Peruzzu, P.; Farace, C.; Pala, S.; Fois, A.G.; Pirina, P.; et al. Level of neurotoxic metals in amyotrophic lateral sclerosis: A population-based case–control study. *J. Neurol. Sci.* **2015**, *359*, 11–17. [CrossRef]

22. IARC, International Agency for Research on Cancer. *Chromium, Nickel and Welding*; IARC: Lyon, France, 1990; Volume 49.
23. Yao, H.; Guo, L.; Jiang, B.-H.; Luo, J.; Shi, X. Oxidative stress and chromium (VI) carcinogenesis. *J. Environ. Pathol. Toxicol. Oncol.* **2008**, *27*, 77–88. [CrossRef]
24. Henkler, F.; Brinkmann, J.; Luch, A. The role of oxidative stress in carcinogenesis induced by metals and xenobiotics. *Cancers* **2010**, *2*, 376–396. [CrossRef]
25. Al-Fartusie, F.; Farhan, M.; Al-Bairmani, H.; Nabil, N.; Aldhaheri, M.; Al-Temimi, R. Estimation of some vital trace elements in patients with acute pancreatitis: A case-control study. *Braz. J. Pharm. Sci.* **2022**, *58*, e20639. [CrossRef]
26. Antwi, S.O.; Eckert, E.C.; Sabaque, C.V.; Leof, E.R.; Hawthorne, K.M.; Bamlet, W.R.; Chaffee, K.G.; Oberg, A.L.; Petersen, G.M. Exposure to environmental chemicals and heavy metals, and risk of pancreatic cancer. *Cancer Causes Control* **2015**, *26*, 1583–1591. [CrossRef]
27. Amaral, A.F.S.; Porta, M.; Silverman, D.T.; Milne, R.L.; Kogevinas, M.; Rothman, N.; Cantor, K.P.; Jackson, B.P.; Pumarega, J.A.; López, T.; et al. Pancreatic cancer risk and levels of trace elements. *Gut* **2012**, *61*, 1583–1588. [CrossRef]
28. Gómez-Tomás, A.; Pumarega, J.; Alguacil, J.; Amaral, A.F.S.; Malats, N.; Pallarès, N.; Gasull, M.; Porta, M. for the PANKRAS II Study Group. Concentrations of trace elements and KRAS mutations in pancreatic ductal adenocarcinoma. *Environ. Mol. Mutagen.* **2019**, *60*, 693–703. [CrossRef]
29. Schilling, K.; Larner, F.; Saad, A.; Roberts, R.; Kocher, H.M.; Blyuss, O.; Halliday, A.N.; Crnogorac-Jurcevicg, T. Urine metallomics signature as an indicator of pancreatic cancer. *Metallomics* **2020**, *12*, 752–757. [CrossRef]
30. Lener, M.R.; Scott, R.J.; Wiechowska-Kozłowska, A.; Serrano-Fernández, P.; Baszuk, P.; Jaworska-Bieniek, K.; Sukiennicki, G.; Marciniak, W.; Muszyńska, M.; Kładny, J.; et al. Serum concentrations of selenium and copper in patients diagnosed with pancreatic cancer. *Cancer Res. Treat.* **2016**, *48*, 1056–1064. [CrossRef]
31. Guo, Q.; Li, L.; Hou, S.; Yuan, Z.; Li, C.; Zhang, W.; Zheng, L.; Li, X. The role of iron in cancer progression. *Front. Oncol.* **2021**, *11*, 778492. [CrossRef]
32. Al-Ostoot, F.H.; Salah, S.; Khamees, H.A.; Khanum, S.A. Tumor angiogenesis: Current challenges and therapeutic opportunities. *Cancer. Treat. Res. Commun.* **2021**, *28*, 100422. [CrossRef]
33. Lugano, R.; Ramachandran, M.; Dimberg, A. Tumor angiogenesis: Causes, consequences, challenges and opportunities. *Cell. Mol. Life Sci.* **2020**, *77*, 1745–1770. [CrossRef]
34. Backe, M.B.; Moen, I.W.; Ellervik, C.; Hansen, J.B.; Mandrup-Poulsen, T. Iron regulation of pancreatic beta-cell functions and oxidative stress. *Annu. Rev. Nutr.* **2016**, *36*, 241–273. [CrossRef]
35. Friedman, G.D.; van den Eeden, S.K. Risk factors for pancreatic cancer: An exploratory study. *Int. J. Epidemiol.* **1993**, *22*, 30–37. [CrossRef]
36. Gaur, A.; Collins, H.; Wulaningsih, W.; Holmberg, L.; Garmo, H.; Hammar, N.; Walldius, G.; Jungner, I.; Van Hemelrijck, M. Iron metabolism and risk of cancer in the Swedish AMORIS study. *Cancer Causes Control* **2013**, *24*, 1393–1402. [CrossRef]
37. Zhao, Z.; Li, C.; Hu, M.; Li, J.; Liu, R. Plasma ferritin levels, HFE polymorphismsm, and risk of pancreatic cancer among Chinese Han population. *Tumour Biol.* **2014**, *35*, 7629–7633. [CrossRef]
38. Zanini, S.; Renzi, S.; Limongi, A.R.; Bellavite, P.; Giovinazzo, F.; Bermano, G. A review of lifestyle and environment risk factors for pancreatic cancer. *Eur. J. Cancer* **2021**, *145*, 53–70. [CrossRef]
39. IARC, International Agency for Research on Cancer. *Inorganic and Organic Lead Compounds*; IARC: Lyon, France, 2006; Volume 87.
40. Sani, A.H.; Amanabo, M. Lead: A concise review of its toxicity, mechanism and health effect. *GSC Biol. Pharm. Sci.* **2021**, *15*, 055–062. [CrossRef]
41. Forte, G.; Madeddu, R.; Tolu, P.; Asara, Y.; Marchal, J.A.; Bocca, B. Reference intervals for blood Cd and Pb in the general population of Sardinia (Italy). *Int. J. Hyg. Environ. Health* **2011**, *214*, 102–109. [CrossRef]
42. Barbosa, F., Jr.; Tanus-Santos, J.E.; Gerlach, R.F.; Parsons, P.J. A critical review of biomarkers used for monitoring human exposure to lead: Advantages, limitations, and future needs. *Environ. Health Perspect.* **2005**, *113*, 1669–1674. [CrossRef] [PubMed]
43. Farzin, L.; Moassesi, M.E.; Sajadi, F.; Ahmadi Faghih, M.A. Evaluation of trace elements in pancreatic cancer patients in Iran. *Middle East J. Cancer* **2013**, *4*, 79–86.
44. Chatterjee, S.; Combs, G.F., Jr.; Chattopadhyay, A.; Stolzenberg-Solomon, R. Serum selenium and pancreatic cancer: A prospective study in the prostate, lung, ovarian, and colorectal cancer cohort. *Cancer Causes Control* **2019**, *30*, 457–464. [CrossRef] [PubMed]
45. Li, M.; Zhang, Y.; Bharadwaj, U.; Zhai, Q.J.; Ahern, C.H.; Fisher, W.E.; Brunicardi, F.C.; Logsdon, C.D.; Chen, C.; Yao, Q. Down-regulation of ZIP4 by RNA interference inhibits pancreatic cancer growth and increases the survival of nude mice with pancreatic cancer xenografts. *Clin. Cancer Res.* **2009**, *15*, 5993–6001. [CrossRef]
46. Yu, Z.; Yu, Z.; Chen, Z.B.; Yang, L.; Ma, M.J.; Lu, S.N.; Wang, C.S.; Teng, C.B.; Nie, Y.Z. Zinc chelator TPEN induces pancreatic cancer cell death through causing oxidative stress and inhibiting cell autophagy. *Cell Physiol.* **2019**, *234*, 20648–20661. [CrossRef] [PubMed]
47. Pungpapong, S.; Scolapio, J.S.; Woodward, T.A.; Wallace, M.B.; Raimondo, M. Is zinc concentration in pancreatic fluid a marker for pancreatic diseases? *JOP* **2005**, *6*, 425–430. [PubMed]
48. Callejón-Leblic, B.; Gómez-Ariza, J.L.; Pereira-Vega, A.; García-Barrera, T. Metal dyshomeostasis based biomarkers of lung cancer using human biofluids. *Metallomics* **2018**, *10*, 1444–1451. [CrossRef]

49. Hohberger, B.; Chaudhri, M.A.; Michalke, B.; Lucio, M.; Nowomiejska, K.; Schlötzer-Schrehardt, U.; Grieb, P.; Rejdak, R.; Jünemann, A.G.M. Levels of aqueous humor trace elements in patients with open-angle glaucoma. *J. Trace Elem. Med. Biol.* **2018**, *45*, 150–155. [CrossRef] [PubMed]
50. Forte, G.; Bocca, B.; Pisano, A.; Collu, C.; Farace, C.; Sabalic, A.; Senofonte, M.; Fois, A.G.; Mazzarello, V.L.; Pirina, P.; et al. The levels of trace elements in sputum as biomarkers for idiopathic pulmonary fibrosis. *Chemosphere* **2021**, *271*, 129514. [CrossRef]

toxics

MDPI

Article

Accumulation and Release of Cadmium Ions in the Lichen *Evernia prunastri* (L.) Ach. and Wood-Derived Biochar: Implication for the Use of Biochar for Environmental Biomonitoring

Andrea Vannini [1,*], Luca Pagano [1,2], Marco Bartoli [1], Riccardo Fedeli [3], Alessio Malcevschi [1], Michele Sidoli [4], Giacomo Magnani [4], Daniele Pontiroli [4], Mauro Riccò [4], Marta Marmiroli [1,†], Alessandro Petraglia [1,†] and Stefano Loppi [3,5,†]

1 Department of Chemistry, Life Sciences, and Environmental Sustainability, University of Parma, Parco Area delle Scienze 11/a, 43124 Parma, Italy; luca.pagano@unipr.it (L.P.); marco.bartoli@unipr.it (M.B.); alessio.malcevschi@unipr.it (A.M.); marta.marmiroli@unipr.it (M.M.); alessandro.petraglia@unipr.it (A.P.)
2 National Interuniveritary Consortium for Environmental (CINSA), University of Parma, Parco Area delle Scienze 95, 43124 Parma, Italy
3 Department of Life Sciences, University of Siena, Via PA Mattioli 4, 53100 Siena, Italy; riccardo.fedeli@student.unisi.it (R.F.); stefano.loppi@unisi.it (S.L.)
4 Department of Mathematical, Physical and Computer Sciences, University of Parma, Parco Area delle Scienze 7/a, 43124 Parma, Italy; michele.sidoli@unipr.it (M.S.); giacomo.magnani@unipr.it (G.M.); daniele.pontiroli@unipr.it (D.P.); mauro.ricco@unipr.it (M.R.)
5 BAT Center-Interuniversity Center for Studies on Bioinspired Agro-Environmental Technology, University of Naples 'Federico II', 80138 Napoli, Italy
* Correspondence: andrea.vannini@unipr.it
† These authors contributed equally to the work.

Abstract: Biochar (BC) boasts diverse environmental applications. However, its potential for environmental biomonitoring has, surprisingly, remained largely unexplored. This study presents a preliminary analysis of BC's potential as a biomonitor for the environmental availability of ionic Cd, utilizing the lichen *Evernia prunastri* (L.) Ach. as a reference organism. For this purpose, the lichen *E. prunastri* and two types of wood-derived biochar, biochar 1 (BC1) and biochar 2 (BC2), obtained from two anonymous producers, were investigated for their ability to accumulate, or sequester and subsequently release, Cd when exposed to Cd-depleted conditions. Samples of lichen and biochar (fractions between 2 and 4 mm) were soaked for 1 h in a solution containing deionized water (control), 10 μM, and 100 μM Cd^{2+} (accumulation phase). Then, 50% of the treated samples were soaked for 24 h in deionized water (depuration phase). The lichen showed a very good ability to adsorb ionic Cd, higher than the two biochar samples (more than 46.5%), and a weak ability to release the metal (ca. 6%). As compared to the lichen, BC2 showed a lower capacity for Cd accumulation (−48%) and release (ca. 3%). BC1, on the other hand, showed a slightly higher Cd accumulation capacity than BC2 (+3.6%), but a release capacity similar to that of the lichen (ca. 5%). The surface area and the cation exchange capacity of the organism and the tested materials seem to play a key role in their ability to accumulate and sequester Cd, respectively. This study suggests the potential use of BC as a (bio)monitor for the presence of PTEs in atmospheric depositions and, perhaps, water bodies.

Keywords: biomonitoring; cadmium accumulation; cadmium release; cadmium removal; cation exchange capacity; surface area

1. Introduction

Cadmium (Cd) is a relatively abundant metal in the Earth's crust (about 0.1 mg/kg; [1]). However, its continuous release from anthropogenic activities [2] can increase its concentration in soils, atmospheric depositions, and water bodies [3]. Cadmium is a non-essential metal for human health. However, its chronic presence at concentrations above environmental background levels can cause several adverse human health effects, including, in the worst case, the development of various types of cancer [4–6]. Due to its high chemical and physical similarity to calcium (Ca; [7], Cd is easily translocated in the human body, where it can remain for decades ($t_{1/2}$ = 10–35 years; [8]. For this reason, Cd is recognized as a global priority pollutant. Daily consumption of contaminated vegetables is the main route by which the human body is exposed to the metal [9]. Therefore, its threshold concentration in vegetables is set at 0.1 mg/kg fresh weight, at the European level [10]. Cases of poisoning and serious illness resulting from the consumption of Cd-contaminated plant material due to Cd contamination of air and water (e.g., Itai-Itai disease) are well known. Cadmium emissions must therefore be controlled, and, in this scenario, Europe has set limits on particulate emissions from industrial activities [11], which has led to a decrease in Cd emissions [12]. The reduction in emissions has also helped to reduce the atmospheric deposition of Cd [13], although some countries, such as Poland, Slovakia, and Hungary, still have high levels [14,15].

Epiphytic lichens are an effective way to assess the presence of potentially toxic elements (PTEs) in atmospheric depositions [16–19]. Due to the lack of roots, protective structures such as plant cuticles, and gas flow regulating structures such as stomata, lichens can efficiently accumulate airborne PTEs far beyond their physiological needs and without regard to their phytotoxicity [20–22]. Accumulation occurs when PTE levels in atmospheric deposition exceed their natural background concentrations but ends when equilibrium with their new environmental availability is reached [23]. On the other hand, when the environmental availability of PTEs decreases, as can occur after environmental improvements [24], PTE releases become the dominant process and thalli experience a decrease in their concentrations [23]. Lichens require long periods of time to release the accumulated metal [22,24], but in some cases, i.e., for some specific chemical species of metals such as mercury, the release appears to be irreversible [25]. There is extensive research in the literature on the ability of lichens to accumulate Cd (see [20,26] for examples), but there is no information on the fate of this metal after its environmental availability has been reduced.

Biochar (BC), the by-product of the pyrolysis of organic materials (often wood) for bioenergy production, is now attracting considerable attention, from an environmental perspective, as a strategy to stabilize PTEs in soils [27–29], thereby enabling their remediation [30]; as a tool for filtering contaminated water and air [31–33]; and as a sustainable strategy for agricultural purposes [34,35]. Despite these diverse applications, the potential of BC for environmental biomonitoring has been surprisingly underexplored.

To assess the suitability of BC for biomonitoring, akin to assessments conducted with lichens, it is essential to understand its capacity to reflect environmental concentrations of PTEs, including Cd. While BC's documented ability to accumulate Cd through various mechanisms is extensive [36], its desorption capacity remains less explored [37]. This study presents a preliminary analysis of BC's potential as a biomonitor for the environmental availability of ionic Cd, utilizing the lichen *Evernia prunastri* as a reference organism. To this end, the dynamics of the accumulation and release of Cd^{2+} in the lichen *E. prunastri* (a well-known biomonitor of environmental PTE availability) and in two types of biochar were compared through an experiment conducted in a controlled environment.

2. Materials and Methods

2.1. BC Samples Description and Pretreatment

Samples of both biochar 1 (BC1) and biochar 2 (BC2) BC were provided by the respective (anonymous) producers. Both materials were produced from woody biomass of forestry origin pyrolyzed at 500–650 °C. BC1 was produced using a 125 kWe industrial gasifier while BC2 was produced using a Cryos Gas Unit plant. BC1 corresponded to material produced and disposed (i.e., waiting for disposal) in non-waterproof plastic bales exposed to the elements (i.e., rain, sun, and temperature variation) for 5 years.

BC1 and BC2 samples were first sieved at 4 mm and then at 2 mm to obtain materials with a flake size between 2 and 4 mm (Figure 1). The samples were then separately subjected to a preliminary wash with deionized water (1.1 μS/cm^2) to remove the first soluble component adsorbed on the BC, such as salts, followed by a wash with warm 1M HCl maintaining a sample mass (g)/solution volume (mL) ratio of 1:250; this procedure was considered to remove most of the metals, including Cd, from the BC surface. The BC samples were then air-dried at 80 °C overnight and stored in polypropylene tubes. The pool of BC1 and BC2 material (60 g each) was divided into 30 samples, each weighing 2 g. Each sample was subsequently enclosed in a nylon mesh with a 0.2 mm mesh size and closed at one end using nylon thread.

Figure 1. Sieved (2–4 mm) samples of BC1 (**left**); and BC2 (**right**).

2.2. Lichen Sample Collection and Preparation

Thalli of the lichen *Evernia prunastri* (L.) Ach. were collected from branches of *Quercus* spp. located in a remote forest area in southwestern Emilia-Romagna (October 2022). For this study, only adult thalli (laciniae > 5 cm) growing on branches more than 1.5 m above the ground were collected. In the laboratory, samples were cleaned of any foreign material, such as bark debris, spider nests, and other lichen species.

Lichen samples were slowly hydrated overnight in a climatic chamber at 90% and then washed with deionized running water (1.1 μS/cm^2) to remove adhering dust and to hypothetically create a pool of lichen material with similar elemental concentrations [18]. The lichen material (60 g) was divided into 30 samples, each weighing 2 g and then underwent the same procedure as those of the BC.

2.3. Surface Area of Lichen and Biochar Samples

The specific surface area (SSA) measurement was conducted on both BCs and *E. prunastri* using the methylene blue (MB) adsorption method. This measurement involves determining the SSA of the materials under investigation by quantifying the amount of MB molecules adsorbed on the surface of the analyte through UV-Vis spectroscopy in a water environment [38–40]. After proper calibration to establish the correlation between absorbance and dye concentration, the analysis was performed by mixing 10 mg of the analyte (samples not treated with Cd) with 10 mL of MB solutions at various concentrations (ranging from 1×10^{-6} M to 3×10^{-4} M) and stirring for 24 h. Subsequently, the suspensions were centrifuged for 30 min at 5000 rpm to separate the solid fraction. The supernatant solutions were then measured using a Jasco V-550 UV–Visible spectrophotometer to assess the differences in MB concentration after coming into contact with the analytes, thereby determining the SSA of the samples under investigation (refer to the Supporting Information for more details).

2.4. Cation Exchange Capacity Measurement

Samples not treated with Cd (2 g) were transferred to 50 mL centrifuge tubes with 30 mL of a pH 8.2 buffered barium chloride ($BaCl_2$) solution. Subsequently, the tubes were centrifuged at 3000× *g* rpm for 1 h. The resulting supernatant was decanted into a 100 mL tared flask. This process was iterated twice, with the clear supernatants being collected in the same 100 mL flask. The final volume was adjusted using the pH 8.2 buffered $BaCl_2$ solution. A comprehensive sample wash was conducted by adding 30 mL of water, followed by centrifugation. Using a precision burette, 30 mL of a 50 mM magnesium sulfate ($MgSO_4$) solution was carefully introduced into the centrifuge tubes. The tubes were subsequently manually agitated to ensure complete dispersion of the sample. Continuous agitation was maintained for 2 h, followed by centrifugation. Ten milliliters of the resulting clear solution were extracted and transferred into a 250 mL flask. To this, 100 mL of distilled water, 10 mL of an ammonium chloride (NH_4Cl) buffer solution, and a minimal amount of indicator were added to create the sample solution. Concurrently, a blank test solution was prepared by transferring 100 mL of distilled water, 10 mL of the 50 mM $MgSO_4$ solution, 10 mL of an NH_4Cl buffer solution, and a small quantity of indicator into a separate 250 mL flask. Both the blank test solution and the sample solution were titrated using a 2.5 cmol/L ethylendiaminotetracetyc acid (EDTA) solution until the appearance of a distinct blue color, signifying the titration endpoint. Results were expressed as meq/100 g. See the Gazzetta ufficiale della Repubblica Italiana, Serie Generale No. 248 of 21 October 1999 of the Ministero delle Politiche Agricole e Forestali for full details [41].

2.5. Treatments with Cd^{2+} and Depuration Phase

Samples were separately immersed for 1 h in solutions containing Cd^{2+} ($CdCl_2$) at concentrations of 0 (control), 10, and 100 μM (accumulation phase), maintaining a sample mass (g)/solution volume (mL) ratio of 1:250. In one hour, both BCs and lichens show (approximately) their maximum Cd^{2+} accumulation capacity [42,43]. After their exposure, all samples were blotted with paper towels to remove the excess water and then left overnight in ambient conditions. Half of each set of the treated samples were then separately soaked in deionized water for 24 h (depuration phase) maintaining a sample mass (g)/solution volume (mL) ratio of 1:250. The samples were blotted with paper towels and dried overnight at 40 °C. All the samples were then stored in vacuum bags at −15 °C until the analysis.

2.6. Sample Analysis

Sample analyses were performed according to [44], with modifications. Lichen samples (2 g) were solubilized with 9 mL HNO_3 (65%) and 1 mL of H_2O_2 at 165 °C for 20 min using a heated digester thermoblock (DK20, Velp Scientifica, Usmate Velate, MB, Italy). Specifically, samples were first solubilized for 10 min with 9 mL of HNO_3 and then with the addition of 1 mL of H_2O_2 for a further 10 min. The solubilized samples were collected in 15 mL tubes and left at room temperature until the analysis. BC samples (2 g) were first incinerated at 450 °C for 24 h and then solubilized as described above; samples were considered ready for acid solubilization when they no longer showed a black discoloration. All digested samples were filtered at 0.22 μm. Cadmium concentrations were quantified by flame atomic absorption spectrometry (AA240FS, Agilent Technologies, Santa Clara, CA, USA). The Cd calibration curve was prepared by diluting 1000 ppm certified standard solutions (Agilent Technologies, Santa Clara, CA, USA). Cadmium concentrations were expressed as mg/kg. Three technical replicates were performed for each sample. Samples with concentrations below the limit of quantification (0.05 mg/kg), such as Cd-untreated samples, were analyzed by ICP-MS (Perkin Elmer-Sciex, Elan 6100; Cd detection limit <0.001 mg/kg). The analytical quality was evaluated using the certified reference material NCS DC73350 "Leaves of Poplar". An average accuracy higher than 95% and a precision of 98% were obtained.

Losses of Cd^{2+} resulting from the incineration step were assessed by comparing the Cd content of BC samples analyzed after the 450 °C process with their direct (albeit partial) chemical digestion using HNO_3 and H_2O_2 (9/1 v/v); see Section 2.6 for details. The recovery of Cd was always higher than 94%.

2.7. Statistics

Due to the low number of statistical replications ($n = 6$), the search for statistically significant differences between the treatments was carried out by means of permutation tests for multiple comparisons, using the Benjamini–Hochberg correction. All calculations were performed through R software [45].

3. Results and Discussion

The concentration of Cd measured in samples of *E. prunastri* before (background concentrations) and after their exposure to two different concentrations of Cd (accumulation phase) as well as after their immersion in a Cd-free solution (depuration phase) are reported in Figure 2.

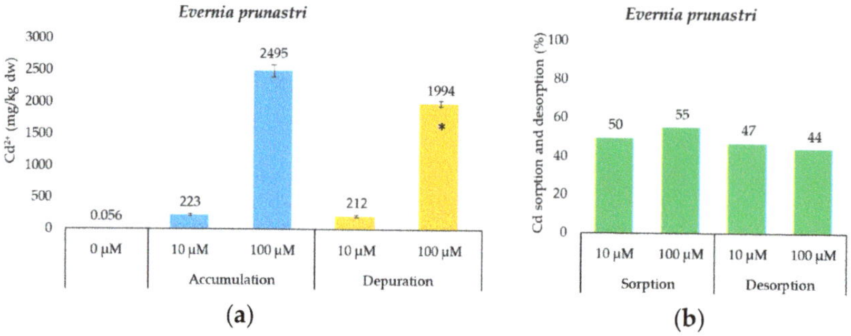

Figure 2. (**a**) concentration of Cd (mean ± standard deviation) in samples of the lichen species *Evernia prunastri* exposed to 10 and 100 μM Cd^{2+} solutions (accumulation phase) and then subjected to a depuration phase (only deionized water) for 24 h; the values above the bars are the averages of the six replicates analyzed. Legend: * indicates statistically significant differences in Cd concentration (for each exposure concentration) between the 'accumulation' and 'depuration' phases ($p < 0.05$; N = 6); and (**b**) cadmium sorption and desorption (%) in lichen samples first exposed to Cd and then exposed to the purification step, respectively. The values above the bars represent the percentages of Cd sequestered and therefore present on the surface of lichen after the accumulation and purification treatments.

Unexposed lichen samples (control samples) showed Cd concentrations consistent with those commonly found in lichens from remote areas (i.e., 0.14 ± 0.03 mg/kg_{dw} from Cecconi et al. [46] and 0.21 ± 0.17 mg/kg_{dw} from Vannini et al. [47]). On the other hand, Cd-exposed samples showed concentrations several orders of magnitude higher than those commonly found in unpolluted thalli, thus confirming the efficiency of lichens to accumulate ionic Cd in proportion to its simulated environmental availability. Specifically, the concentrations of Cd measured within the lichen *E. prunastri* treated with 10 uM Cd solutions are approximately 32 times higher than those detectable in thalli of the fruticose lichen *Ramalina farinacea* exposed to emissions from a lignite power plant [48] and many orders of magnitude higher than those measured in thalli of *Parmelia sulcata* native to a mining area in Ghana [49]. Nevertheless, Cd concentrations measured in treated thalli of *E. prunastri* are much higher than those measured in thalli of the same lichen species treated with the same concentrations of Cd used in this experiment, albeit after a shorter exposure time (about 30 min; [50]). In detail, Sujetovienė and Šliumpaitė [50] measured Cd concentrations ca. 2.5 times lower than those measured in this study after the treatment with 10 μM Cd^{2+}. These differences are probably due to the different treatment the samples underwent before total chemical analysis, which in the case of Sujetovienė and Šliumpaitė [50] was a rapid washing of the samples with deionized water immediately after the accumulation phase. This step, which was not carried out in this study, likely hindered lichen samples from accumulating the metal more efficiently. This is because the part of the metal not yet bound to the active sites of the organism would have been immediately removed by washing. After all, the lichen exhibits an excellent ability to accumulate the metal from the exposure solution, and this capacity seems to be independent of the provided metal concentration. In fact, a calculation aimed at estimating the percentage of metal sequestered by the lichen following its exposure to the lowest (10 μM) and highest (100 μM) treatment concentrations indicates a metal sequestration capacity of 50% and 55%, respectively. These results are quite consistent with those reported by Paoli et al. [26] for the lichen *Xanthoria parietina*.

The accumulation of ionic forms of elements within lichens is the result of both intracellular and extracellular processes [51]. Intracellular accumulation occurs when an element has lower affinity for extracellular exchange groups (i.e., carboxylic and hydrocarboxylic groups and chitin; [52]) compared to intracellular ones. On the other hand, extracellular accumulation occurs through the chelating action of oxalates produced by the mycobiont, along with ionic bonding with the cell wall of the fungal and algal partners [53]. When compared to other essential elements for metabolism such as Cu, for example, Cd exhibits higher rates of extracellular accumulation than intracellular rates [26,54].

The accumulation of metals within lichens is, however, a partially reversible process [55]; in fact, compared to lichen samples exposed to the accumulation phase, samples exposed to the purification phase (24 h) showed a non-significant (about 3%; $p > 0.05$) as well as a significant Cd release (i.e., about 9%) when exposed to the lowest and highest Cd concentrations, respectively, with an average Cd release of about 6%. The absence of a significant release of ionic Cd^{2+} from samples exposed to the lowest treatment concentration may indicate an excellent ability of the lichen to accumulate the metal irreversibly, at least over the 24 h period tested; this result could be attributed to the incomplete saturation of the lichen's binding sites for Cd. Instead, the statistically significant reduction in the Cd concentration from thalli treated with the highest treatment concentration (100 μM) may be due to the release of Cd that was not effectively immobilized in the thallus. In addition, however, the reduction in Cd content after exposure to the purification phase could also be due to partial release of the metal temporarily accumulated intracellularly due to an increase in its extracellular concentration, a process observed in both *Peltigera rufescens* and *Cladina arbuscula* [20]. In fact, Loppi et al. [56] clearly reported that the reduction in PTE content 24 h after the purification phase was mainly due to the release of the metal from the intracellular level of the lichen, while its content at the extracellular level remained unchanged. Cadmium exposure easily activates defense responses at the photobiont level (such as phytochelatin activation and protein synthesis; [57]) to avoid physiological impairment [58], mechanisms that are probably involved in the dynamics assumed above. Cadmium is, in fact, a non-essential element for the metabolism of lichens (or living organisms in general), which easily justifies the low rate of intracellular uptake (μmol/g/h) in lichen thalli as compared to their extracellular uptake [42].

Figure 3 shows the concentration of Cd measured in BC1 and BC2 before (background concentrations) and after their exposure to two different concentrations of Cd (accumulation phase) as well as after their immersion in a Cd-free solution (depuration phase).

Both unexposed biochar samples showed Cd concentrations consistent with each other and with unexposed lichen samples. In particular, the Cd concentration in both unexposed BC samples was below the maximum acceptable level for organic certification (0.7 mg/kg; [59]), the most stringent level for the use of BC in agriculture. After exposure to the accumulation phase, BC1 showed Cd accumulation proportional to the treatment concentration, whereas BC2 did not. In fact, the Cd concentration measured in BC2 samples after the treatment with the highest Cd concentration was only six times higher than that measured after the treatment with the 10 μM Cd^{2+} solution. The Cd content measured in BC samples treated with 10 μM Cd^{2+} solution aligns with the adsorption capacity of the material for the metal studied (0.3–39.1 mg/kg; [60]), thus confirming the ability of BC as a sorbent material for the removal of PTEs from water bodies. On the other hand, BCs exposed to the highest Cd concentration (100 μM; i.e., not environmentally relevant) easily exceeded this range.

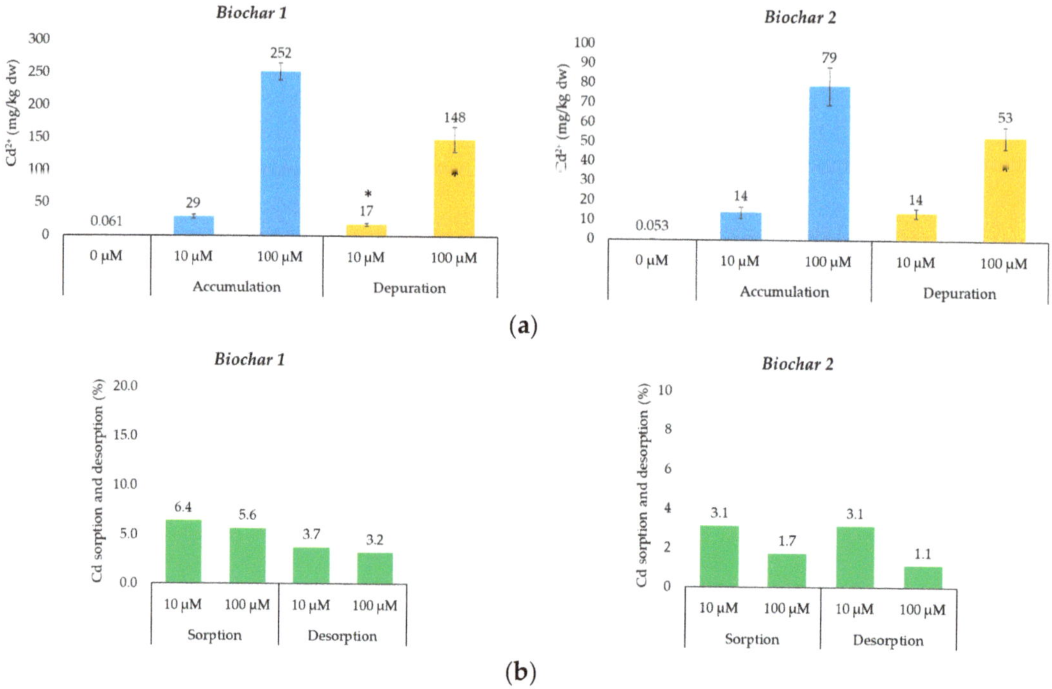

Figure 3. (**a**) concentration of Cd (mean ± standard deviation) in sample of the aged (BC1) and new (BC2) wood-derived biochar exposed to 10 and 100 μM Cd^{2+} solutions (accumulation phase) and then subjected to depuration phase (only deionized water) for 24 h; the values above the bars are the averages of the six replicates analyzed. Legend: * indicates statistically significant differences in Cd concentration (for each exposure concentration) between the 'accumulation' and 'depuration' phases ($p < 0.05$; N = 6); and (**b**) cadmium sorption and desorption (%) in in BC1 and BC2 samples first exposed to Cd and then exposed to the purification step, respectively. The values above the bars represent the percentages of Cd sequestered and therefore present on the surface of the BC after the accumulation and purification treatments.

After being subjected to the purification step, BC1 and BC2 showed a different ability to retain the accumulated metal. Specifically, BC1 showed a constant (ca. 40%) capacity to release Cd, regardless of its initial ionic concentration, whereas BC2 released the metal significantly ($p < 0.01$) only after treatment with 100 μM Cd; no Cd release was measured in BC2 samples treated with 10 μM Cd^{2+} and exposed to the depuration phase ($p > 0.05$). These results suggest that BC2 seems to be able to accumulate the metal stably (but not permanently), at least when adsorbed up to a concentration of about 14 mg/kg. Since BC1 showed a Cd concentration of 17 mg/kg after the purification step, this value could perhaps be considered as the Cd concentration that biochar (in a general sense) would be able to sequester stably after 24 h, provided there is no competition for exchange sites with other elements [26]. The presence of other metals in the solution could reduce the accumulation of Cd on the surface of BC, thereby diminishing its ability to sequester the metal [61].

Looking at Figure 2, both BCs showed a different ability to accumulate and release Cd, with the former (BC1) appearing to accumulate it more efficiently than the latter (BC2), although it has little capacity to retain it once adsorbed. Specifically, BC1 showed an ability to accumulate Cd in an ionic form that was approximately three times greater ($p < 0.001$; Student's t-test) than that of BC2 when exposed to high Cd concentrations (100 μM) and approximately two times greater ($p < 0.001$; Student's t-test) when exposed to environmentally relevant concentrations (10 μM). However, when the concentrations compared were those measured after the depuration phase, BC2 showed a better ability to retain Cd than BC1 only when exposed to very high concentrations of the metal. Therefore, BC1, having both a greater capacity to accumulate Cd and to release it once its environmental concentration decreases compared to BC2, could be tested in the future as a biomonitor, for the environmental availability of heavy metals such as Cd, dissolved in aquatic environments. Comparing the results of Cd accumulation in BC1 with those obtained from *E. prunastri* lichens, the latter have a greater capacity to both accumulate and retain the metal in ionic form ($p < 0.001$; Student's t-test).

Table 1 summarizes the results of the surface area analysis (m^2/g) calculated for each of the materials studied before their exposure to Cd, see the corresponding curves (Figures S1–S3) in the Supplementary Materials File.

Table 1. Estimated surface area (m^2/g) and relative standard error as well as, cation exchange capacity (CEC), and density, of the lichen *Evernia prunastri*, biochar 1 (BC1), and biochar 2 (BC2).

Samples	Surface Area (m^2/g) ± Std. Dev.	Cation Exchange Capacity (CEC; meq/100 g)	Density (g/cm^3)
Lichen (*E. prunastri*)	356 ± 26	37.7	-
Biochar 1 (BC1)	78 ± 10	28.0	0.21
Biochar 2 (BC2)	28 ± 2	14.8	0.44

It is clear from the table that the three materials studied in terms of surface area follow the hierarchy: lichen > BC1 > BC2, results that are clearly consistent with the hierarchy described above for Cd accumulation capacity: i.e., lichen > BC1 > BC2. Therefore, the parameters Cd accumulation and surface area appear to be positively correlated, albeit in a power rather than a linear relationship (Figure 4).

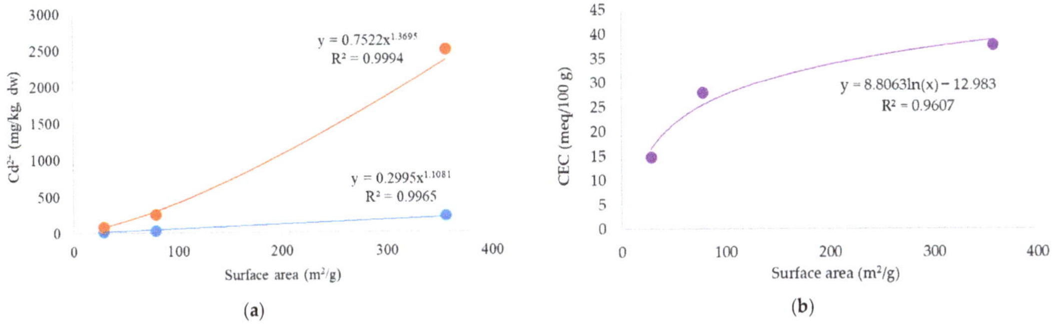

Figure 4. (**a**) correlation between surface area (m^2/g) and Cd content (mg/kg) for the environmental matrices tested exposed to the lower (10 μM; blue line and dots) and higher (100 μM; brown line and dots) Cd concentration; and (**b**) correlation between surface area (m^2/g) and CEC (meq/100 g) for the environmental matrices tested before their exposure to Cd.

The greater capacity of BC1 than BC2 to accumulate Cd could be due to its 'seniority'. Ageing increases the water-holding capacity of BC [62] and therefore its ability to quantitatively sequester elements such as Cd [63]. Consistent with this, BC1 has a lower density than BC2, confirming the higher surface area measured compared to that of BC2. Biochemical ageing can also increase the amount of oxygenated functional groups on the BC surface, although without increasing the surface area [64], thus providing an additional sequestration surface for Cd. Indeed, the importance of these groups for the removal of Cd from contaminated sites is well known [65,66]. However, it is important to specify that, since these two types of BC came from different biomass sources, the differences in Cd accumulation between the two investigated BCs may simply be due to variations in their chemical and physical characteristics [67,68], such as surface area [69] and cation exchange capacity (CEC, i.e., the substrate's ability to exchange ions), parameters that are closely related. This relationship was observed among the three materials analyzed, although it follows a logarithmic trend rather than a linear one. This observation confirms what was previously noted: as the surface area increases, so does the material's capacity to adsorb (and subsequently exchange) PTEs such as Cd. It is reasonable to speculate that the activation of BC, in addition to increasing the sample's surface area [70], may also enhance its CEC [71], thereby improving its ability to adsorb PTEs. Comparing the PTE-accumulation capacity between lichen and activated BC would certainly yield intriguing results, given that activation can significantly increase BC's surface area. As a result, all parameters related to Cd sequestration, such as complexation, physical absorption, cation exchange capacity, precipitation, and electrostatic absorption [60], would be amplified. Future indications regarding the differential capacity for the accumulation of PTEs in atmospheric depositions between the lichen *E. prunastri* and activated BC could shed light on the potential use of BC to assess the environmental availability of PTEs in atmospheric depositions primarily associated with particulate matter (PM) [18]. These findings could be of considerable interest in testing other valid surrogates for this purpose [72]. Indeed, as far as we know, research on BC's ability to trap PM from atmospheric depositions seems to be an underexplored topic.

Currently, the use of biochar in the environmental field concerns both the reduction of PTE mobility in soils to increase the edibility of crop products [28,29] and the removal of heavy metals from wastewater [31,32]. However, based on the results obtained in this study, its ability to provide information on the environmental availability of heavy metals such as Cd should not be ruled out, especially when it comes to water bodies. Its use for soil monitoring would, therefore, be unnecessary, as this matrix can be easily sampled using traditional methodologies. By exposing biochar to a water body for a defined period (i.e., one month), it would be possible (this is purely speculative for now) to obtain integrated information on the availability of the chemical elements in the water body under investigation, without having to resort to numerous point source analyses of the water body, which may provide an accurate indication of element concentrations, but require numerous monitoring campaigns. Using this approach, biochar could be used in a very similar way to lichens, i.e., to provide information on the environmental availability of PTEs during the exposure period, but in the aquatic rather than the atmospheric compartment. Specific work would therefore be useful to clarify the feasibility of using this substrate to monitor the integrated availability of PTEs in streams of environmental interest.

4. Conclusions

Biochar has various environmental applications. However, its largely unexplored potential for environmental biomonitoring has prompted us to fill this gap. This study provides an initial analysis of how BC could serve as a biomonitor for the environmental availability of ionic Cd, using the lichen *Evernia prunastri* (L.) Ach. as a reference organism. The lichen showed a very good ability to adsorb ionic Cd—higher than the two biochar samples (more than 46.5%)—and a weak ability to release the metal (ca. 6%). As compared to the lichen, BC2 had a lower capacity for Cd accumulation (−48%) and release (ca. 3%). BC1, on the other hand, showed a slightly higher Cd accumulation capacity than BC2 (+3.6%), but a release capacity similar to that of the lichen (ca. 5%). Specific information regarding the dynamics of PTE accumulation and release in BC requires further investigation. The surface area and the CEC of the organisms/materials being tested seem to play a key role in their ability to accumulate Cd. Based on the results obtained, this work suggests the potential use of biochar as a (bio)monitor for the presence of PTEs in atmospheric depositions as well as in water bodies, when activated by, for example, increasing their surface area to enhance the material's capacity to accumulate the target metal.

Supplementary Materials: The following supporting information can be downloaded at: https://www.mdpi.com/article/10.3390/toxics12010066/s1, Figure S1: Langmuir isotherm fit for BC1 (new) sample; Figure S2: Langmuir isotherm fit for BC2 (aged) sample; Figure S3: Langmuir isotherm fit for *Evernia prunastri* (L.) Ach. sample.

Author Contributions: Conceptualization: A.V. and S.L.; methodology: A.V., L.P, M.M., R.F., M.B., M.S., G.M., D.P. and M.R.; validation: A.V.; formal analysis: A.V. and L.P.; investigation: A.V. and L.P.; resources: A.V., M.M., S.L. and A.P.; data curation: A.V.; writing—original draft preparation: A.V.; writing—review and editing: L.P, M.B., R.F., A.M., M.S., D.P., M.R., M.M., A.P. and S.L.; supervision: A.V. and S.L. All authors have read and agreed to the published version of the manuscript.

Funding: This research received no external funding.

Institutional Review Board Statement: Not applicable.

Informed Consent Statement: Not applicable.

Data Availability Statement: The raw data presented in this study are available on request from the corresponding author.

Acknowledgments: LP and MM thanks the support of FIL ("Fondi Locali per la Ricerca"), the Director of Consorzio Interuniveristario Nazionale per le Scienze Ambientali (CINSA), and Nelson Marmiroli. AV thanks the REI Laboratory in Parma for the analysis of the cation exchange capacity of the samples. This work has benefited from the equipment and framework of the COMP-HUB and COMP-R Initiatives, funded by the 'Departments of Excellence' program of the Italian Ministry for University and Research (MIUR, 2018–2022 and MUR, 2023–2027). The publication was made by a researcher with a research contract co-funded by the European Union—PON Research and Innovation 2014–2020 in accordance with Article 24, paragraph 3, letter a), of Law no. 240 of December 30, 2010, as amended, and of Ministerial Decree No. 1062 of August 10, 2021.

Conflicts of Interest: The authors declare no conflicts of interest.

References

1. Wedepohl, K.H. The composition of the continental crust. *Geochim. Cosmochim. Acta* **1995**, *59*, 1217–1232. [CrossRef]
2. WHO, World Health Organization, 2000. Air Quality Guidelines—Second Edition. Available online: https://www.who.int/publications/i/item/9789289013581 (accessed on 23 December 2023).
3. Pan, J.; Plant, J.A.; Voulvoulis, N.; Oates, C.J.; Ihlenfeld, C. Cadmium levels in Europe: Implications for human health. *Environ. Geochem. Health* **2010**, *32*, 1–12. [CrossRef] [PubMed]
4. IARC International Agency for Research on Cancer. Available online: https://monographs.iarc.who.int/wp-content/uploads/2018/06/mono100C-8.pdf (accessed on 7 May 2023).
5. Cho, Y.A.; Kim, J.; Woo, H.D.; Kang, M. Dietary Cadmium Intake and the Risk of Cancer: A Meta-Analysis. *PLoS ONE* **2013**, *8*, e75087. [CrossRef] [PubMed]
6. Genchi, G.; Sinicropi, M.S.; Lauria, G.; Carocci, A.; Catalano, A. The Effects of Cadmium Toxicity. *Int. J. Environ. Res. Public Health* **2020**, *17*, 3782. [CrossRef] [PubMed]
7. Thornton, I. Geochemistry of cadmium. Cadmium in the Environment. In *Experientia Supplementum*; Mislin, H., Ravera, O., Eds.; Birkhäuser: Basel, Switzerland, 1986; Volume 50. [CrossRef]
8. WHO, World Health Organization, 2011. Cadmium in Drinking-Water. Available online: https://www.who.int/publications/m/item/background-documents-for-development-of-who-guidelines-for-drinking-water-quality-and-guidelines-for-safe-recreational-water-environments (accessed on 23 December 2023).
9. Kubier, A.; Wilkin, R.T.; Pichler, T. Cadmium in soils and groundwater: A review. *Appl. Geochem.* **2019**, *108*, 104388. [CrossRef]
10. European Regulation, 1881/2006. Available online: https://eur-lex.europa.eu/LexUriServ/LexUriServ.do?uri=OJ:L:2006:364:0005:0024:EN:PDF (accessed on 20 September 2023).
11. Aarhus Protocol on Heavy Metals. 1998. Available online: https://unece.org/environment-policy/air/protocol-heavy-metals#:~:text=The%201998%20Aarhus%20Protocol%20on%20Heavy%20Metals&text=The%20Protocol%20aims%20to%20cut,road%20transport)%20and%20waste%20incineration (accessed on 3 June 2023).
12. EEA European Environmental Agency. 2023. Available online: https://www.eea.europa.eu/ims/heavy-metal-emissions-in-europe (accessed on 23 June 2023).
13. ICP Vegetation, 2023. Heavy Metals, Nitrogen and Pops in European Mosses: 2020 Survey. Available online: https://icpvegetation.ceh.ac.uk/get-involved/manuals/moss-survey (accessed on 27 November 2023).

14. Pacyna, J.M.; Pacyna, E.G.; Aas, W. Changes of Emissions and Atmospheric Deposition of Mercury, Lead, and Cadmium. *Atmos. Environ.* **2009**, *43*, 117–127. [CrossRef]
15. Ballabio, C.; Jones, A.; Panagos, P. Cadmium in Topsoils of the European Union—An Analysis Based on LUCAS Topsoil Database. *Sci. Total Environ.* **2024**, *912*, 168710. [CrossRef]
16. Bargagli, R. Moss and lichen biomonitoring of atmospheric mercury: A review. *Sci. Total Environ.* **2016**, *572*, 216–231. [CrossRef]
17. Vannini, A.; Paoli, L.; Nicolardi, V.; Di Lella, L.A.; Loppi, S. Seasonal variations in intracellular trace element content and physiological parameters in the lichen Evernia prunastri transplanted to an urban environment. *Acta Bot. Croat.* **2017**, *76*, 171–176. [CrossRef]
18. Vannini, A.; Paoli, L.; Russo, A.; Loppi, S. Contribution of submicronic (PM1) and coarse (PM > 1) particulate matter deposition to the heavy metal load of lichens transplanted along a busy road. *Chemosphere* **2019**, *231*, 121–125. [CrossRef]
19. Contardo, T.; Vannini, A.; Sharma, K.; Giordani, P.; Loppi, S. Disentangling sources of trace element air pollution in complex urban areas by lichen biomonitoring. A case study in Milan (Italy). *Chemosphere* **2020**, *256*, 127155. [CrossRef] [PubMed]
20. Bačkor, M.; Kováčik, J.; Piovár, J.; Pisani, T.; Loppi, S. Physiological Aspects of Cadmium and Nickel Toxicity in the Lichens Peltigera rufescens and Cladina arbuscula Subsp. mitis. *Water Air Soil Pollut.* **2010**, *207*, 253–262. [CrossRef]
21. Pisani, T.; Munzi, S.; Paoli, L.; Bačkor, M.; Kováčik, J.; Piovár, J.; Loppi, S. Physiological effects of mercury in the lichens Cladonia arbuscula subsp. mitis (Sandst.) Ruoss and Peltigera rufescens (Weiss) Humb. *Chemosphere* **2011**, *82*, 1030–1037. [CrossRef] [PubMed]
22. Vannini, A.; Tedesco, R.; Loppi, S.; Di Cecco, V.; Di Martino, L.; Nascimbene, J.; Dallo, F.; Barbante, C. Lichens as monitors of the atmospheric deposition of potentially toxic elements in high elevation Mediterranean ecosystems. *Sci. Total Environ.* **2021**, *798*, 149369. [CrossRef]
23. Godinho, R.M.; Verburg, T.G.; Freitas, M.D.C.; Wolterbeek, H.T. Dynamics of element accumulation and release of Flavoparmelia caperata during a long-term field transplant experiment. *Int. J. Environ. Health* **2011**, *5*, 49. [CrossRef]
24. Walther, D.A.; Ramelow, G.J.; Beck, J.N.; Young, J.C.; Callahan, J.D.; Maroon, M.F. Temporal changes in metal levels of the lichens *Parmotrema praesorediosum* and *Ramalina stenospora*, southwest Louisiana. *Water Air Soil Pollut.* **1990**, *53*, 189–200. [CrossRef]
25. Vannini, A.; Nicolardi, V.; Bargagli, R.; Loppi, S. Estimating atmospheric mercury concentrations with lichens. *Environmental* **2014**, *48*, 8754–8759. [CrossRef]
26. Paoli, L.; Vannini, A.; Monaci, F.; Loppi, S. Competition between heavy metal ions for binding sites in lichens: Implications for biomonitoring studies. *Chemosphere* **2018**, *199*, 655–660. [CrossRef]
27. Servin, A.D.; De la Torre-Roche, R.; Castillo-Michel, H.; Pagano, L.; Hawthorne, J.; Musante, C.; White, J.C. Exposure of agricultural crops to nanoparticle CeO_2 in biochar-amended soil. *Plant Physiol. Biochem.* **2017**, *110*, 147–157. [CrossRef]
28. He, L.; Zhong, H.; Liu, G.; Dai, Z.; Brookes, P.C.; Xu, J. Remediation of heavy metal contaminated soils by biochar: Mechanisms, potential risks and applications in China. *Environ. Pollut.* **2019**, *252*, 846–855. [CrossRef]
29. Vannini, A.; Jamal, M.B.; Gramigni, M.; Fedeli, R.; Ancora, S.; Monaci, F.; Loppi, S. Accumulation and release of mercury in the lichen Evernia prunastri (L.) Ach. *Biology* **2021**, *10*, 1198. [CrossRef] [PubMed]
30. Wang, Y.; Liu, Y.; Zhan, W.; Zheng, K.; Wang, J.; Zhang, C.; Chen, R. Stabilization of heavy metal-contaminated soils by biochar: Challenges and recommendations. *Sci. Total Environ.* **2020**, *729*, 139060. [CrossRef] [PubMed]
31. Kumar, A.; Singh, E.; Mishra, R.; Kumar, S. Biochar as environmental armour and its diverse role towards protecting soil, water and air. *Sci. Total Environ.* **2022**, *806*, 150444. [CrossRef]
32. Zhao, Z.; Wang, B.; Theng, B.K.G.; Lee, X.; Zhang, X.; Chen, M.; Xu, P. Removal performance, mechanisms, and influencing factors of biochar for air pollutants: A critical review. *Biochar* **2022**, *4*, 30. [CrossRef]
33. Gwenzi, W.; Chaukura, N.; Wenga, T.; Mtisi, M. Biochars as media for air pollution control systems: Contaminant removal, applications and future research directions. *Sci. Total Environ.* **2021**, *753*, 142249. [CrossRef]
34. Mukherjee, A.; Majumdar, S.; Servin, A.D.; Pagano, L.; Dhankher, O.P.; White, J.C. Carbon Nanomaterials in Agriculture: A Critical Review. *Front. Plant Sci.* **2016**, *7*, 172. [CrossRef] [PubMed]
35. Marmiroli, M.; Caldara, M.; Pantalone, S.; Malcevschi, A.; Maestri, E.; Keller, A.A.; Marmiroli, N. Building a risk matrix for the safety assessment of wood derived biochars. *Sci. Total Environ.* **2022**, *839*, 156265. [CrossRef] [PubMed]
36. Rahim, H.U.; Akbar, W.A.; Alatalo, J.M. A Comprehensive Literature Review on Cadmium (Cd) Status in the Soil Environment and Its Immobilization by Biochar-Based Materials. *Agronomy* **2022**, *12*, 877. [CrossRef]
37. Beesley, L.; Marmiroli, M. The immobilisation and retention of soluble arsenic, cadmium and zinc by biochar. *Environ. Pollut.* **2011**, *159*, 474–480. [CrossRef]
38. Hegyesi, N.; Vad, R.T.; Pukánszky, B. Determination of the specific surface area of layered silicates by methylene blue adsorption: The role of structure, pH and layer charge. *Appl. Clay Sci.* **2017**, *146*, 50–55. [CrossRef]
39. Nunes, C.A.; Guerreiro, M.C. Estimation of surface area and pore volume of activated carbons by methylene blue and iodine numbers. *Quím. Nova* **2011**, *34*, 472–476. [CrossRef]
40. Fornasini, L.; Scaravonati, S.; Magnani, G.; Morenghi, A.; Sidoli, M.; Bersani, D.; Pontiroli, D. In Situ decoration of laser-scribed graphene with TiO_2 nanoparticles for scalable high-performance micro-supercapacitors. *Carbon* **2021**, *176*, 296–306. [CrossRef]
41. Gazzetta Ufficiale Della Repubblica Italiana. Available online: https://www.gazzettaufficiale.it/eli/gu/1999/10/21/248/so/185/sg/pdf (accessed on 4 February 2023).

42. Beckett, R.P.; Brown, D.H. The Control of Cadmium Uptake in the Lichen Genus Peltigera. *J. Exp. Bot.* **1984**, *35*, 1071–1082. [CrossRef]
43. Xiang, A.; Gao, Z.; Zhang, K.; Jiang, E.; Ren, Y.; Wang, M. Study on the Cd (II) adsorption of biochar based carbon fertilizer. *Ind. Crops Prod.* **2021**, *174*, 114213. [CrossRef]
44. Marmiroli, M.; Bonas, U.; Imperiale, D.; Lencioni, G.; Mussi, F.; Marmiroli, N.; Maestri, E. Structural and functional features of chars from different biomasses as potential plant amendments. *Front. Plant Sci.* **2018**, *9*, 1119. [CrossRef] [PubMed]
45. R Development Core Team. Available online: https://www.R-project.org/ (accessed on 9 March 2022).
46. Cecconi, E.; Fortuna, L.; Benesperi, R.; Bianchi, E.; Brunialti, G.; Contardo, T.; Di Nuzzo, L.; Frati, L.; Monaci, F.; Munzi, S.; et al. New Interpretative Scales for Lichen Bioaccumulation Data: The Italian Proposal. *Atmosphere* **2019**, *10*, 136. [CrossRef]
47. Vannini, A.; Bianchi, E.; Avi, D.; Damaggio, N.; Di Lella, L.; Nannoni, F.; Protano, G.; Loppi, S. Biochar Amendment Reduces the Availability of Pb in the Soil and Its Uptake in Lettuce. *Toxics* **2021**, *9*, 268. [CrossRef]
48. Riga-Karandinos, A.N.; Karandinos, M.G. Assessment of Air Pollution from a Lignite Power Plant in the Plain of Megalopolis Ž Greece. Using as Biomonitors Three Species of Lichens; Impacts on Some Biochemical Parameters of Lichens. *Sci. Total Environ.* **1998**, *215*, 167–183. [CrossRef]
49. Boamponsem, L.K.; Adam, J.I.; Dampare, S.B.; Nyarko, B.J.B.; Essumang, D.K. Assessment of Atmospheric Heavy Metal Deposition in the Tarkwa Gold Mining Area of Ghana Using Epiphytic Lichens. *Nucl. Instrum. Methods Phys. Res. Sect. B Beam Interact. Mater. At.* **2010**, *268*, 1492–1501. [CrossRef]
50. Sujetovienė, G.; Šliumpaitė, I. Effects of Cadmium on Physiological Parameters of the Lichen Evernia Prunastri and Ramalina Fastigiata. *E3S Web Conf.* **2013**, *1*, 29007. [CrossRef]
51. Bargagli, R.; Mikhailova, I. Accumulation of Inorganic Contaminants. In *Monitoring with Lichens—Monitoring Lichens*; Nimis, P.L., Scheidegger, C., Wolseley, P.A., Eds.; NATO Science Series (Series IV: Earth and Environmental Sciences); Springer: Dordrecht, The Netherlands, 2002; Volume 7, pp. 65–84.
52. Galun, M.; Keller, P.; Malki, D.; Feldstein, H.; Galun, E.; Siegel, S.M.; Siegel, B.Z. Removal of Uranium(VI) from Solution by Fungal Biomass and Fungal Wall-Related Biopolymers. *Science* **1983**, *219*, 285–286. [CrossRef]
53. Bačkor, M.; Loppi, S. Interactions of Lichens with Heavy Metals. *Biol. Plant.* **2009**, *53*, 214–222. [CrossRef]
54. Rola, K. Insight into the Pattern of Heavy-Metal Accumulation in Lichen Thalli. *J. Trace Elem. Med. Biol.* **2020**, *61*, 126512. [CrossRef] [PubMed]
55. Nieboer, E.; Richardson, D.H.S.; Tomassini, F.D. Mineral Uptake and Release by Lichens: An Overview. *Bryologist* **1978**, *81*, 226. [CrossRef]
56. Loppi, S.; Di Lucia, A.; Vannini, A.; Ancora, S.; Monaci, F.; Paoli, L. Uptake and release of copper ions in epiphytic lichens. *Biologia* **2020**, *75*, 1547–1552. [CrossRef]
57. Sanità di Toppi, L.; Pawlik-Skowrońska, B.; Vurro, E.; Vattuone, Z.; Kalinowska, R.; Restivo, F.M.; Musetti, R.; Skowroński, T. First and second line mechanisms of cadmium detoxification in the lichen photobiont Trebouxia impressa (Chlorophyta). *Environ. Pollut.* **2008**, *151*, 280–286. [CrossRef]
58. Kováčik, J.; Dresler, S.; Babula, P.; Hladký, J.; Sowa, I. Calcium has protective impact on cadmium-induced toxicity in lichens. *Plant Physiol. Biochem.* **2020**, *156*, 591–599. [CrossRef] [PubMed]
59. EBC European Biochar Certificate. 2023. Available online: https://www.european-biochar.org/en (accessed on 23 June 2023).
60. Zhang, K.; Yi, Y.; Fang, Z. Remediation of cadmium or arsenic contaminated water and soil by modified biochar: A review. *Chemosphere* **2023**, *311*, 136914. [CrossRef]
61. Meng, Z.; Xu, T.; Huang, S.; Ge, H.; Mu, W.; Lin, Z. Effects of Competitive Adsorption with Ni(II) and Cu(II) on the Adsorption of Cd(II) by Modified Biochar Co-Aged with Acidic Soil. *Chemosphere* **2022**, *293*, 133621. [CrossRef]
62. Xing, D.; Cheng, H.; Ning, Z.; Liu, Y.; Lin, S.; Li, Y.; Wang, X.; Hill, P.; Chadwick, D.; Jones, D.L. Field aging declines the regulatory effects of biochar on cadmium uptake by pepper in the soil. *J. Environ. Manag.* **2022**, *321*, 115832. [CrossRef]
63. Li, K.; Yin, G.; Xu, Q.; Yan, J.; Hseu, Z.Y.; Zhu, L.; Lin, Q. Influence of Aged Biochar Modified by Cd^{2+} on Soil Properties and Microbial Community. *Sustainability* **2020**, *12*, 4868. [CrossRef]
64. Goñi-Urtiaga, A.; Courtier-Murias, D.; Picca, G.; Valentín, J.L.; Plaza, C.; Panettieri, M. Response of water-biochar interactions to physical and biochemical aging. *Chemosphere* **2022**, *307*, 136071. [CrossRef] [PubMed]
65. Bian, Y.; Bian, Z.; Zhang, J.; Ding, A.; Liu, S.; Zheng, L.; Wang, H. Adsorption of cadmium ions from aqueous solutions by activated carbon with oxygen-containing functional groups. *Chin. J. Chem. Eng.* **2015**, *23*, 1705–1711. [CrossRef]
66. Navas-Cárdenas, C.; Caetano, M.; Endara, D.; Jiménez, R.; Lozada, A.B.; Manangón, L.E.; Navarrete, A.; Reinoso, C.; Sommer-Márquez, A.E.; Villasana, Y. The Role of Oxygenated Functional Groups on Cadmium Removal using Pyrochar and Hydrochar Derived from Guadua angustifolia Residues. *Water* **2023**, *15*, 525. [CrossRef]
67. Khan, M.A.; Khan, S.; Ding, X.; Khan, A.; Alam, M. The effects of biochar and rice husk on adsorption and desorption of cadmium on to soils with different water conditions (upland and saturated). *Chemosphere* **2018**, *193*, 1120–1126. [CrossRef]
68. Carnier, R.; Coscione, A.R.; Abreu, C.A.D.; Melo, L.C.A.; Silva, A.F.D. Cadmium and lead adsorption and desorption by coffee waste-derived biochars. *Bragantia* **2022**, *81*, e0622. [CrossRef]
69. Yuan, S.; Hong, M.; Li, H.; Ye, Z.; Gong, H.; Zhang, J.; Huang, Q.; Tan, Z. Contributions and Mechanisms of Components in Modified Biochar to Adsorb Cadmium in Aqueous Solution. *Sci. Total Environ.* **2020**, *733*, 139320. [CrossRef]

70. Leng, L.; Xiong, Q.; Yang, L.; Li, H.; Zhou, Y.; Zhang, W.; Huang, H. An overview on engineering the surface area and porosity of biochar. *Sci. Total Environ.* **2021**, *763*, 144204. [CrossRef]
71. Kharel, G.; Sacko, O.; Feng, X.; Morris, J.R.; Phillips, C.L.; Trippe, K.; Lee, J.W. Biochar surface oxygenation by ozonization for super high cation exchange capacity. *ACS Sustain. Chem. Eng.* **2019**, *7*, 16410–16418. [CrossRef]
72. Loppi, S.; Vannini, A.; Monaci, F.; Dagodzo, D.; Blind, F.; Erler, M.; Fränzle, S. Can Chitin and Chitosan Replace the Lichen Evernia prunastri for Environmental Biomonitoring of Cu and Zn Air Contamination? *Biology* **2020**, *9*, 301. [CrossRef]

toxics

MDPI

Article

Hardness-Dependent Freshwater Quality Criteria for the Protection of Aquatic Organisms for Cadmium in China

Zeya Zhang [1], Rui Huang [1], Zhongjie Shen [1,2], Yili Fan [1,3], Chenglian Feng [1] and Yingchen Bai [1,*]

[1] State Key Laboratory of Environmental Criteria and Risk Assessment, Chinese Research Academy of Environmental Sciences, Beijing 100012, China; z.craes@163.com (Z.Z.); huangrui233@mails.ucas.ac.cn (R.H.); 17851550286@163.com (Z.S.); fanyl@craes.org.cn (Y.F.); fengcl@craes.org.cn (C.F.)

[2] School of Environmental Science and Engineering, Changzhou University, Changzhou 213164, China

[3] College of Water Science, Beijing Normal University, Beijing 100875, China

[*] Correspondence: baiyc@craes.org.cn

Abstract: Cadmium poses a significant threat to freshwater aquatic organisms and ecosystems, making it essential to establish regional freshwater quality criteria (FWQC) in China to safeguard these organisms. The toxicity database for cadmium covered 249 acute toxicity data from 52 species (seven phyla and 27 families) and 62 chronic toxicity data from 21 species (four phyla and 12 families). During short-term exposure, *Morone saxatilis* displayed the most sensitivity to cadmium, whereas *Daphnia magna* showed the most sensitivity in long-term exposure scenarios. Significant correlations were identified between water hardness and the toxicity data for cadmium, with the acute toxicity coefficient (K_{ATD}) at 1.0227 (n = 52, $p < 0.05$) and the chronic toxicity coefficient (K_{CTD}) at 0.4983 (n = 21, $p < 0.05$). With the species sensitivity distribution method, the short-term freshwater quality criteria (S-FWQC) were derived with a normal distribution as the best fit (R^2 0.9793), while the long-term freshwater quality criteria (L-FWQC) were calculated using a logistic distribution as the best fit (R^2 0.9686). The formulas for the S-FWQC and L-FWQC were represented as $10^{(1.0227 \times lg(H) - 1.5444)}$ and $10^{(0.4983 \times lg(H) - 1.7549)}$, respectively, with water hardness serving as an independent variable. This study offers valuable insights for improving the management of cadmium to protect freshwater aquatic organisms in China.

Keywords: cadmium; freshwater quality criteria; water hardness; species sensitivity distribution

1. Introduction

Due to its high energy density, corrosion resistance, and optoelectronic properties, cadmium, as a non-essential element and common pollutant of heavy metal, was widely utilized in industries such as batteries, pigments, solar cells, etc. [1–3]. With large-scale industrial applications, cadmium inevitably entered natural water systems during production, usage, and disposal [4–6]. Previous studies showed that cadmium was a prevalent environmental contaminant with both acute and chronic toxicity to aquatic organisms [7,8]. For example, long-term exposure to low environmental doses of cadmium of 0.1–1 μg/L was demonstrated to cause damage to the gills, muscles, brain, and intestines of *Oreochromis mossambicus* at 28–32 °C and dissolved O_2 of 5 mg/L [9]. Concurrently, cadmium can affect metal ion binding, oxidative stress, and energy metabolism within *Pectinidae* tissues in seawater equipped with an air pump [10]. Histopathological changes in the liver and kidney tissues of *Morone saxatilis* were induced by following 96 h exposure to cadmium at a concentration of 20 μg/L [11]. Furthermore, in the 30 d chronic toxicity tests of cadmium (7.5 μg/L) on *Daphnia magna*, tissue damage was induced, leading to stunted growth and reduced reproductive capacity in *Daphnia magna* [12]. Due to its hazardous properties, cadmium was classified as a priority pollutant to freshwater aquatic organisms and human health in several countries including the USA, Canada, Australia, New Zealand, etc., as well as international organizations including the European Union (EU) [13–16]. The freshwater quality criteria (FWQC) for the protection of aquatic organisms were defined as the maximum allowable concentrations of pollutants in water to ensure the health and safety of aquatic organisms [17–19]. As the scientific foundation of water quality standards (WQSs), the FWQC play an important role in environmental management. The FWQC for the protection of aquatic organisms for cadmium were investigated and established in several countries, including the USA, Canada, Australia, and New Zealand [20–22].

WQSs for freshwater in China were formulated by adopting and referring to the FWQC of developed countries/international organizations [17,23,24]. For instance, 0.001 mg/L for Class I, 0.005 mg/L for Class II-IV, and 0.01 mg/L for Class V were set in environmental quality standard for surface water (GB3838-2002) [25] based on the criteria and WQSs of the USA (FWQC for aquatic life of 0.001 mg/L, human health FWQC of 0.005 mg/L, WQS for drinking water sources of 0.01 mg/L, and agricultural irrigation WQS of 0.005 mg/L), Japan (drinking WQS of 0.01 mg/L and WQS for drinking water sources of 0.01 mg/L), the UK (drinking WQS of 0.01 mg/L), etc. [26]. However, the applicability and scientific accuracy of the FWQC have been questioned in protecting the bio-environmental system because of the significant differences among other countries and the geographical and climatic conditions, aquatic biota, etc. in China [18,23,27,28]. For instance, the families of *Salmonidae* and *Cyprinidae* were proposed for the FWQC for the protection of aquatic organisms in the USA and China, respectively, because of the variations in freshwater biota among different nations [29,30]. The FWQC for the protection of aquatic organisms for cadmium, ammonia, and phenol were issued by the Ministry of Ecology and Environment (MEE) in China in 2020 [31–33]. With the processes of research, many research studies indicated that invasive species and internationally common species without distribution in China should be ruled out when deriving FWQC. Therefore, the FWQC for the protection of aquatic organisms from cadmium should be revisited based on the current research results and current guidelines to ensure the scientific and effective management of cadmium in freshwater environments in China [30].

The species sensitivity distribution (SSD) method was successfully applied to derive the FWQC of various pollutants in several countries, including the USA and Canada [20,21]. The SSD method assumed that the tested species were collected randomly and representative of all aquatic organisms within the freshwater ecosystem. The SSD method had clear advantages due to its support by statistical theory and its ability to utilize a full array of toxicological data of freshwater aquatic organisms [34,35]. Therefore, the SSD method might serve as a valuable tool for deriving the FWQC for the protection of aquatic organisms for cadmium in China [36–39]. In addition, the toxicity of cadmium to aquatic organisms was affected by water quality parameters, such as hardness, temperature, pH, etc. [32,40,41]. As water hardness increased, the toxicity of cadmium decreased in aquatic organisms, which was attributed to competition between heavy metals and Ca^{2+} and Mg^{2+} ions for binding sites on the cell membrane [29,42,43]. The median effect concentration (EC_{50}) of cadmium to *Morone saxatilis* in soft water (3.7 μg/L, 40 mg/L as $CaCO_3$) was 86.3% less than that in hard water (27.0 μg/L, 285 mg/L as $CaCO_3$), demonstrated by 96 h acute toxicity tests [44]. For instance, previous studies indicated that higher calcium concentrations can reduce the acute toxicity and uptake of cadmium in *Daphnia magna* with calcium concentrations of 0.5–200 mg/L at pH 8.00–8.20 [45]. Significantly positive relationships have also been documented between the water hardness of experimental water and the toxicity of cadmium to aquatic organisms over the USA by the USEPA [29]. Therefore, it is necessary to derive the hardness-dependent FWQC for the protection of aquatic organisms for cadmium in China.

The objectives of this research included the following: (1) to compare the sensitivities of various species by establishing a toxicity database for cadmium to aquatic organisms, (2) to determine the quantitative relationship between the toxicity data (TD) of cadmium to aquatic organisms and water hardness, and (3) to derive the FWQC for the protection of aquatic organisms for cadmium in China.

2. Materials and Methods

2.1. Toxicity Data Collection and Screen

With the development of environmental criteria research and the increasing amount of acute toxicity data (ATD) and chronic toxicity data (CTD) for pollutants, stricter requirements on the selection of aquatic organisms were imposed by the technical guidelines for deriving water quality criteria for freshwater organisms. The guideline highlighted that tested species should reflect the characteristics of freshwater biota in China, and invasive species such as *Oreochromis niloticus*, *Gambusia affinis*, *Procambarus clarkia*, and others should not be considered as test species. To derive the FWQC for cadmium, aquatic species in China were screened based on documents such as the Chinese biodiversity catalog, the recommended species list for deriving freshwater quality criteria in China, and high-quality scientific papers published on China's national knowledge infrastructure and science citation index. During the derivation of FWQC, the short-term freshwater quality criteria (S-FWQC) and long-term freshwater quality criteria (L-FWQC) were established to provide an appropriate level of protection for aquatic organisms with ATD and CTD, respectively. Short-term exposure was associated with acute toxicity, typically characterized by exposure durations spanning from 24 to 96 h. In the derivation of FWQC for cadmium, the preferred times of short-term exposure are 24, 48, and 96 h for Rotifera, Daphnia/midges, and other organisms, respectively [31]. The median lethal concentration (LC_{50}) and EC_{50} were chosen as endpoints for short-term exposure experiments for cadmium to freshwater aquatic organisms [18]. Long-term exposure, on the other hand, refers to chronic toxicity, with exposure durations designed to be at least 21 days or spanning across one or more generations. The no observed effect concentration (NOEC), the lowest observed effect concentration (LOEC), the maximum acceptable toxicant concentration (MATC), the 10% effect concentration (EC_{10}), and the 20% effect concentration (EC_{20}) were typically chosen as endpoints for the long-term exposure of chronic toxicity during FWQC derivation [34]. Due to their rapid cell division rates, an exposure time of four days was selected as chronic toxicity for algae, including *Chlorella vulgaris*, *Scenedesmus acutus*, etc. [18]. When the LOEC and NOEC were obtained under the same experimental conditions, the MATC was calculated as the geometric mean of the NOEC and LOEC to derive the FWQC for cadmium in our investigation. A list of abbreviations can be found in the Supplementary Materials (Table S1).

During the derivation of FWQC, both ATD and CTD of cadmium to aquatic organisms were collected from the literature and toxicity databases. In detail, the retrieval strategy of "TI = (Cd ion or Cadmium) AND TS = (toxicity or LC_{50} or EC_{50} or EC_{20} or EC_{10} or NOEC or LOEC or MATC)" was applied to the Chinese knowledge resource integrated database (http://www.cnki.net/, accessed on 4 August 2024), Elsevier (http://www.sciencedirect.com, accessed on 13 August 2024), and Web of Science (http://www.webofscience.com, accessed on 15 August 2024), respectively, to research articles related to ATD and CTD of cadmium. For the ECOTOX database (http://cfpub.epa.gov/ecotox, accessed on 20 August 2024), the retrieval strategy "Chemicals = (cadmium) and Effects = (all) and Endpoints = (LC_{50} and EC_{50} and NOEC and LOEC and MATC) and Species = (both animals and plants) and Test condition = (fresh water)" was employed to collect the ATD and CTD for cadmium to protect aquatic organisms. Additionally, both ATD and CTD without water hardness were excluded due to the potential effect of water hardness on the toxicity of cadmium to aquatic organisms.

According to the technical guidelines for deriving FWQC for the protection of aquatic organisms [25], toxicity data for the derivation of FWQC were required to cover species from three trophic levels, and the primary producers must be included among the freshwater aquatic species. Both ATD and CTD were expected to cover at least ten species from specified taxonomic groups, including one Cyprinidae fish, one non-Cyprinidae fish, one zooplankton, one benthic non-fish animal, as well as one phytoplankton or aquatic vascular plant [30].

2.2. Hardness Adjustment for TD

An analysis of covariance should be utilized to account for the relationship between water quality characteristics and toxicity of contaminants during the derivation of the FWQC for the protection of aquatic organisms [29,30,46]. ATD and CTD of cadmium to aquatic organisms were determined to meet the requirements for analysis of covariance according to technical guidelines for deriving water quality criteria for aquatic organisms [29]. A least-squares regression analysis was conducted on the logarithms of ATD and CTD in relation to the logarithms of water hardness for H_A and H_C, resulting in the pooled slope (K_{ATD} and K_{CTD}) as expressed in Equations (1) and (2). The water hardness adjustment of cadmium *ATD* and *CTD* for aquatic organisms was calculated in Equations (3) and (4) with K_{ATD} and K_{CTD}, respectively.

$$\lg(ATD) = K_{ATD}\lg(H_A) + C_A \tag{1}$$

$$\lg(CTD) = K_{CTD}\lg(H_C) + C_C \tag{2}$$

$$ATD_H = 10^{k_{ATD}lg(ATD) - k_{ATD}lg(H_A)} \tag{3}$$

$$CTD_H = 10^{k_{CTD}lg(CTD) - k_{CTD}lg(HC)} \tag{4}$$

where H was represented as the water hardness, with values of 50, 100, 150, 200, 250, 300, 350, and 450 mg/L. H_A was the original water hardness value for ATD, and H_C was the original water hardness value for CTD. The ATD_H and CTD_H were the TD adjusted to the corresponding water hardness, respectively. The C_A and C_C were the acute and chronic toxicity constants, respectively.

2.3. Statistical Analysis and the S-FWQC/L-FWQC Derivation by SSD Method

The SSD model was employed to calculate the S-FWQC and L-FWQC for the protection of aquatic organisms based on ATD and CTD of cadmium, respectively. The geometric mean of the available ATD and CTD for each species were designed as the species geomean acute toxicity data (SMAD) and species geomean chronic toxicity data (SMCD) as Equations (5) and (6), respectively.

$$SMAD_{H,i} = \sqrt[m]{(ATD_H)_{i,1} \times (ATD_H)_{i,2} \times \ldots \times (ATD_H)_{i,m}} \tag{5}$$

$$SMCD_{H,i} = \sqrt[n]{(CTD_H)_{i,1} \times (CTD_H)_{i,2} \times \ldots \times (CTD_H)_{i,n}} \tag{6}$$

where m and n were the total number of *ATD* and *CTD* for certain species.

Next, the $SMAD_{H,i}/SMCD_{H,i}$ for all species were sorted in ascending order, with ranks assigned from the smallest (R = 1) to the largest (R = N). The cumulative probability (F_R) for each species was then determined using Equation (7):

$$F_R = \frac{\sum_{1}^{R} f}{\sum f + 1} \times 100\% \tag{7}$$

where f was the number of species corresponding to the toxicity data rank R.

Four different models of the normal distribution model, log-normal distribution model, logistic distribution model, and log-logistic distribution model were employed to derive the S-FWQC/L-FWQC by plotting the logarithms of $SMAD_{H.i}/SMCD_{H.i}$ as the independent variable and the cumulative probability of species as the dependent variable. Determined by the coefficient of determination R^2, the most suitable model was applied to estimate the hazardous concentration of 5% (HC_5) of freshwater species to protect the remaining 95% of species in the freshwater ecosystem. The S-FWQC and L-FWQC were calculated by dividing the acute and chronic HC_5 values by an assessment factor (AF) ranging from 2 to 5 depending on the quantity and quality of the ATD and CTD.

Typically, when the number of species exceeded 15 and covered a sufficient range of trophic levels for ATD and CTD, an AF of 2 was recommended according to the technical guideline for deriving water quality criteria for aquatic organisms [29]. Data analysis was conducted using EEC-SSD (Version 1.0, Ministry of Ecology and Environment, Beijing, China) and Origin 2022 for deriving the S-FWQC and L-FWQC (OriginLab, Northampton, MA, USA). The flowchart of FWQC for cadmium was shown in Figure 1.

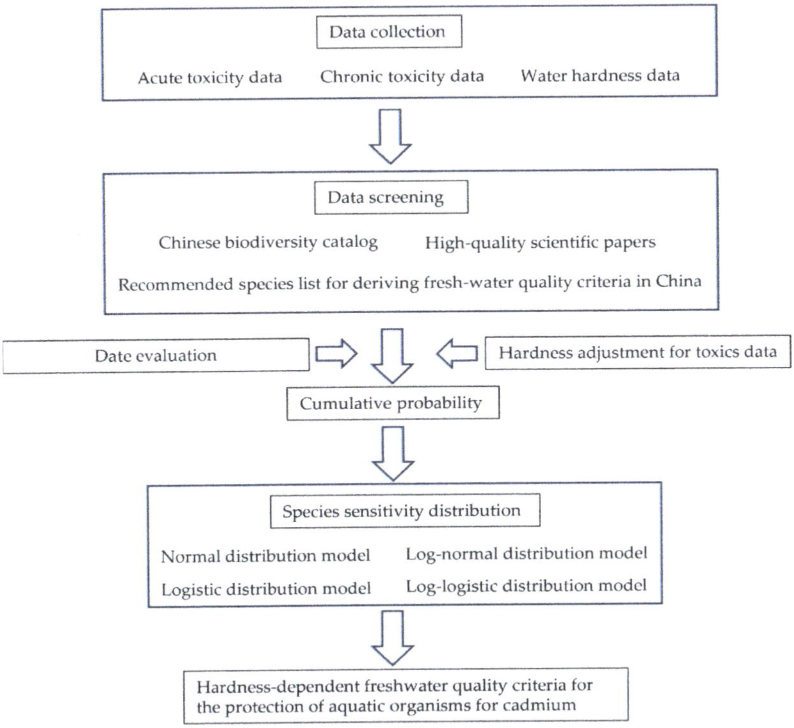

Figure 1. Flowchart of deriving hardness-dependent freshwater quality criteria for the protection of aquatic organisms for cadmium.

3. Results and Discussion

3.1. The Database of ATD and CTD of Cadmium to Aquatic Organisms

The toxicity databases for cadmium were compiled from the literature and existing toxicity databases, involving 249 ATD for 52 species with the corresponding water hardness of aquatic organisms in China, as presented in Table 1. For acute toxicity, ATD were available for one freshwater plant, 30 invertebrate species (including nine planktonic and 21 benthic species), and 21 vertebrate species (including five cyprinid and 16 non-cyprinid teleost fishes) (Figure 2a). Regarding chronic toxicity, 62 CTD were identified for 21 species with the corresponding water hardness, which included one freshwater plant, five invertebrate species (three planktonic and two benthic species), and 12 vertebrate species (two cyprinid and 10 non-cyprinid teleost fishes) in China (Table 2 and Figure 2b). The data toxicity of cadmium to aquatic organisms adequately met the requirements for the derivation of ATD and CTD according to the technical guideline for deriving water quality criteria for freshwater organisms [29].

Table 1. Ranked freshwater aquatic organisms of the species geomean acute toxicity data with water hardness of 100 mg/L as $CaCO_3$ ($SMAD_{100,i}$) and acute toxicity data (ATD) of cadmium to freshwater aquatic organisms in China.

Rank	Species	N	Hardness (mg/L)	ATD (μg/L)	$SMAD_{100,i}$
1	*Morone saxatilis*	3	40–475	4–10	3.05
2	*Oncorhynchus mykiss*	7	20–427	2.07–7.56	3.21
3	*Salvelinus confluentus*	13	29.3–31.7	0.9–6.6	5.14
4	*Salmo trutta*	13	29.2–151	1.23–15.1	7.34
5	*Oncorhynchus tshawytscha*	6	21–343	1.1–57	9.61
6	*Oncorhynchus kisutch*	5	22–90	2–17.5	15.73
7	*Gammarus pulex*	2	94.6–117.4	20–50	29.97
8	*Duphnia magna*	6	30–250	30–244	38.24
9	*Hydra viridissima*	2	19.5–210	3–210	39.62
10	*Cherax quadricarinatus*	3	43.79	8.48–44.8	43.18
11	*Ceriodaphnia dubia*	8	40–172	31.47–361.1	79.28
12	*Gammarus pseudolimnaeus*	5	43.5–76.8	22–68.3	81.23
13	*Ceriodaphnia rericulata*	6	45–240	66–184	82.20
14	*Simocephalus vetulus*	2	45–67	24–89.3	85.47
15	*Daphnia pulex*	9	40–240	44.96–99	103.16
16	*Simocephalus serrulatus*	9	9.7–67	3.5–123	105.69
17	*Moina macrocopa*	5	82.00	71.25–412	133.82
18	*lemra minor*	1	39.00	650.00	141.45
19	*Hydra oligactis*	1	210.00	320.00	149.84
20	*Hydra vulgaris*	5	19.5–210	82.5–520	167.75
21	*Aplexa hypnorum*	2	44.4–44.8	93.00	212.38
22	*Neocaridina denticulate*	4	30–400	230–2592	299.04
23	*Oryzias latipes*	2	50–100	130–350	304.04
24	*Diaphanosoma brachyurum*	2	67.1–93	69.8–1060	346.18
25	*Nais elinguis*	5	17.89–18.72	27–158	347.60
26	*Lumbriculus variegatus*	5	10–290	74–780	408.34
27	*Lymnaea stagnalis*	3	250.00	752–1585	477.07
28	*Chydorus sphaericus*	6	10.5–83.6	149–560	1419.18
29	*Limnodrilus hoffmeisteri*	4	5.3–152	170–2400	1542.59
30	*Brachionus calyciflorus*	1	36.20	650.00	1837.48
31	*Anguilla rostrata*	3	55.00	820–1500	2038.44
32	*Procambarus acutus*	11	85.5–262.5	1390–7160	2397.29
33	*Hyriopsis cumingii*	3	51.43	388–6346	2669.07
34	*Prychocheilus oregonensis*	4	25–347	1092–5555	2715.64
35	*Bufo gargarizans*	1	90.00	2592.00	2886.90
36	*Xenopus laevis*	4	85–116	1600–4000	3064.99
37	*Poecilia reticulata*	9	18.72–209.2	170–16,000	4691.95
38	*Tubifex tubifex*	6	5.3–305	320–56,000	6150.98
39	*Gasterosteus aculeatus*	2	107.15–115	6500–23,000	10,988.70
40	*Tanichthys albonubes*	2	39.16–44.5	4610–4447	11,063.56
41	*Ctenopharyngod on idella*	5	42.72–210.1	3490–24,500	11,348.44
42	*Oreochromis mossambica*	2	17–28.4	1000–6000	11,538.02
43	*Ictalurus punctatus*	5	44.4–67	4610–10,200	12,849.68
44	*Carassius auratus*	4	20–144	2130–46,800	14,828.01
45	*Lepomis macrochirus*	16	20–350	1700–48,200	15,968.42
46	*Cyprinus carpio*	8	100–312.5	6500–220,770	18,273.53
47	*Lepomis cyanellus*	7	20–360	2840–88,600	19,967.71
48	*Cirrhinus mrigala*	2	19.5–72	5300–13,700	23,253.70
49	*Aristichthys nobilis*	3	2.50	245–2250	29,040.73
50	*Branchiura sowerbyi*	10	5.3–195	240–88,780	29,461.18
51	*Pseudorasbora parva*	1	5.80	5170.00	95,089.54
52	*Chironomus riparius*	4	10–170	128,840–1,106,000	389,447.18

N is the number of ATD collected in the literature and toxicity databases.

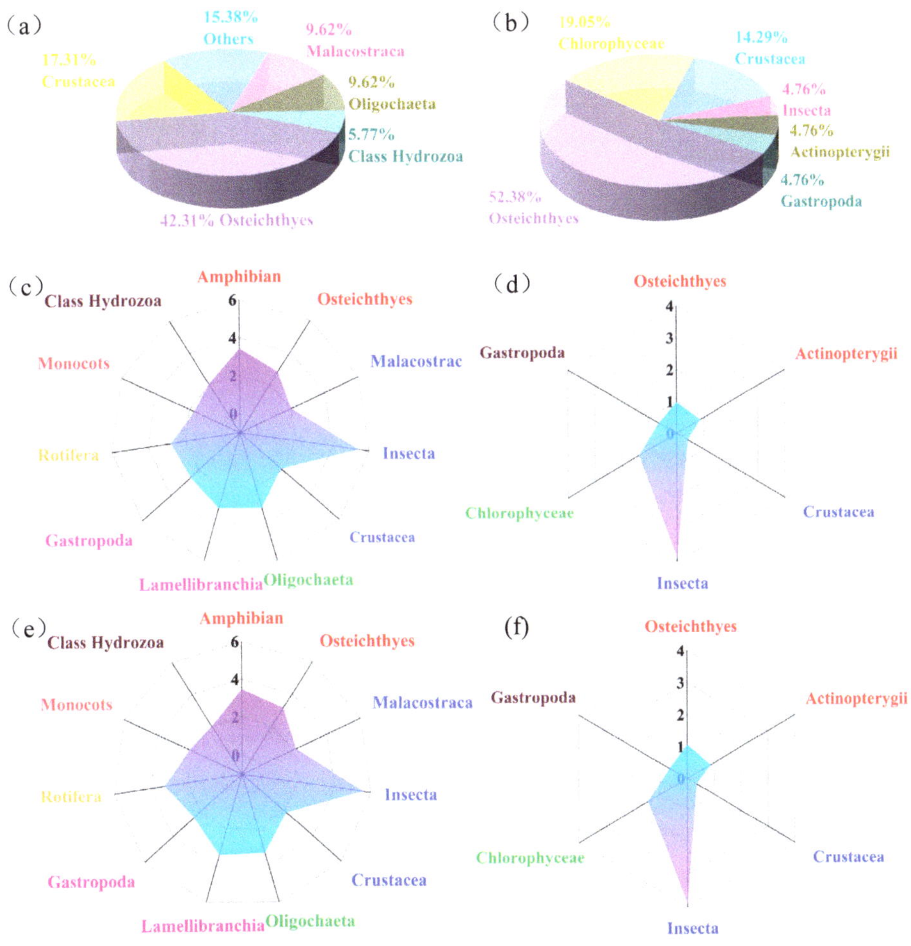

Figure 2. The proportion of freshwater aquatic organisms of different classes in acute toxicity data (**a**) and chronic toxicity data in China (**b**), the species geomean acute toxicity data (**c**), the species geomean chronic toxicity data (**d**), the species geomean acute toxicity data with water hardness of 100 mg/L as $CaCO_3$ (**e**), and the species geomean chronic toxicity data with water hardness of 100 mg/L as $CaCO_3$ (**f**) of cadmium to various classes groups of freshwater aquatic organisms (concentrations in μg/L) in China. The same-colored classes in (**c**–**f**) represent the same phylum.

The percentages of Osteichthyes, Crustacea, Malacostraca, Oligochaeta, Hydrozoa, and other classes related to the ATD of cadmium to aquatic organisms were 42.31%, 17.31%, 9.62%, 9.62%, 5.77%, and 15.38%, respectively (Figure 2a). The SMAD for cadmium were ranked in ascending order as follows: Monocots < Crustacea < Malacostraca < Class Hydrozoa < Gastropoda < Osteichthyes < Rotifera < Oligochaeta < Lamellibranchia < Amphibian < Insecta (Figure 2c). Similarly, the percentages of Osteichthyes, Chlorophyceae, Crustacea, Insecta, Actinopterygii, and Gastropoda related to the CTD of cadmium to aquatic organisms were 52.38%, 19.05%, 14.29%, 4.76%, 4.76%, and 4.76%, respectively (Figure 2b). The SMCD for cadmium were ranked in ascending order as follows: Crustacea < Gastropoda < Actinopterygii < Osteichthyes < Chlorophyceae < Insecta (Figure 2d).

Table 2. Ranked freshwater aquatic organisms of the species geomean chronic toxicity data with water hardness of 100 mg/L ($SMCD_{100,i}$) as $CaCO_3$ and chronic toxicity data (CTD) of cadmium to freshwater aquatic organisms in China.

Rank	Species	N	Hardness (mg/L)	CTD (μg/L)	$SMCD_{100,i}$
1	*Daphnia magna*	5	99–200	0.3–2.39	0.79
2	*Oncorhynchus mykiss*	17	6.8–413.8	0.4–4.31	1.94
3	*Ceriodaphnia dubia*	7	100–270	1.602–6.257	2.04
4	*Oncorhynchus kisutch*	1	44.00	2.10	3.16
5	*Oryzias latipes*	2	340.00	50.00	3.17
6	*Oncorhynchus tshawytscha*	2	25.00	1.57–1.88	3.43
7	*Salvelinus fontinalis*	4	37–188	2.045–9.165	3.82
8	*Salmo trutta*	7	30.6–250	0.4–16.49	5.01
9	*Scenedesmus acutus*	1	90.00	5.00	5.27
10	*Chlorella vulgaris*	1	90.00	5.00	5.27
11	*Aplexa hypnorum*	2	45.30	3.46–5.801	6.65
12	*Daphnia pulex*	2	65–106	5–7.49	6.71
13	*Salmo salar*	1	23.50	4.53	9.32
14	*Esox lucius*	1	44.00	7.36	11.08
15	*Lepomis macrochirus*	3	147–207	4.167–49.8	15.00
16	*Oreochromisaurea*	1	145.00	52.00	43.21
17	*Pseudokirchneri ella subcapitata*	1	171.00	120.00	91.85
18	*Cirrhinus mrigala*	2	71.50	98–132	134.43
19	*Chlamydomonas Reinhardii*	1	24.00	99.00	201.59
20	*Cyprinus carpio*	1	188.50	650.00	473.94
21	*Pachydiplax longipennie*	1	120.00	8249.00	7532.61

N is the number of CTD collected in the literature and toxicity databases.

3.2. Derivation of the FWQC for Cadmium

3.2.1. Correlations Between Water Hardness and Toxicity of Cadmium

A significant correlation was established between water hardness and the ATD of cadmium, with a slope (K_{ATD}) of 1.0227 and a coefficient of determination (R^2) of 0.1031 ($p < 0.05$) (Figure 3a). Additionally, a significant correlation was observed between water hardness and the CTD of cadmium, with a slope (K_{CTD}) of 0.4983 and a coefficient of determination (R^2) of 0.0448 ($p < 0.05$) (Figure 3b).

The TD of cadmium to freshwater aquatic organisms were influenced by the presence of calcium and magnesium ions because of the same valence and similar biological targets of aquatic organisms, including both ATD and CTD as mentioned above [39,44,47–49]. Consequently, cadmium exhibited greater toxicity in soft water than that in hard water for both short-term and long-term exposures to aquatic organisms. This result was also reported by the aquatic life ambient water quality criteria for cadmium published by the USEPA in 2016 [28]. Furthermore, the water hardness affected cadmium toxicity significantly in aquatic environments, which was consistent with the FWQC for the protection of aquatic organisms for cadmium released by the MEE in China in 2020 based on the characteristics of aquatic biota in China [49].

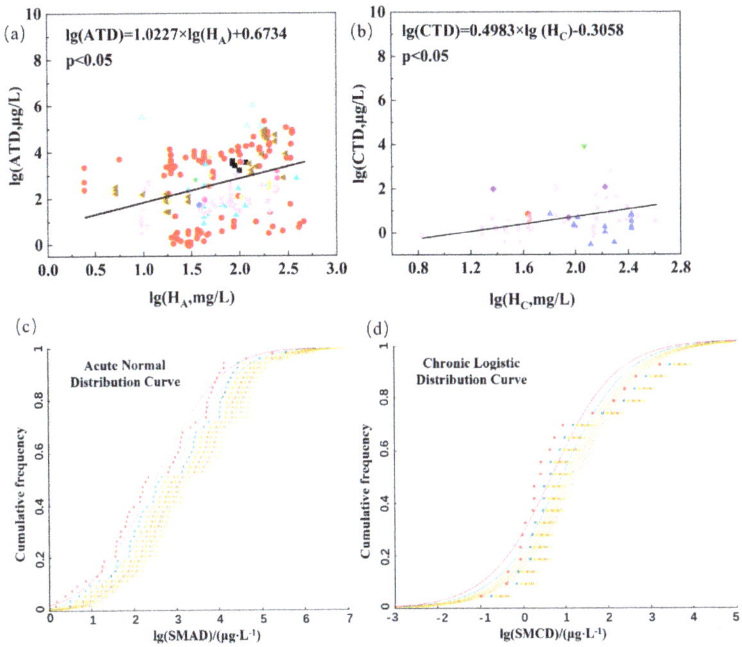

Figure 3. The linear relationship between the water hardness and (**a**) acute toxicity data (ATD) of 249 and (**b**) chronic toxicity data (CTD) of 62 of cadmium in China; different colors represent different classes. (**c**) the species sensitivity distribution (SSD) curves of cadmium ions of the logarithms of species geomean acute toxicity data [lg(SMAD)]; and (**d**) the logarithms of species geomean chronic toxicity data [lg(SMCD)] with different water hardness levels of 50, 100, 150, 200, 250, 300, 350, and 450 mg/L as $CaCO_3$ from left to right.

3.2.2. Hardness-Dependent FWQC for Cadmium

According to the third national evaluation on surface water quality in China, the distribution of water hardness across the total surface water area was characterized by proportions of 42% for a water hardness below 150 mg/L, 34% for levels between 150 and 300 mg/L, 11% for levels between 300 and 450 mg/L, and 13% for levels above 450 mg/L, respectively [30,50]. Utilizing the principle of equal data distribution, ATD and CTD were adjusted to eight distinct water hardness using the correction equation (Equation (2)) along with the K_{ATD} and K_{CTD} coefficients, respectively. The range of ATD for cadmium was found to be from 0.9 μg/L to 1,106,000 μg/L with water hardness levels varying from 2.5 mg/L to 475 mg/L as $CaCO_3$ (Table 1). The range of CTD for cadmium was identified as from 0.3 μg/L to 8,249 μg/L with water hardness values from 6.8 mg/L to 340 mg/L as $CaCO_3$ (Table 2). Following the water hardness adjustment, the $SMAD_{100,i}$ ranged from 3.05 μg/L to 389,447.18 μg/L (Table 1), and the $SMCD_{100,i}$ varied from 0.79 μg/L to 7532.61 μg/L with a water hardness reference of 100 mg/L as $CaCO_3$ (Table 2). The percentages of species and the $SMAD_{100,i}/SMAD_{100,i}$ of cadmium were similar to those before water hardness correction (Figure 2e,f).

Following the adjustment of the $SMAD_{H,i}$ for water hardness, the three species identified as most sensitive to cadmium among the 52 species were non-cyprinid teleost fish: *Morone saxatilis*, *Oncorhynchus mykiss*, and *Salvelinus confluentus*. Conversely, the three species exhibiting the most tolerance to cadmium were *Branchiura sowerbyi*, *Pseudorasbora parva*, and *Chironomus riparius* based on short-term exposure (Table 1). After the adjustment of $SMCD_{H,i}$, the three species most sensitive to cadmium among the 21 species were *Daphnia magna*, *Oncorhynchus mykiss*, and *Ceriodaphnia dubia* based on long-term exposure.

The three species demonstrating the most tolerance to cadmium were *Chlamydomonas Reinhardtii, Cyprinus carpio*, and *Pachydiplax longipennis* as a comparison of the SMCD in China (Table 2). Therefore, *Daphnia magna* might be used as an indicator species for cadmium pollution in the aquatic environment in China.

The most sensitive species identified were *Salvelinus confluentus, Cottus bairdii*, and *Salmo trutta*, while the most tolerant species were *Chironomus plumosus* and *Cyprinus carpio* for short-term exposure according to the aquatic life ambient water quality criteria for cadmium published by the USEPA in 2016 (Table 3) [19,28]. For long-term exposure, the most sensitive species included *Hyalella azteca, Ceriodaphnia dubia*, and *Cottus bairdii*, whereas the most tolerant species were *Aeolosoma headleyi* and *Oreochromis aureus* in the USA based on the aquatic life ambient water quality criteria for cadmium (Table 3) [19, 28]. Therefore, the different sensitive and tolerant species for cadmium in the freshwater ecosystem indicated the difference in aquatic biota between China and the USA.

Table 3. Three sensitive species of cadmium in different countries (China, Canada, USA) in short-term exposure and long-term exposure.

Nation	Short-Term Exposure	Long-Term Exposure
China	*Morone Saxatilis;* *Oncorhynchus mykiss* *Salvelinus confluentus;*	*Daphnia magna;* *Oncorhynchus mykiss* *Ceriodaphnia dubia;*
Canada	*Oncorhynchus mykiss;* *Hyalella Azteca;* *Daphnia magna*	*Daphnia magna;* *Ceriodaphnia reticulata;* *Hyalella Azteca;*
USA	*Salvelinus confluentus;* *Cottus bairdii;* *Salmo trutta*	*Hyalella Azteca;* *Ceriodaphnia dubia;* *Cottus bairdii*

By using the $SMAD_{H,i}$ and cumulative probability as independent and dependent variables, the R^2 values for the normal, log-normal, logistic, and log-logistic distribution models were calculated as 0.9793, 0.9786, 0.9785, and 0.9747, respectively, for the $SMAD_{100,i}$ of aquatic organisms in China (Table S2). Similarly, the SSD curves were applied to fit $SMCD_{100,i}$, resulting in R^2 values of 0.9644 and 0.9686 for the normal and logistic distribution models, respectively (Table S2). Therefore, the normal distribution model and logistic distribution model were the best models fit for the $SMAD_{H,i}$ and $SMCD_{H,i}$ of cadmium, respectively, to derive the FWQC. The SSD curves and HC_5 showed a rightward shift as water hardness increased for both the $SMAD_{H,i}$ and $SMCD_{H,i}$, suggesting that both the ATD and CTD of cadmium decreased with increasing water hardness (Figure 3c,d).

The acute HC_5 value of cadmium ranged from 3.12 to 29.52 μg/L, while the chronic HC_5 value ranged from 0.25 to 0.74 μg/L with water hardness levels varying from 50 to 450 mg/L as $CaCO_3$ (Tables 4 and S3). An assessment factor of 2 was established for the toxicity data covering 15 species for both ATD and CTD, which was in accordance with the technical guidelines for deriving water quality criteria in China. This assessment factor aligns with the FWQC for the protection of aquatic organisms for cadmium published by the MEE in China in 2020 [30].

Consequently, the S-FWQC for cadmium were calculated in a range of 1.56 to 14.76 μg/L at a water hardness of 50–450 mg/L as $CaCO_3$ (Table 4). The L-FWQC for cadmium ranged from 0.12 to 0.37 μg/L at a water hardness of 50–450 mg/L as $CaCO_3$ (Table 4). Notably, it was observed that the ratio of S-FWQC to L-FWQC at the lowest water hardness level (50 mg/L) was nearly 9/3 times greater than that at the highest water hardness level (450 mg/L) in China. This highlights the importance of considering water hardness during deriving both S-FWQC and L-FWQC to protect aquatic organisms from cadmium exposure. The S-FWQC and L-FWQC of cadmium can also be expressed with the equation $10^{(1.0227 \times lg(H) - 1.5444)}$ and $10^{(0.4983 \times lg(H) - 1.7549)}$, respectively, in China. In detail, the derived S-FWQC (3.17 μg/L) in this research was 24.5% less than that (4.2 μg/L) published by MEE in China, while L-FWQC (0.17 μg/L) in this research was 26.1% less than that (0.23 μg/L) published by MEE in China at the water hardness of 100 mg/L as $CaCO_3$ [30]. Therefore, it is necessary to eliminate invasive species and internationally common species without distribution in China during the derivation of FWQC for the protection of aquatic organisms.

Table 4. Hazardous concentration of 5% (HC_5) and short-term freshwater quality criteria (S-FWQC)/long-term freshwater quality criteria (L-FWQC) of cadmium for different water hardness deriving with the species sensitivity distribution model.

	H ($CaCO_3$, mg/L)	HC_5 (μg/L)	FWQC (μg/L)
S-FWQC	50	3.12	1.56
	100	6.34	3.17
	150	9.60	4.80
	200	12.88	6.44
	250	16.18	8.09
	300	19.50	9.75
	350	22.83	11.41
	450	29.52	14.76
L-FWQC	50	0.25	0.12
	100	0.35	0.17
	150	0.43	0.21
	200	0.49	0.25
	250	0.55	0.28
	300	0.60	0.30
	350	0.65	0.33
	450	0.74	0.37

3.3. The Comparison with Other FWQC and WQSs for Cadmium

The FWQC for the protection of aquatic organisms for cadmium were investigated and published by several countries, including the USA, Canada, Australia, New Zealand, EU, etc. The criterion maximum concentration (CMC) and criterion continuous concentration (CCC) for cadmium were expressed as $e^{(0.9789\times\ln(H)-3.866)}$ and $e^{(0.7977\times\ln(H)-3.909)}$, respectively, with water hardness serving as the independent variable for the protection of aquatic organisms issued by the USEPA [28]. In Canada, the S-FWQC and L-FWQC were given by $10^{(1.016\times log(H))}$ and $10^{(0.83\times log(H)-2.46)}$ for the water quality guidelines of cadmium with water hardness as the independent variable [21]. With a water hardness of 100 mg/L as $CaCO_3$, S-FWQC (3.17 μg/L) in the present research was 76.1% greater than the CMC (1.8 μg/L) in the USA and was 26.8% greater than that in Canada (2.5 μg/L) (Figure 4c). The L-FWQC (0.17 μg/L) in the present research was 76.4% less than the CMC (0.72 μg/L) in the USA and was similar to that in Canada (0.18 μg/L), with a water hardness of 100 mg/L as $CaCO_3$ (Figure 4d). According to the freshwater quality guidelines in Australia and New Zealand, the trigger value for cadmium was set at 0.2 μg/L, with a chosen protection level of 95% (30 mg/L), applicable to moderately hard water at 90 mg/L (as $CaCO_3$) [22], which was greater than the L-FWQC of 0.17 μg/L determined in this research at a water hardness of 100 mg/L. Environmental quality standards (EQSs) were established to protect aquatic organisms by the EU, including the annual average concentration (AA-EQS) and the maximum allowable concentration (MAC-EQS) [16]. Both the AA-EQS and MAC-EQS increased with the increase in water hardness from 40 mg/L to 200 mg/L, which was similar to the variation in the CMC and CCC in the USA and the S-FWQC and L-FWQC in China in this study. The differences in water hardness range and classification are primarily due to the distribution range of water hardness in China and the EU. In detail, according to the third national surface water quality assessment in China, a surface water hardness <150 mg/L, 150 mg/L~<300 mg/L, 300 mg/L~≤450 mg/L, and >450 mg/L accounted for 42%, 34%, 11%, and 13% in China, respectively. The S-FWQC (3.17 μg/L) in the present research was 2.5 times greater than the MAC-EQS (0.9 μg/L) in the EU, and the L-FWQC (0.17 μg/L) was approximately comparable to the AA-EQS (0.15 μg/L) with a water hardness of 100 mg/L as $CaCO_3$. The variations in water quality standards/criteria among different countries and international organizations might be attributed to differences in aquatic biota, freshwater conditions, and toxicity data. Therefore, it is essential to derive the FWQC from the toxicity data of native species to minimize the potential influence of water characteristics and specific taxonomic groups of species in different countries and regions.

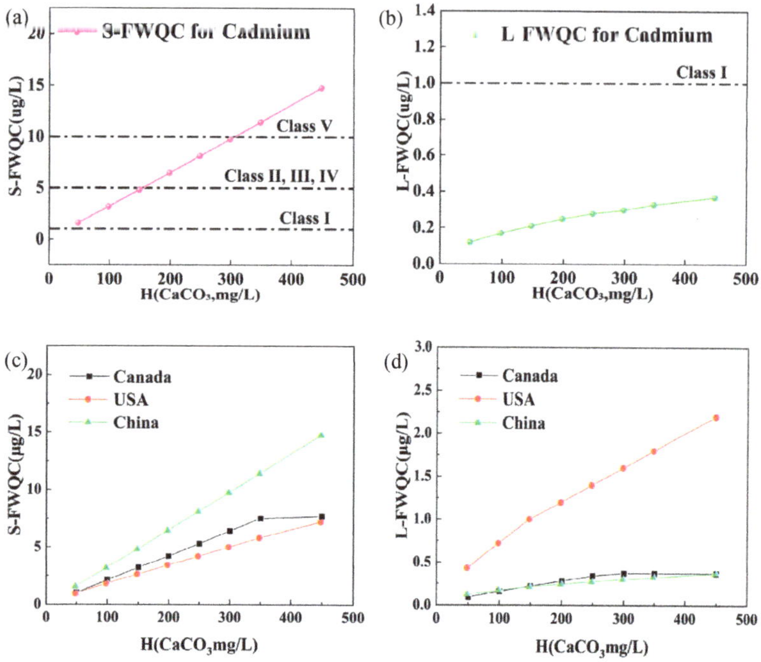

Figure 4. Comparison of the limits of GB3838-2002 and the short-term freshwater quality criteria (S-FWQC) (**a**) and long-term freshwater quality criteria (L-FWQC) (**b**) of cadmium for the protection of aquatic organisms. The change in S-FWQC (**c**) and L-FWQC (**d**) in China, Canada, and USA with water hardness.

At present, the permissible limits for cadmium were set at 1 μg/L for Class I, 5 μg/L for Classes II, III, and IV, and 10 μg/L for Class VI for the GB3838-2002 [25]. Under short-term exposure at a certain water hardness of 100 mg/L of $CaCO_3$, the S-FWQC of 3.17 μg/L was 3.17 times greater than Class I (1 μg/L) of GB3838-2002, 36.6% less than Classes II-IV (5 μg/L) of GB3838-2002, and 68.3% times less than Class V (10 μg/L) of GB3838-2002. Under Class II-IV of GB3838-2002, *Morone saxatilis* and *Oncorhynchus mykiss* might experience adverse effects by cadmium to aquatic organisms with short-term exposure at an arbitrary water hardness of 100 mg/L of $CaCO_3$. Under Class V of GB3838-2002, *Morone saxatilis*, *Oncorhynchus mykiss*, *Salvelinus confluentus*, *Salmo trutta*, and *Oncorhynchus tshawytscha* might be affected by cadmium to aquatic organisms with short-term exposure at an arbitrary water hardness of 100 mg/L of $CaCO_3$ at least. In long-term exposure at a certain water hardness of 100 mg/L of $CaCO_3$, the L-FWQC of 0.17 μg/L was 83% less than Class I (1 μg/L) of GB3838-2002, 96.6% less than Classes II-IV (5 μg/L) of GB3838-2002, and 98.3% less than Class V (10 μg/L) of GB3838-2002. Under Class I of GB3838-2002, *Daphnia magna* may be harmed by cadmium to aquatic organisms with long-term exposure at an arbitrary water hardness of 100 mg/L of $CaCO_3$. Under Class II-IV of GB3838-2002, *Daphnia magna*, *Oncorhynchus mykiss*, *Ceriodaphnia dubia*, *Oncorhynchus kisutch*, *Oryzias latipes*, *Oncorhynchus tshawytscha*, and *Salvelinus fontinalis* were likely to be adversely affected by cadmium to aquatic organisms with long-term exposure at an arbitrary water hardness of 100 mg/L of $CaCO_3$ at least. Under Class V of GB3838-2002, 13 species might suffer harm by cadmium to aquatic organisms with long-term exposure at an arbitrary water hardness of 100 mg/L of $CaCO_3$, including *Aplexa hypnorum*, *Daphnia pulex*, *Salmo salar*, etc., shown as rank1-13 in Table 2. The limit of cadmium of GB3838-2002 and the environmental risks of cadmium require careful consideration in the freshwater ecosystems in China based on the L-FWQC.

Water hardness was considered a key factor in deriving the FWQC to support the scientific management and assessment of water quality. The S-FWQC at a hardness of 500 mg/L (14.76 μg/L) was about 10 times greater than at a hardness of 50 mg/L (1.56 μg/L), while the L-FWQC at the hardness of 500 mg/L (0.37 μg/L) was about 3 times greater than at a hardness of 50 mg/L (0.12 μg/L) in China. Additionally, invasive species and internationally common species without distribution in China were excluded from the FWQC for cadmium to ensure its greater applicability to unique ecosystems. The toxicity database for cadmium covered 249 ATD from 52 species and 62 CTD from 21 species to derive the FWQC in China. Abundant aquatic species exist in the surface water in China (more than 20,000 species), and less than 100 species were applied to derive the FWQC to protect the aquatic organisms in China. Therefore, there was an urgent need to supplement the TD of cadmium for native species in China, particularly the CTD, to improve the accuracy and applicability of the FWQC. Water hardness was considered the primary factor in this study because the cadmium toxicity varied with the changes in water hardness for both the ATD and CTD in China. However, other water quality parameters, such as pH, dissolved oxygen, temperature, organic matter, and other metal ions, might affect the toxicity of cadmium to protect aquatic organisms. Due to the lack of TD for cadmium under controlled experimental conditions, such as pH, dissolved oxygen, temperature, organic matter, and other metal ions, more research studies are needed to investigate the effects of these additional factors on aquatic organisms. By utilizing the hardness-based FWQC, a nationwide ecological risk assessment for cadmium can be conducted to protect aquatic organisms in surface water in China. The re-evaluation of cadmium remediation technologies, such as adsorption technology, precipitation technology, and microbial remediation, should also be conducted for potential application in the treatment of environmental pollution in surface waters of rivers, lakes, and reservoirs based on the FWQC in China.

4. Conclusions

The ATD of 249 from 52 species and the CTD of 62 from 21 species were applied to establish the FWQC for the protection of aquatic organisms for cadmium in China, excluding invasive species and internationally common species without distribution in China. Significant correlations between the water hardness and toxicity databases of cadmium were obtained with a K_{ATD} of 1.0227 (n = 52, $p < 0.05$) for the ATD and a K_{CTD} of 0.4983 (n = 21, $p < 0.05$) for the CTD, respectively, in China. Among the studied species, *Morone saxatilis* exhibited the most sensitivity for cadmium during short-term exposure, while *Daphnia magna* demonstrated the most sensitivity for cadmium during long-term exposure, which made it a potential indicator of cadmium contamination in the freshwater ecosystems under long-term exposure. The S-FWQC and L-FWQC can be expressed using the equation $10^{(1.0227 \times lg(H) - 1.5444)}$ and $10^{(0.4983 \times lg(H) - 1.7549)}$, respectively, with water hardness as $CaCO_3$ as an independent variable. The best fitting models of the normal distribution model and logistic distribution model were selected to derive the S-FWQC and L-FWQC, respectively, and were calculated to be 1.56–14.76 μg/L and 0.12–0.37 μg/L at a water hardness of 50–450 mg/L as $CaCO_3$. The possible environmental risk and cadmium remediation technologies should be established and given more concern regarding the FWQC of cadmium in freshwater to protect the aquatic organisms in China, especially with the L-FWQC.

Supplementary Materials: The following supporting information can be downloaded at https://www.mdpi.com/article/10.3390/toxics12120892/s1, Table S1. List of Abbreviations; Table S2. Fitting results of short-term freshwater quality criteria and long-term freshwater quality criteria species sensitivity distribution model with water hardness of 100 mg/L as $CaCO_3$; Table S3. Hazard concentration of short-term species and long-term species.

Author Contributions: Z.Z.: methodology, validation, investigation, and writing—original draft; R.H.: methodology, formal analysis, and writing; Z.S.: conceptualization, visualization, and writing—review and editing; Y.F.: conceptualization, visualization, and writing—review and editing; C.F.: conceptualization, writing—review and editing, funding acquisition, and project administration; Y.D.: conceptualization, writing—review and editing, funding acquisition, and project administration. All authors have read and agreed to the published version of the manuscript.

Funding: This research was funded by the National Key Research and Development Program of China (No. 2022YFC3204103–01, No. 2023YFC3208401, and No. 2021YFC3200801) and the Department Budget Project of the Ministry of Ecology and Environment (No. 22110105012001).

Institutional Review Board Statement: Not applicable.

Informed Consent Statement: Not applicable.

Data Availability Statement: The data will be made available on request.

Conflicts of Interest: The authors declare no conflicts of interests.

References

1. Chang, Y.; Jiang, X.; Dou, J.; Xie, R.; Zhao, W.; Cao, Y.; Gao, J.; Yao, F.; Wu, D.; Mei, H.; et al. Investigating the Potential Risk of Cadmium Exposure on Seizure Severity and Anxiety-Like Behaviors through the Ferroptosis Pathway in Epileptic Mice: An Integrated Multi-Omics Approach. *J. Hazard. Mater.* **2024**, *480*, 120964.
2. Lanctot, C.; Callaghan, P.; Cresswell, T. Cadmium Bioaccumulation Dynamics During Amphibian Development and Metamorphosis. *J. Hazard. Mater.* **2024**, *474*, 1465–1474. [CrossRef] [PubMed]
3. Wang, R.; Sang, P.; Guo, Y.; Jin, P.; Cheng, Y.; Yu, H.; Xie, Y.; Yao, W.; Qian, H. Cadmium in Food: Source, Distribution and Removal. *Food Chem.* **2023**, *405*, 127135. [CrossRef] [PubMed]
4. Jannetto, P.J.; Cowl, C.T. Elementary Overview of Heavy Metals. *Clin. Chem.* **2023**, *69*, 336–349. [CrossRef] [PubMed]
5. Li, F.; Yang, X.; Zhang, Z.; Jiang, Y.; Gong, Y. Behaviour, Ecological Impacts of Microplastics and Cadmium on Soil Systems: A Systematic Review. *Environ. Technol. Innov.* **2024**, *35*, 116643. [CrossRef]
6. Zhang, H.; Reynolds, M. Cadmium Exposure in Living Organisms: A Short Review. *Sci. Total Environ.* **2019**, *678*, 761–767. [CrossRef]
7. Lee, J.W.; Jo, A.H.; Lee, D.C.; Choi, C.Y.; Kang, J.C.; Kim, J.H. Review of Cadmium Toxicity Effects on Fish: Oxidative Stress and Immune Responses. *Environ. Res.* **2023**, *236*, 105525. [CrossRef]
8. Saidon, N.B.; Szabo, R.; Budai, P.; Lehel, J. Trophic Transfer and Biomagnification Potential of Environmental Contaminants (Heavy Metals) in Aquatic Ecosystems. *Environ. Pollut.* **2024**, *340*, 345–360. [CrossRef]
9. Yang, Z.; Wong, J.; Wang, L.; Sun, F.; Lee, M.; Yue, G.H. Unveiling the Underwater Threat: Exploring Cadmium's Adverse Effects on Tilapia. *Sci. Total Environ.* **2024**, *912*, 119000. [CrossRef]
10. Bach, L.; Sonne, C.; Riget, F.F.; Dietz, R.; Asmund, G. A Simple Method to Reduce the Risk of Cadmium Exposure from Consumption of Iceland Scallops (*Chlamys islandica*) Fished in Greenland. *Environ. Int.* **2014**, *69*, 100–103. [CrossRef]
11. Robohm, R.A. Paradoxical Effects of Cadmium Exposure on Antibacterial Antibody-Responses in 2 Fish Species—Inhibition in Cunners (Tautoglabrus-Adspersus) and Enhancement in Striped Bass (*Morone-Saxatilis*). *Vet. Immunol. Immunopathol.* **1986**, *12*, 251–262. [CrossRef] [PubMed]
12. Santojanni, A.; Gorbi, G.; Sartore, F. Prediction of Fecundity in Chronic Toxicity Tests on *Daphnia magna*. *Water Res.* **1998**, *32*, 3146–3156. [CrossRef]
13. Irfan, M.; Liu, X.; Hussain, K.; Mushtaq, S.; Cabrera, J.; Zhang, P. The Global Research Trend on Cadmium in Freshwater: A Bibliometric Review. *Environ. Sci. Pollut. Res.* **2023**, *30*, 71585–71598. [CrossRef] [PubMed]
14. El Rasafi, T.; Oukarroum, A.; Haddioui, A.; Song, H.; Kwon, E.E.; Bolan, N.; Tack, F.M.G.; Sebastian, A.; Prasad, M.N.V.; Rinklebe, J. Cadmium Stress in Plants: A Critical Review of the Effects, Mechanisms, and Tolerance Strategies. *Crit. Rev. Environ. Sci. Technol.* **2022**, *52*, 675–726. [CrossRef]
15. Peana, M.; Pelucelli, A.; Chasapis, C.T.; Perlepes, S.P.; Bekiari, V.; Medici, S.; Zoroddu, M.A. Biological Effects of Human Exposure to Environmental Cadmium. *Biomolecules* **2023**, *13*, 36. [CrossRef]
16. European Parliament and Council. Directive 2013/39/EU of 12 August 2013 Amending Directives 2000/60/EC and 2008/105/EC as Regards Priority Substances in the Field of Water Policy. *Off. J. Eur. Union* **2013**. Available online: https://eur-lex.europa.eu/LexUriServ/LexUriServ.do?uri=OJ:L:2013:226:0001:0017:en:PDF (accessed on 4 August 2024).

17. Wu, F.; Meng, W.; Zhao, X.; Li, H.; Zhang, R.; Cao, Y.; Liao, H. China Embarking on Development of Its Own National Water Quality Criteria System. *Environ. Sci. Technol.* **2010**, *44*, 7992–7993. [CrossRef]
18. Feng, C.; Li, H.; Yan, Z.; Wang, Y.; Wang, C.; Fu, Z.; Liao, W.; Giesy, J.P.; Bai, Y. Technical Study on National Mandatory Guideline for Deriving Water Quality Criteria for the Protection of Freshwater Aquatic Organisms in China. *J. Environ. Manag.* **2019**, *250*, 109539. [CrossRef]
19. Wu, F.; Meng, W.; Cao, Y.; Li, H.; Zhang, R.; Feng, C.; Yan, Z. Derivation of Aquatic Life Water Quality Criteria for Cadmium in Freshwater in China. *Res. Environ. Sci.* **2011**, *24*, 172–184.
20. EPA. *National Recommended Water Ouality Criteria*; U.S. Environmental Protection Agency: Washington, DC, USA, 2009.
21. CCME. *Canadian Tissue Residue Guidelines for the Protection of Wildlifeconsumers of Aquatic Biota*; Canadian Council of Ministers of the Environment: Winnepeg, MB, Canada, 2000.
22. ANZECC; ARMCANZ. Australian and New Zealand Guidelines for Fresh and Marine Water Quality. *Canberra* **2000**, *1*, 1–314.
23. Wu, F. *Water Quality Criteria Theory, Methodology and Case Studies*; Science Press: Beijing, China, 2012.
24. Wu, F. *Medium and Long-Term Roadmap for the Development of the Environmental Quality Standard System in China*, 2nd ed.; Science Press: Beijing, China, 2020.
25. *GB3838-2002*; Environmental Quality Standard for Surface Water. State Environmental Protection Administration of China: Beijing, China, 2002.
26. Xia, Q.; Chen, Y.; Liu, X. *Water Quality Criteria and Water Quality Standard*; China Standards Press: Beijing, China, 2004.
27. Karthikeyan, P.; Marigoudar, S.R.; Mohan, D.; Sharma, K.V.; Murthy, M.V.R. Prescribing Sea Water Quality Criteria for Arsenic, Cadmium and Lead through Species Sensitivity Distribution. *Ecotoxicol. Environ. Saf.* **2021**, *208*, 117249. [CrossRef] [PubMed]
28. Wu, F.; Feng, C.; Zhang, R.; Li, Y.; Du, D. Derivation of Water Quality Criteria for Representative Water-Body Pollutants in China. *Sci. China Earth Sci.* **2012**, *55*, 900–906. [CrossRef]
29. EPA. *Aquatic Life Ambient Water Quality Criteria Cadmium*; U.S. Environmental Protection Agency: Washington, DC, USA, 2016.
30. MEEC. *Technical Guideline for Deriving Water Quality Criteria for Freshwater Organisms (HJ 831-2022)*; Minisitry of Ecology and Environment of China: Beijing, China, 2022.
31. MEEC. *Water Quality Criteria for Freshwater Aquatic Organisms-Cadmium*; Ministry of Ecology and Environment of the People's Republic of China: Beijing, China, 2020.
32. MEEC. *Water Quality Criteria for Freshwater Aquatic Organisms-Ammonia Nitrogen*; Ministry of Ecology and Environment of the People's Republic of China: Beijing, China, 2020.
33. MEEC. *Water Quality Criteria for Freshwater Aquatic Organisms-Phenol*; Ministry of Ecology and Environment of the People's Republic of China: Beijing, China, 2020.
34. Xu, Y.Q.; Huang, P.; Li, X.W.; Liu, S.S.; Lu, B.Q. Derivation of Water Quality Criteria for Paraquat, Bisphenol a and Carbamazepine Using Quantitative Structure-Activity Relationship and Species Sensitivity Distribution (QSAR-SSD). *Sci. Total Environ.* **2024**, *948*, 114539. [CrossRef] [PubMed]
35. Wheeler, J.R.; Grist, E.P.M.; Leung, K.M.Y.; Morritt, D.; Crane, M. Species Sensitivity Distributions: Data and Model Choice. *Mar. Pollut. Bull.* **2002**, *45*, 192–202. [CrossRef] [PubMed]
36. Hong, Y.; Xiao, S.; Naraginti, S.; Liao, W.; Feng, C.; Xu, D.; Guo, C.; Jin, X.; Xie, F. Freshwater Water Quality Criteria for Phthalate Esters and Recommendations for the Revision of the Water Quality Standards. *Ecotoxicol. Environ. Saf.* **2024**, *279*, 116517. [CrossRef] [PubMed]
37. Lou, Y.; Bian, J.; Sun, X.; Wang, F.; Xu, L.; Sun, G. Optimization of Ammonia Nitrogen Benchmarks and Ecological Risk Assessment in Monsoon Freezing Lakes Based on Species Sensitivity Distribution with Lake Chagan in Northeastern China as an Example. *Ecol. Indic.* **2024**, *166*, 18734–18748. [CrossRef]
38. Qin, L.; Yu, L.; Wang, M.; Sun, X.; Wang, J.; Liu, J.; Wang, Y.; White, J.C.; Chen, S. The Environmental Risk Threshold (HC_5) for Cd Remediation in Chinese Agricultural Soils. *J. Environ. Manag.* **2024**, *362*, 119885. [CrossRef]
39. Sun, X.; Qin, L.; Yu, L.; Wang, J.; Liu, J.; Wang, M.; Chen, S. Ecological Risk Threshold for Chromium in Chinese Soils and Its Prediction Models. *Environ. Res.* **2024**, *262*, 120498. [CrossRef]
40. Ding, T.; Du, S.; Zhang, Y.; Wang, H.; Zhang, Y.; Cao, Y.; Zhang, J.; He, L. Hardness-Dependent Water Quality Criteria for Cadmium and an Ecological Risk Assessment of the Shaying River Basin, China. *Ecotoxicol. Environ. Saf.* **2020**, *198*, 110666. [CrossRef]
41. Xie, D.; Wei, H.; Huang, Y.; Qian, J.; Zhang, Y.; Wang, M. Elevated Temperature as a Dominant Driver to Aggravate Cadmium Toxicity: Investigations through Toxicokinetics and Omics. *J. Hazard. Mater.* **2024**, *474*, 116614. [CrossRef]
42. Davies, P.H.; Gorman, W.C.; Carlson, C.A.; Brinkman, S.F. Effect of Hardness on Bioavailability and Toxicity of Cadmium to Rainbow-Trout. *Chem. Spec. Bioavailab.* **1993**, *5*, 67–77. [CrossRef]
43. Yim, J.H.; Kim, K.W.; Kim, S.D. Effect of Hardness on Acute Toxicity of Metal Mixtures Using *Daphnia magna*: Prediction of Acid Mine Drainage Toxicity. *J. Hazard. Mater.* **2006**, *138*, 16–21. [CrossRef] [PubMed]
44. Palawski, D.; Hunn, J.B.; Dwyer, F.J. Sensitivity of Young Striped Bass to Organic and Inorganic Contaminants in Fresh and Saline Waters. *Trans. Am. Fish. Soc.* **1985**, *114*, 748–753. [CrossRef]
45. Pilehvar, A.; Cordery, K.I.; Town, R.M.; Blust, R. The Synergistic Toxicity of Cd (II) and Cu (II) to Zebrafish (*Danio rerio*): Effect of Water Hardness. *Chemosphere* **2020**, *247*, 113759. [CrossRef] [PubMed]

46. Jin, Q.; Feng, C.; Xia, P.; Bai, Y. Hardness-Dependent Water Quality Criteria for Protection of Freshwater Aquatic Organisms for Silver in China. *Int. J. Environ. Res. Public Health* **2022**, *19*, 6067. [CrossRef]
47. Ding, R.; Wei, D.; Wu, Y.; Liao, Z.; Lu, Y.; Chen, Z.; Gao, H.; Xu, H.; Hu, H. Profound Regional Disparities Shaping the Ecological Risk in Surface Waters: A Case Study on Cadmium across China. *J. Hazard. Mater.* **2024**, *465*, 121054. [CrossRef]
48. Canli, E.G.; Canli, M. Calcium Reduces the Effects of Cadmium on the Responses of Biomarkers in Freshwater Fish (*Oreochromis niloticus*). *Chem. Ecol.* **2024**, *40*, 664–677. [CrossRef]
49. Adams, W.; Blust, R.; Dwyer, R.; Mount, D.; Nordheim, E.; Rodriguez, P.H.; Spry, D. Bioavailability Assessment of Metals in Freshwater Environments: A Historical Review. *Environ. Toxicol. Chem.* **2020**, *39*, 48–59. [CrossRef]
50. Zhou, H.; Peng, W.; Du, X.; Huang, H. Evaluation of Surface Water Quality in China. *Water Resour.* **2004**, *24*, 21–30.

toxics

Article

Exogenously Applied Triacontanol Mitigates Cadmium Toxicity in *Vigna radiata* L. by Optimizing Growth, Nutritional Orchestration, and Metal Accumulation

Saba Mudassar [1], Shakil Ahmed [1], Rehana Sardar [2,*], Nasim Ahmad Yasin [3], Muhammad Jabbar [4] and Maximilian Lackner [5,*]

1 Institute of Botany, University of the Punjab, Lahore 54590, Pakistan
2 Department of Biological and Environmental Sciences, Emerson University, Multan 60000, Pakistan
3 Faculty of Agricultural Sciences, University of the Punjab, Lahore 54590, Pakistan
4 Faculty of Bio-Sciences, Cholistan University of Veterinary and Animal Sciences (CUVAS), Bahawalpur 63100, Pakistan
5 Department of Industrial Engineering, University of Applied Sciences Technikum Wien, 17 Hoechstaedtplatz 6, 1200 Vienna, Austria
* Correspondence: rehana.sardar@eum.edu.pk (R.S.); maximilian.lackner@technikum-wien.at (M.L.)

Abstract: Cadmium (Cd) is one of the foremost phytotoxic elements. Its proportion in agricultural soil is increasing critically due to anthropogenic activities. Cd stress is a major crop production threat affecting food security globally. Triacontanol (TRIA) is a phytohormone that promotes growth, development, and metabolic processes in plants. The current study explicates the mitigation of Cd toxicity in *Vigna radiata* L. (mung bean) seedlings through the application of TRIA by a seed priming technique under Cd stress. The role of TRIA in improving metabolic processes to promote *Vigna radiata* (mung bean, green gram) vegetative growth and performance under both stressed and unstressed conditions was examined during this study. To accomplish this, three doses of TRIA (10, 20, and 30 µmol L^{-1}) were used to pretreat *V. radiata* seeds before they were allowed to grow for 40 days in soil contaminated with 20 mg kg^{-1} Cd. Cd stress lowered seed germination, morphological growth, and biomass in *V. radiata* plants. The maximum root and shoot lengths, fresh and dry weights of roots, and shoot and seed germination rates were recorded for TRIA2 compared with those of TRIA1 and TRIA3 under Cd stress. In Cd-stressed *V. radiata* plants, TRIA2 increased the content of chlorophyll *a* (2.1-fold) and *b* (3.1-fold), carotenoid (4.3-fold), total chlorophyll (3.1-fold), and gas exchange attributes, such as the photosynthetic rate (2.9-fold), stomatal conductance (6.0-fold), and transpiration rate (3.5-fold), compared with those in plants treated with only Cd. TRIA seed priming increased nutrient uptake (K^{1+}, Na^{1+}, Mg^{2+}, and Zn^{2+}), total phenolic content, total soluble protein content, and DPPH (2,2-diphenyl-1-picrylhydrazyl) activity. Additionally, TRIA2 significantly reduced the quantity of Cd in the plants (3.0-fold) and increased the metal tolerance index (6.6-fold) in plants contrasted with those in the Cd-treated plants. However, TRIA2 promoted plant growth and biomass production by lowering Cd-induced stress through modifying the plant antioxidant machinery and reducing oxidative stress. The improved yield characteristics of *V. radiata* seedlings treated with TRIA suggest that exogenous TRIA may be used to increase plant tolerance to Cd stress.

Keywords: cadmium; growth; mung bean; proline; TRIA; *Vigna radiata*

1. Introduction

Potentially toxic elements poisoning soil is one of the foremost important abiotic stresses in the world [1]. Because of its nonbiodegradable and persistent characteristics in the ecosystem [2], metal contamination in soil is a serious concern. The main cause of increasing heavy metal pollution is human activity. The use of industrial effluents for irrigation methods has the potential to increase the content of heavy metals in soil, which can also affect the electrical properties, salinization, chloride content, total dissolved solids, pH, and levels of dissolved oxygen and phosphate [3]. Owing to the global freshwater shortage, irrigation by using wastewater containing potentially hazardous heavy metals is a common practice in some countries, although this is illegal [4]. Owing to Pakistan's water deficit, the use of wastewater in agriculture will likely increase quickly [5]. Wastewater and floods result in sedimentation on residential and agricultural lands [6]. Floodplain soils and polluted sediments are damaging the environment, the safety of agricultural products, and, consequently, human and animal health [6]. In farmed areas, heavy metal pollution prevents plants from growing and lowers food quality [1].

Cadmium (Cd), along with other polluting metals, is frequent in arable areas worldwide due to natural weathering of rocks and agricultural and industrial practices resulting in agricultural yield loss [7]. Cd is listed as the seventh most hazardous element out of 20 [8] and is considered a main cause of concern because it is poisonous and mobile even at low concentrations [9]. The primary sources of Cd in the environment are agricultural sources, including pesticides and chemical fertilizers, as well as those associated with mining and industry [10]. The permitted limit for Cd^{2+} is 0.005–0.02 mg/L for plants and 1 mg/L for animals according to USEPA [11]. Cadmium harms plants physiologically, morphologically, molecularly, and biochemically [12]. Crops that receive wastewater irrigation are more likely to accumulate Cd, which inhibits seed germination, slows down root and shoot growth, decreases the plant height overall, reduces the number of leaves per plant, and, ultimately, kills plants [10]. Plants exposed to Cd exhibit distorted chloroplasts, reduced water and nutrient absorption and retention, and poor transpiration and photosynthesis [13]. Plants contain more reactive oxygen species (ROS), including H_2O_2, OH, and ${O_2}^{1-}$, under Cd stress, which also increases oxidative stress [14]. Cd can disrupt the photosynthetic system, resulting in reduced chlorophyll synthesis, and induce leaf chlorosis, necrosis, reduced plant growth, and other effects [11].

Priming enhances the physiochemical and metabolic activities required for appropriate seed germination. A higher germination percentage, improved radicle and plumule growth, uniform crop stand, and higher productivity are salient outcomes of seed priming [15]. Furthermore, seed priming increases nutrient uptake, activates enzymes, enhances plant stress resilience, and improves biomass production in the stressed environment [16–18].

Several endogenous and exogenous growth regulators enable plants to mitigate environmental stresses [19]. Triacontanol (TRIA, $C_{30}H_{61}OH$, CAS no. 595-50-0), a saturated primary alcohol that was initially discovered in alfalfa [20], is considered a novel plant growth regulator (PGR) that can modify numerous physio-biochemical phenomena in agricultural plants [21]. Plant species including *Medicago sativa*, *Oryza sativa*, and *Jatropha curcas*. naturally contain TRIA in their epicuticular wax. TRIA promotes plant development and increases plant growth when given exogenously to most crops at low concentrations [22]. When applied exogenously, it improves plant biomass, photosynthetic pigments, gas exchange parameters, mineral nutrient uptake, antioxidant enzyme activities, yield, and quality features [23]. Furthermore, TRIA alters the anatomy of leaf, stem, and vascular tissue systems in plants [24]. TRIA controls the expression of genes to suppress or amplify stress responses. The critical functions of TRIA in plant responses to abiotic stressors such as acid mist, cold, drought, heavy metals, and salt stress have been well characterized [25]. By increasing biomass production, chlorophyll content, gas exchange attributes, mineral nutrient uptake, and the antioxidant defense system, its application reduces the harmful effects of these stressors on plants [26].

Vigna radiata L. (mung bean) is the chief nutritious and economical grain legume belonging to the Leguminosae family. It offers basic nutritional protein benefits, making it a great choice for vegetarians. Despite its short lifespan, it contributes significantly to major agricultural systems and increases soil fertility. Although *V. radiata* can withstand different environmental conditions, its yield is still declining [27]. In addition to some regions of Africa and Australia, *V. radiata* crops are commonly grown in Asia. Currently, Asia produces 90% of the world's *V. radiata*, with China, Pakistan, India, and Thailand being the continent's top producers [28]. *V. radiata* is mostly cultivated as a summer crop in Pakistan; however, its average production is quite poor compared with other countries. Pakistan imports pulses in significant quantities to manage its rising protein demand [29]. Recent advancements in agriculture, molecular breeding, and integrated multiomics have promoted heavy metal resistance in many crop plants. However, these sophisticated methods are merely applicable in laboratory settings and are not very adaptable by farmers [30]. Moreover, there is a lack of research concerning the effects of seed priming with TRIA on the germination, growth, and maturity of *V. radiata* seeds under Cd stress. The present study aimed to investigate the impact of TRIA priming on seed germination, various physiochemical traits, and growth in Cd-exposed *V. radiata* plants.

2. Materials and Methods

The Roshan Seed Centre in Lahore provided certified *V. radiata* seeds of the Azri variety. *V. radiata* seeds were sterilized for 3–5 min by soaking them in 0.5% sodium hypochlorite solution and thoroughly rinsing them three times with distilled water. *V. radiata* seeds were primed for 8 h at room temperature with various concentrations of Triacontanol (TRIA) (10, 20, and 30 μM), referred to as TRIA1, TRIA2, and TRIA3, respectively. Following the priming process, the primed seeds were carefully washed thoroughly before being spread on blotting paper to air dry them at room temperature. As a control, seeds that had not been primed were used. Thirty-two clay pots were used for the experiments. With cadmium chloride ($CdCl_2$), the soil (7.4 kg) in each of the allocated clay pots was contaminated with 20 mg kg^{-1} Cd. The soil that contained Cd spikes was subjected to Cd treatment. Both the Cd-contaminated and non-Cd-contaminated soils were placed in their designated pots. A randomized complete block design (RCBD) was used for this experiment to decrease systematic error, and after the pots had been filled, they were arranged according to their treatments. Additionally, four replicates of each treatment were set up. The pots were sorted and labeled based on their respective treatments and replication counts, and after 15 days, the soil was pretreated. Different levels of TRIA-primed seeds were sown in the appropriate pots.

2.1. Measurement of Growth Parameters

The harvest of plants was taken 40 days after sowing (DAS). Roots of plants were carefully taken out from the pots; they were cleaned properly. All the plants were subjected to measurements of several morphological parameters, including root length (cm), shoot length (cm), plant length (cm), leaf area (cm^2), and number of leaves.

2.2. Biomass Production Assessment

V. radiata plants were harvested, and their fresh weights, such as shoot fresh weight (g), root fresh weight (g), and total plant fresh weight (g), were measured via an electrical balance (Sartorius GmbH, model 1216MP 6E, Goettingen, Germany). The plants were dried for 72 h at 70 °C in an oven (Wiseven, type WOF-105, Republic of Korea) to determine the dry weight, including root dry weight (g), shoot dry weight (g), and total plant dry weight (g).

2.3. Plant Photosynthetic Pigment Determination

Chlorophyll *a*, chlorophyll *b*, and total chlorophyll content were determined via the Arnon [31] method, whereas the concentration of carotenoids was determined via the Davis [32] method. Fresh *V. radiata* leaves were manually crushed with a mortar and pestle, extracted with 10 mL of an 80% acetone solution, and then kept at a low temperature. The mixture was subsequently centrifuged at 10,000 rpm for 5 min. On a spectrophotometer (Uv-1800 240V. CAT. No. 206-25400-38 SHIMZDZU Corporation, Kyoto, Japan), measurements were taken at 480 nm, 645 nm, and 663 nm to determine the optical density of the supernatant.

2.4. Assessment of Cd Accumulation

V. radiata shoots were cleaned with water and dried in a drying oven for two days at 65 °C (Wiseven, Model WOF-105, Korea) to determine the concentration of Cd by making acid-digested samples. The dried shoots were mashed with a mortar and pestle and passed through 60-screen mesh. A total of 1.5 mL of $HClO_4$ (60%) and 5 mL of HNO_3 (70%) were applied to a 0.5 g sample of plant shoots. The solution was heated continuously until the brown color disappeared. By adopting the procedure of Moseley and Jones [33], the solution was cooled and diluted (1:1) with approximately 5 mL of HCl (density of 1.18 g mL^{-1}) in a 25 mL solution of water. An atomic absorption spectrophotometer (GBC XPLOR AA-Dual) was employed to quantify the Cd accumulation concentration factor [34]. The Al-Farraj et al. [35] technique was used to find the accumulation coefficient (*AC*).

$$AC\ factor = \frac{Concentration\ (roots)}{Concentration\ (soil)} \tag{1}$$

The %MTI (metal tolerance index) was determined according to Balint et al. [36], as described below.

$$\%MTI = \frac{Dry\ weight\ of\ treated\ plants}{Dry\ weight\ of\ untreated\ plants} \times 100 \tag{2}$$

2.5. Estimation of Mineral Content

The mineral content from the digested samples of the *V. radiata* plants was analyzed. A flame photometer (Model 410, Corning) was used to evaluate Na^+ and K^+ content according to Sagner et al. [37], whereas an atomic absorption spectrophotometer (GBC XPLOR AA-Dual) was employed to quantify the concentrations of Zn^{2+} and Mg^{2+}, as suggested by Chapman and Dale [34].

2.6. Assessment of Soluble Protein

For estimating total soluble protein, a 2 g plant sample was homogenized with 4 mL of 1 N phosphate buffer (34 g K_2HPO_4 in 2 mL of distilled water) with the help of an ice-chilled mortar and pestle. The mixture was centrifuged at 6000 rpm for 15 min. The homogenization of 0.8 mL of the supernatant was carried out with Folin solution (4 mL) for 15 min by keeping it at room temperature. Each sample received Folin reagent (0.5 mL), which was then shaken and left at room temperature for approximately 45 min. Using a UV-1800 Shimadzu spectrophotometer (Shimadzu, Kyoto, Japan), optical density measurements were made at 715 nm. The quantity of total soluble protein was estimated via a bovine serum albumin (BSA) standard curve.

2.7. Total Proline Content Determination

The proline concentration was determined using the protocol of Bates et al. [38]. A 0.25 g ice-chilled sample of plant leaves was extensively vortexed after adding 10 mL of 3% sulfosalicylic acid. The mixture underwent filtering. First, 2.8 g of ninhydrin, 48.16 mL of 85% phosphoric acid, 72.8 mL of acetic acid, and 2 mL of glacial acetic acid were added to 2 mL of filtrate to prepare a homogenous mixture. The homogenate was kept in a water bath at 100 °C (N.S Engineering concern XMTG-9000). After one hour, the solution was chilled to stop the process. Toluene (4 mL) was vortexed vigorously for 30 s in the mixture to draw off the upper toluene phase that contained the proline–ninhydrin chromophore complex. The mixture was incubated at 25 °C for approximately 0.5 h. In a Shimadzu UV-1800 spectrophotometer, the absorbance was measured at 520 nm and contrasted with the proline standard curve [38].

$$Proline\left(\mu\frac{\text{mol}}{\text{g}}\text{FW}\right) = \left[\frac{\mu\text{g proline/mlx ml of toluene}}{115.5}\right] / \left[\frac{\text{g of sample}}{10}\right] \tag{3}$$
$$FW = fresh\ weight\ (g)$$

2.8. Analyzing the Gas Exchange Parameters

The transpiration rate (*E*), net photosynthetic rate (*A*), and stomatal conductance (*gs*) were determined via the IRGA (infrared gas analyzer) LCA-4 system of Analytical Development Co. Ltd. (ADC, Ltd. 12 Spurling works, Pindar Road, Hoddesdon, Herts, UK). Between 10:30 and 11:30 am, the highest completely extended leaf was used to take readings for several parameters.

The activity of 2,2-diphenyl-1-picrylhydrazyl (DPPH) was measured for the scavenging of free radicals. In total, 1 g of *V. radiata* sample was obtained, and 10 mL of methanol was used to create a methanolic extract via the Chen et al. [39] technique. For this purpose, 5 mL of 0.1 mM methanolic extract and 1 mL of complete mixture were combined. Freshly made 2,2-diphenyl-1-picrylhydrazyl (DPPH) methanolic solution was stored in the dark. After one hour, the absorbance was observed at 517 nm through a Shimadzu UV-1800 spectrophotometer. In total, 1 mL of methanol was used to prepare the blank.

The percentage (%) of free radical scavenging activity was determined as follows.

$$\text{Activity}(\%) = \text{Scavenging}\left[1 - \left(\frac{A517\text{nm}, Sample}{A517\text{nm}, Blank}\right)\right] \times 100 \quad (4)$$

A = Absorbance

2.9. Assessment of Total Phenolic Content

To estimate the total phenolic content, 2 g of fresh leaves and 10 mL of 80% aqueous methanol were heated at 65 °C for approximately 15 min. A total of 250 µL of Folin–Ciocalteu reagent (1 N), 5 mL of sterilized distilled water, and 1 mL of the extract were combined and then stored at 30 °C. To determine the total amount of phenols, the absorbance of the blue hue emerging in the reaction mixture was observed at 725 nm and compared to the standard curve of gallic acid [40,41].

2.10. Statistical Analysis

The gathered data were examined via one-way ANOVA and SPSS software (IBM-SPSS Statistics 20). Duncan's multiple range test was used to segregate means for significant treatment at 0.05 *p*-value. The data represent the average of four replicate ± S.E. Additionally, principal component analysis (PCA) and correlation were performed via RStudio software 2023.09.1+494.

3. Results

3.1. Growth Parameters of V. radiata

Compared with the control plants, the plants whose seeds were primed with TRIA2 showed a 1.41-fold increase in shoot length. Compared with those of the Cd-only treated plants, the shoot length of the TIA2+Cd-treated plants increased 2.5-fold. The plant seeds treated with TRIA3+Cd also had a 2.3-fold greater length than those grown only in Cd-spiked soil without TRIA treatment, as shown in Table 1.

Table 1. Impact of TRIA on growth parameters and germination percentage of *V. radiata* under Cd stress.

Treatments	Growth Parameters					
	Shoot Length (cm)	Root Length (cm)	Total Length (cm)	Leaf Area (cm^2)	No. of Leaves	Germination %
C	5.44 ± 0.10 ^{b}	3.12 ± 0.11 ^{b}	9.57 ± 0.24 ^{b}	3.94 ± 0.21 ^{b}	5.50 ± 0.46 ^{b}	84 ± 4.00 ^{b}
Cd	3.84 ± 0.15 ^{a}	2.5 ± 0.09 ^{a}	6.24 ± 0.25 ^{a}	1.57 ± 0.10 ^{a}	2.52 ± 0.40 ^{a}	66 ± 3.00 ^{a}
TRIA1	7.75 ±0.76 ^{d}	4.65 ± 0.30 ^{d}	12.4 ± 0.43 ^{d}	5.93 ± 0.20 ^{cd}	7.25 ± 0.27 ^{d}	90 ± 2.00 ^{bc}
TRIA2	9.11 ± 0.45 ^{g}	5.76 ± 0.27 ^{h}	14.87 ± 0.05 ^{f}	7.98 ± 0.28 ^{e}	10.5 ± 0.50 ^{f}	100 ± 0.90 ^{d}

Table 1. *Cont.*

Treatments	Growth Parameters					
	Shoot Length (cm)	Root Length (cm)	Total Length (cm)	Leaf Area (cm^2)	No. of Leaves	Germination %
TRIA3	8.54 ± 0.27 ^{f}	5.01 ± 0.05 ^{g}	13.55 ± 0.08 ^{e}	6.12 ± 0.18 ^{d}	8.25 ± 0.25 ^{e}	95 ± 2.00 ^{c}
TRIA1+Cd	6.92 ± 0.07 ^{c}	4.33 ± 0.20 ^{c}	11.25 ± 0.05 ^{c}	5.01 ± 0.19 ^{c}	6 ± 0.84 ^{c}	85 ± 5.50 ^{b}
TRIA2+Cd	8.12 ±0.75 ^{e}	5.12 ± 0.06 ^{f}	13.24 ± 0.27 ^{e}	6.32 ± 0.34 ^{d}	8.5 ± 0.40 ^{e}	95 ± 3.00 ^{cd}
TRIA3+Cd	7.45 ± 0.36 ^{d}	4.87 ± 0.50 ^{e}	12.32 ± 0.30 ^{d}	5.47 ± 0.07 ^{c}	7 ± 0.42 ^{d}	90 ± 4.00 ^{bc}

Values illustrate means ± SE of four replicates. Dissimilar letters on error bars showed significant differences between treatments at $p \leq 0.05$. C = control, Cd = 20 mg kg^{-1} Cd, TRIA1 = 10 µM L^{-1} TRIA, TRIA2 = 20 µM L^{-1} TRIA, TRIA3 = 30 µM L^{-1} TRIA.

3.2. Assessment of Biomass

The plants subjected to Cd stress exhibited 0.6- and 0.4-fold lower fresh weights of roots and shoots, respectively, as compare to plants under control conditions. Reciprocally, the dry and fresh weights of *V. radiata* seeds increased, and Cd stress was alleviated in *V. radiata* seeds, as shown in Table 2. Moreover, TRIA2+Cd enhanced shoot and root fresh weight as well as total plant fresh weight by 5.7-, 3.8-, and 5.2-fold, respectively, through the amelioration of Cd stress in comparison with that in Cd-only treated plants. There was a slight reduction in the dry and fresh weights of roots and shoots in the case of TRIA3 compared with those in the case of TRIA2.

Table 2. Impact of TRIA on biomass assessment of *V. radiata* under Cd stress.

Treatments	Shoot Fresh Weight (g $plant^{-1}$)	Root Fresh Weight (g $plant^{-1}$)	Total Fresh Weight (g $plant^{-1}$)	Shoot Dry Weight (g $plant^{-1}$)	Root Dry Weight (g $plant^{-1}$)	Total Dry Weight (g $plant^{-1}$)
C (control)	0.94 ± 0.03 ^{bc}	0.25 ± 0.02 ^{b}	1.18 ± 0.02 ^{b}	0.28 ± 0.02 ^{b}	0.08 ± 0.02 ^{b}	0.46 ± 0.01 ^{b}
Cd	0.38 ± 0.01 ^{a}	0.16 ± 0.03 ^{a}	0.53 ± 0.03 ^{a}	0.18 ± 0.04 ^{a}	0.04 ± 0.04 ^{a}	0.30 ± 0.05 ^{a}
TRIA1	1.23 ± 0.09 ^{c}	0.46 ± 0.09 ^{d}	1.69 ± 0.14 ^{d}	0.68 ± 0.01 ^{d}	0.25 ± 0.05 ^{d}	0.93 ± 0.32 ^{d}
TRIA2	2.85 ± 0.10 ^{f}	0.72 ± 0.06 ^{g}	3.57 ± 0.06 ^{g}	1.13 ± 0.26 ^{g}	0.53 ± 0.08 ^{g}	1.66 ± 0.04 ^{h}
TRIA3	2.21 ± 0.12 ^{e}	0.58 ± 0.05 ^{e}	2.79 ± 0.16 ^{f}	0.84 ± 0.09 ^{e}	0.39 ± 0.04 ^{ef}	1.23 ± 0.05 ^{f}
TRIA1+Cd	1.01 ± 0.38 ^{cd}	0.33 ± 0.09 ^{c}	1.34 ± 0.33 ^{c}	0.46 ± 0.32 ^{c}	0.18 ± 0.15 ^{c}	0.64 ± 0.04 ^{c}
TRIA2+Cd	2.17 ± 0.40 ^{e}	0.61 ± 0.14 ^{f}	2.78 ± 0.06 ^{f}	0.91 ± 0.07 ^{f}	0.41 ± 0.17 ^{f}	1.32 ± 0.12 ^{g}
TRIA3+Cd	1.87 ± 0.08 ^{d}	0.49 ± 0.32 ^{d}	2.36 ± 0.09	0.68 ± 0.03 ^{d}	0.34 ⊥ 0.37 ^{e}	1.02 ± 0.23 ^{e}

Values illustrate means ± SE of four replicates. Dissimilar letters on error bars showed significant differences among the treatments at $p \leq 0.05$. C = control, Cd = 20 mg kg^{-1} Cd, TRIA1 = 10 µM L^{-1} TRIA, TRIA2 = 20 µM L^{-1} TRIA, TRIA3 = 30 µM L^{-1} TRIA.

3.3. Estimation of Photosynthetic Pigments and Cadmium Uptake Content

The toxicity of cadmium negatively affects plant pigments in *V. radiata* (Table 3). The plants were harvested on 30 DAS. The plants whose seeds were primed with TRIA2 demonstrated the maximum content of chlorophyll *a*, which was 2.46-fold greater than the control plants, which were not subjected to any treatment. The plants treated with TRIA2+Cd also showed the greatest chlorophyll *a* content among those under Cd stress. Furthermore, TRIA diminished Cd stress and increased chlorophyll (*a*, *b*), total chlorophyll, and carotenoid content in the TRIA1+Cd, TRIA2+Cd, and TRIA3+Cd groups. Compared with the control plants, the Cd-stressed plants exhibited decreases in chlorophyll *a* and *b* content of 1.6- and 1.76-fold, respectively, as depicted in Table 3.

Table 3. Impact of TRIA on plant pigments and Cd uptake, accumulation, and metal tolerance index in *V. radiata* under Cd stress.

Treatments	Chl *a* (mg g^{-1} FW)	Chl *b* (mg g^{-1} FW)	Total Chl (mg g^{-1} FW)	Carotenoids (mg g^{-1} FW)	Cd Uptake in Plant (mg/g)	AC Factor	MTI
C	0.74 ± 0.04 ^{b}	1.43 ± 0.14 ^{b}	2.15 ± 0.15 ^{b}	0.08 ± 0.02 ^{a}	-	-	-
Cd	0.45 ± 0.03 ^{a}	0.81 ± 0.02 ^{a}	1.27 ± 0.03 ^{a}	0.03 ± 0.03 ^{a}	0.84 ± 0.03 ^{d}	42 ± 0.40 ^{d}	55.56 ± 1.56 ^{a}
TRIA 1	1.30 ± 0.03 ^{c}	1.92 ± 0.06 ^{cd}	3.21 ± 0.04 ^{d}	0.13 ± 0.02 ^{a}	-	-	-
TRIA 2	1.82 ± 0.05 ^{e}	3.10 ± 0.06 ^{f}	4.92 ± 0.02 ^{g}	0.17 ± 0.02 ^{a}	-	-	-
TRIA 3	1.42 ± 0.03 ^{d}	2.68 ± 0.13 ^{e}	4.11 ± 0.13 ^{f}	0.12 ± 0.02 ^{a}	-	-	-
TRIA 1+Cd	1.08 ± 0.05 ^{c}	1.72 ± 0.03 ^{c}	2.79 ± 0.04 ^{c}	0.10 ± 0.03 ^{a}	0.67 ± 0.02	33.5 ± 0.57 ^{c}	177.78 ± 1.14 ^{b}
TRIA 2+Cd	1.38 ± 0.07 ^{d}	2.52 ± 0.03d ^{e}	3.90 ± 0.04 ^{e}	0.13 ± 0.02 ^{a}	0.28 ± 0.01 ^{a}	14 ± 0.60 ^{a}	366.67 ± 0.90 ^{d}
TRIA 3+Cd	1.21 ± 0.01 ^{c}	2.19 ± 0.01 ^{d}	3.40 ± 0.01 ^{d}	0.11 ± 0.01 ^{a}	0.45 ± 0.02 ^{b}	22.5 ± 0.27 ^{b}	283.33 ± 0.07 ^{c}

Values illustrate means ± SE of four replicates. Dissimilar letters on error bars showed significant differences among the treatments at $p \leq 0.05$. C = control, Cd = 20 mg kg^{-1} Cd, TRIA1 = 10 µM L^{-1} TRIA, TRIA2 = 20 µM L^{-1} TRIA, TRIA3 = 30 µM L^{-1} TRIA.

The cadmium uptake content reached a maximum in *V. radiata* plants growing in Cd-susceptible soil, which presented 0.84 mg Cd g^{-1} DW. The seeds of the plants that were primed with TRIA alleviated the toxicity of Cd. The maximum alleviation was noted for TRIA2+Cd, which decreased the uptake of Cd by 66.67% compared with that in Cd-contaminated plants that were not treated with TRIA. The highest accumulation factor was calculated for the plants grown under Cd-spiked conditions, and it was alleviated by 20%, 66%, and 46.43%, respectively, in the seeds of the plants treated with TRIA1+Cd, TRIA2+Cd, and TRIA3+Cd. The metal tolerance index was higher in TRIA2+Cd plants than in Cd-only plants by 84.85% (Table 3).

3.4. Determination of Plant Nutrition

The uptake of plant nutrients, including Mg, Zn, K, and Na, is affected by Cd toxicity. However, the nutrient content recovered in control plants and plants under Cd stress primed with TRIA indicated that seed priming with TRIA2+Cd greatly enriched the nutrient uptake of Mg and Zn by 4.5- and 2.9-fold, respectively, compared to that in only the Cd-treated plants. The lowest nutrient concentration was noted in the plants affected with Cd toxicity, which were deprived of TRIA. Additionally, TRIA3+Cd increased Mg and Zn by 3.5- and 2.7-fold, respectively, compared with those in plants treated with only Cd (Table 4).

Table 4. Impact of TRIA on mineral content (mg g $^{-1}$) of *V. radiata* under Cd stress.

Treatments	Mg^{+2} (mg g^{-1})	Zn^{+2} (mg g^{-1})	K^{+} (mg g^{-1})	Na^{+} (mg g^{-1})
C	0.34 ± 0.60 ^{b}	0.34 ±0.20 ^{b}	17.82 ±1.20 ^{b}	1.81 ± 0.04 ^{b}
Cd	0.17 ± 0.27 ^{a}	0.18 ± 0.40 ^{a}	11.43 ± 0.09 ^{a}	0.94 ± 0.08 ^{a}
TRIA1	0.48 ± 0.60 ^{c}	0.44 ± 0.05 ^{c}	20.04 ± 0.43 ^{d}	2.68 ± 0.16 ^{c}
TRIA2	0.84 ± 0.64 ^{g}	0.55 ± 0.08 ^{d}	23.75 ± 0.09 ^{f}	3.64 ± 0.09 ^{d}
TRIA3	0.68 ± 0.16 ^{e}	0.54 ± 0.26 ^{d}	22.57 ± 0.05 ^{e}	3.18 ± 0.07 ^{d}
TRIA1 + Cd	0.38 ± 0.09 ^{b}	0.42 ± 0.31 ^{c}	18.78 ± 0.44 ^{c}	2.26 ± 0.03 ^{c}
TRIA2 + Cd	0.72 ± 0.40 ^{f}	0.5 ± 0.07 ^{cd}	22.49 ± 0.20 ^{e}	3.05 ± 0.05 ^{d}
TRIA3 + Cd	0.56 ± 0.04 ^{d}	0.47 ± 0.05 ^{c}	20.36 ± 0.15 ^{d}	2.72 ± 0.03 ^{c}

Values illustrate means ± SE of four replicates. Dissimilar letters on error bars showed significant differences among the treatments at $p \leq 0.05$. C = control, Cd = 20 mg kg^{-1} Cd, TRIA1 = 10 µM L^{-1} TRIA, TRIA2 = 20 µM L^{-1} TRIA, TRIA3 = 30 µM L^{-1} TRIA.

3.5. Assessment of Soluble Protein and Proline Content

TRIA had positive effects on the soluble protein content of *V. radiata* plants. Compared with the plants under controlled conditions, the Cd-stressed plants presented a decrease in soluble protein content of 32.53%. However, the maximum soluble protein content was observed in the plants obtained from the TRIA2 seed priming. In contrast, the soluble protein content of the plants raised from TRIA seed priming and under Cd stress (TRIA2+Cd) was 20.74% lower than that of the TRIA2 plants (not under Cd stress), as shown in Figure 1.

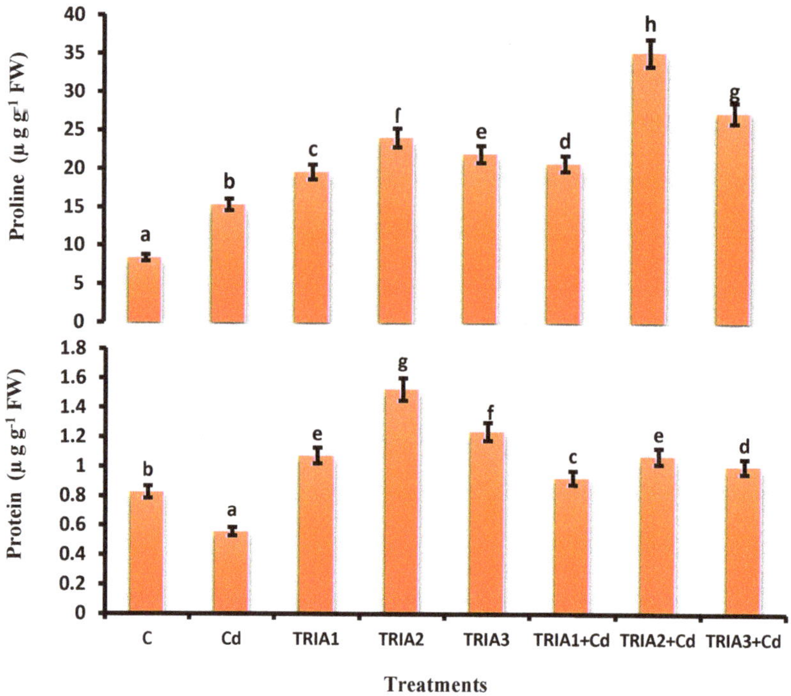

Figure 1. Effect of TRIA and Cd on proline and total soluble protein content of *V. radiata*. Values demonstrate means ± SE of four replicates (n = 4). Non-identical letters over error bars exhibit significant differences between the treatments at $p \leq 0.05$. C = control, Cd = 20 mg kg^{-1} Cd, TRIA1 = 10 µM L^{-1} TRIA, TRIA2 = 20 µM L^{-1} TRIA, TRIA3 = 30 µM L^{-1} TRIA.

3.6. Assessment of Proline Content

In comparison to the plants under controlled conditions, the proline synthesis in the Cd-affected *V. radiata* plants was significantly greater (82.57%). Compared with those of TRIA1, TRIA2+Cd, TRIA2+Cd, and TRIA3+Cd, which were not under Cd stress, the proline content of the plants that were under Cd stress from seeds primed with TRIA increased by 8.36%, 32.08%, and 32.25%, respectively. The plants supplemented with TRIA and not spiked with Cd showed lower proline content (Figure 1).

3.7. Estimation of the Photosynthetic Rate

The pronounced effect of TRIA2+Cd on the photosynthetic rate was 2.8-fold greater than under only Cd conditions. However, the photosynthetic rate decreased by 1.08-fold in the TRIA2+Cd group compared with that in the TRIA2 group. The maximum photosynthetic rate was calculated for TRIA2, i.e., 74.89 µmol/m^3/s, whereas the minimum photosynthetic rate was calculated for the plants under Cd stress, i.e., only 24 µmol/m^3/s (Figure 2).

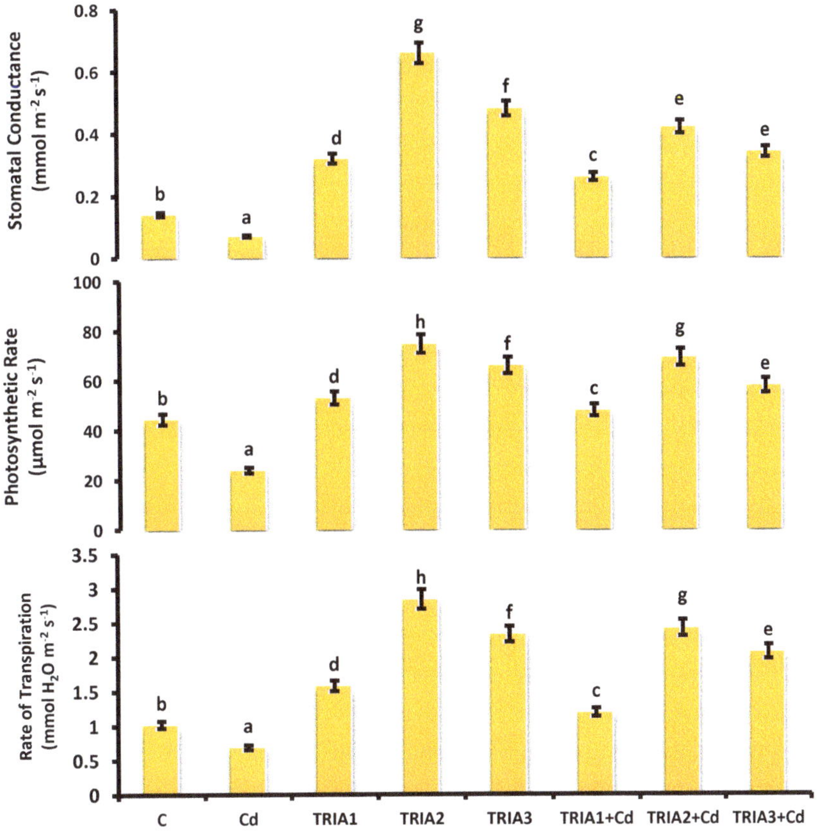

Figure 2. Effect of TRIA and Cd on photosynthetic rate, stomatal conductance, and transpiration rate of *V. radiata*. Values demonstrate means ± SE of four replicates (n = 4). Non-identical letters over error bars exhibit significant differences between the treatments at $p \leq 0.05$. C = control, Cd = 20 mg kg^{-1} Cd, TRIA1 = 10 µM L^{-1} TRIA, TRIA2 = 20 µM L^{-1} TRIA, TRIA3 = 30 µM L^{-1} TRIA.

3.8. Determination of Stomatal Conductance

TRIA alleviated the toxicity of Cd stress and increased stomatal conductance in plants, i.e., TRIA1+Cd, TRIA2+Cd, and TRIA3+Cd, by 3.7-, 6.0-, and 4.9-fold, respectively, compared with that in only Cd-stressed plants. The highest stomatal conductance was recorded in the TRIA2 plants, i.e., 0.64 mmol m^{-2}s^{-1}, which was 4.17-fold greater than the control plants, as shown in Figure 2.

3.9. Estimation of the Transpiration Rate

Cadmium stress decreased the transpiration rate by 32% compared with the control. Compared with the control, seed priming with TRIA1, TRIA2, and TRIA3 increased the transpiration rate by 1.5-, 2.8-, and 2.3-fold, respectively. TRIA2+Cd increased the transpiration rate by 2.5-fold compared with that under only the Cd condition. The maximum transpiration rate was calculated in TRIA2, i.e., 2.81 mmol H_2O m^{-2} s^{-1}, and the minimum transpiration rate was recorded in Cd-affected plants, i.e., 0.69 mmol H_2O m^{-2} s^{-1} (Figure 2).

3.10. Assessment of Total Phenolic Content

The total phenolic content was 41.67% lower in the Cd-affected plants than in the control plants. The maximum phenolic concentration was recorded in TRIA2, i.e., 0.13 mg g^{-1} FW, 2.7-fold greater than that of the control plants, i.e., 0.05 mg g^{-1} FW. The phenolic content in TRIA2+Cd-treated plants was 3.9-fold greater (Figure 3) compared with Cd-affected plants.

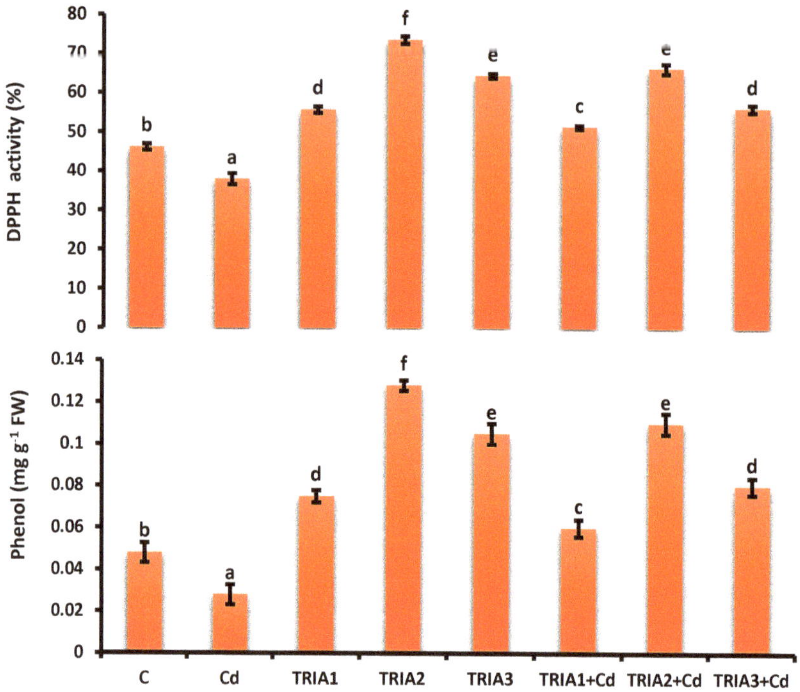

Figure 3. Effect of TRIA and Cd on DDPH and total phenolic content of *V radiata*. Values demonstrate means ± SE of four replicates (n = 4). Non-identical letters over error bars exhibit significant differences between the treatments at $p \leq 0.05$. C = control, Cd = 20 mg kg^{-1} Cd, TRIA1 = 10 μM L^{-1} TRIA, TRIA2 = 20 μM L^{-1} TRIA, TRIA3 = 30 μM L^{-1} TRIA.

3.11. Estimation of 2,2-Diphenyl-1-Picrylhydrazyl (DDPH) Activity

The DDPH activity was 18% lower in the Cd-affected plants than in the control plants. The maximum DPPH activity was recorded for TRIA2, i.e., 72.12%, and for TRIA+Cd, i.e., 63.94%. In TRIA3 and TRIA3+Cd, the DPPH activity was slightly lower, i.e., 66.87% and 56.73%, respectively. Compared with that under Cd-only conditions, DPPH activity under Cd stress was increased by TRIA, as depicted in Figure 3.

3.12. Pearson Correlation

According to the Pearson correlation, there was a strong positive association between the Cd content and the AC factor. Significant negative correlations between the Cd concentration and characteristics such as the germination rate; shoot length; root length; chlorophyll *a*, chlorophyll *b*, total chlorophyll, and carotenoid content; total phenolic content; DPPH, soluble protein, and photosynthetic rates; transpiration rate; stomatal conductance; and Zn, Mg, K, and Na concentrations were detected. Figure 4 shows no positive or negative association between MTI and the other variables.

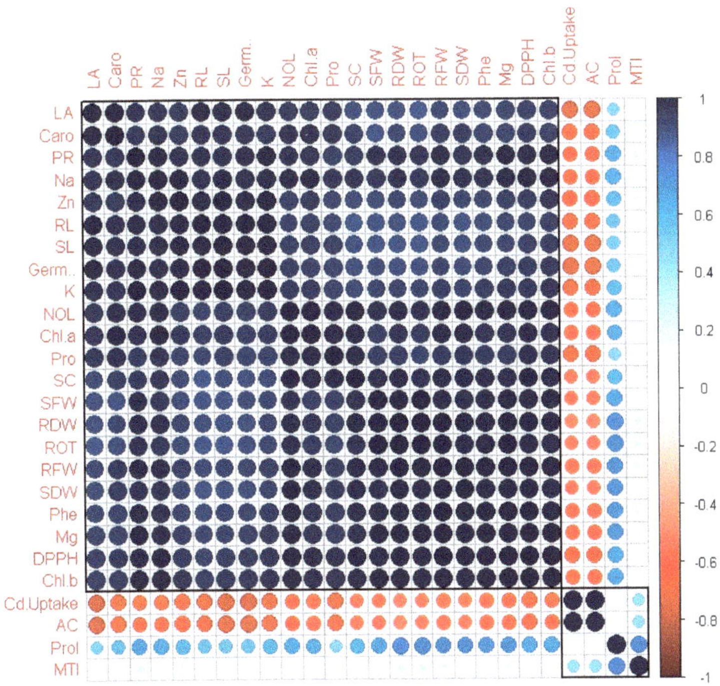

Figure 4. Pearson correlation for *V. radiata* under TRIA and Cd affect. Different abbreviated forms shown in figure as follows: Zn (Zn concentration in shoots), Mg (Mg concentration in shoots), k (K concentration in shoots), Na (Na concentration in shoots), Chl (chlorophyll), RL (root length), SL (shoot length), LA (leaf area), NP (net photosynthesis), SC (stomatal conductance), Caro (carotenoid content), Pro (protein content), Germ (germination percentage), MTI (metal tolerance index), AC (accumulation factor), Prol (proline concentration), Phe (phenolic level), NOL (number of leaves), SFW (shoot fresh weight), RFW (shoot fresh weight), RDW (root dry weight), SDW (shoot dry weight), ROT (rate of transpiration).

3.13. Principal Component Analysis

The relationships between the growth and physio-biochemical features of *V. radiata* grown in soil contaminated with Cd treated with TRIA were further demonstrated via the loading plots of the principal component analysis (PCA). More than 94.8% of the entire database consists of the first two primary components, Dim 1 and Dim 2, creating the largest portion of all the components. All the treatments were successfully dispersed across the entire database. Each contributes a percentage of the overall dataset, with Dim 1 contributing 85.4% and Dim 2 contributing 9.4%. The germination rate, root length, chlorophyll *a* content, chlorophyll *b* content, carotenoid content, total chlorophyll content, stomatal conductance, transpiration rate, photosynthetic rate, proline content, soluble protein content, DPPH content, total phenolic level, and Zn, Na, K, and Mg concentrations are positively related to the PCI. Figure 5 shows the negative correlation between the PCA and the amount of Cd in the shoot.

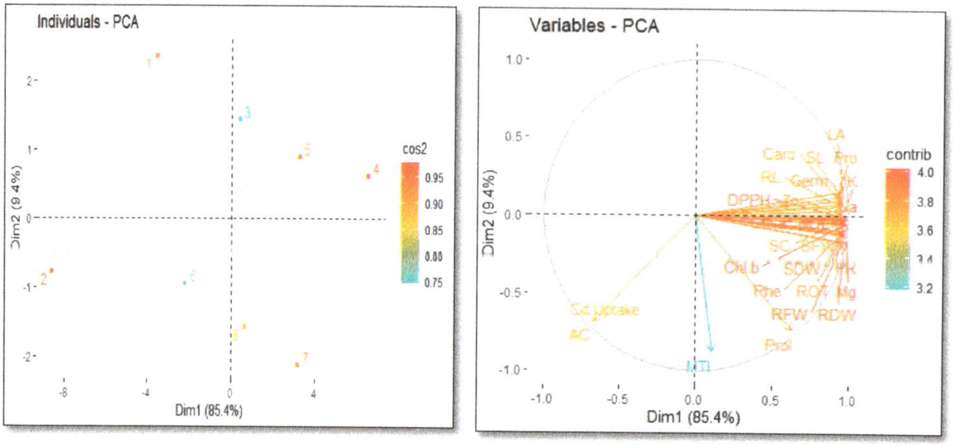

Figure 5. Loading plots of principal component analysis (PCA) demonstrated a relation between physiological parameters and growth under TRIA treatment and Cd on *Vigna radiate* L. Various abbreviations used in the figures are as follows: Zn (Zn amount in shoots), Chl (chlorophyll concentration), RL (length of root), SL (length of shoot), LA (leaf area), NP (net photosynthesis), TR (rate of transpiration), SC (stomatal conductance), Caro (carotenoid concentration), pro (protein), MTI (metal tolerance index), AC (accumulation factor), Prol (proline concentration), Phe (phenolic content), Ger (percentage of germination).

4. Discussion

Triacontanol (TRIA) seed priming increases seed vitality, which helps promote early seedling emergence [42] and increases the ability of plants to tolerate Cd toxicity for better establishment. TRIA priming helps to regulate and activate enzymes involved in stress tolerance and growth promotion in the early stages of embryogenesis, allowing for faster metabolism for germination. Our findings indicated dramatically enhanced growth in the presence of TRIA, which is consistent with the findings of Zaid et al. [23].

Across several plant species, heavy metal stress causes a decrease in root as well as shoot biomass production, which is generally regulated by plant hormones. The exogenously applied growth-promoting hormones mitigate heavy metal toxicity [43,44]. Our results also exhibited that the exogenous application of TRIA enhanced plant growth and biomass production under metal stress. In accordance with our findings, [25] also reported that seedlings treated with TRIA were found to have higher levels of CO_2 and H_2O absorption ability, as well as elevated activity of carbonic anhydrase. Naeem et al. [45] observed that exogenously applied TRIA augmented the biomass production and protein content in *Oryza sativa*. In the same way, Ahmad et al. [46] demonstrated the involvement of TRIA in the growth, stress alleviation, and yield of various crop plants.

Chlorophyll regulates the rate of photosynthesis, which adjusts physiochemical and metabolic activities related to plant growth and biomass production [47]. Our results exhibited decreased chlorophyll content and lower photosynthetic activity, resulting in poor growth in Cd-stressed *V. radiata* plants. Another researcher also found that Cd stress decreased chlorophyll *a*, chlorophyll *b*, and total chlorophyll [48]. Cd-stressed plants exhibit reduced activity of photosystem I (PSI) and photosystem II (PSII), in addition to the rate of the photosynthetic electron transport chain [49]. Cadmium ions hinder the absorption and translocation of Mg^{2+}, leading to chlorophyll degradation [50]. Hayat et al. [51] observed reduced chlorophyll content in Cd-stressed plants, which confers our findings. Correspondingly, Handa et al. [52] noted a reduction in the photosynthetic content of metal-stressed plants. However, TRIA mitigates stress through adjusting stress tolerance processes. Islam and Mohammad [20] revealed that plants treated with exogenous TRIA showed modulated activity of enzymes, improved photosynthetic activity, and productivity under stress and normal circumstances. Nickel stress enhanced the biosynthesis of ROS (reactive oxygen species), augmented protein, DNA, and chlorophyll deprivation in maize plants [53]. Yet, plants treated with TRIA mitigated Ni stress and demonstrated higher enzymatic activity and improved photosynthetic activity and growth. The higher chlorophyll content in TRIA-treated plants during our study increased the rate of photosynthesis, resulting in improved root and shoot length as well as more biomass production. Perhaps TRIA augmented photosynthetic activity by increasing the activity of the Rubisco enzyme and upregulating genes involved in photosynthesis. TRIA improved the morphophysiological features and growth of canola and sunflower plants by increasing chlorophyll synthesis [54]. Similarly, our findings revealed that TRIA enhanced the chlorophyll *a*, chlorophyll *b*, total chlorophyll, and carotenoid content of *V. radiata* plants.

Phenolic compounds can chelate metals during heavy metal stress and directly absorb molecular species of active oxygen. Peroxidase oxidizes phenols, particularly flavonoids, and phenylpropanoids, then function in the phenolic system to scavenge H_2O_2. The chemical makeup of these compounds mostly accounts for their antioxidant effects. Plants can induce their phenolic metabolism in response to stressors (including heavy metal stress) [55]. To further understand how cadmium (Cd) and zinc (Zn) affect the metabolism of phenolic compounds and the mechanisms underlying heavy metal tolerance in *Kandelia obovata*, a pot experiment was carried out [56]. Perhaps the higher level of phenolic content synthesized during this study alleviated metal toxicity by chelating Cd and detoxifying ROS. This study suggests that TRIA-treated seedlings activated the PAL enzyme, which enhanced the synthesis of phenolic compounds. Zoufan et al. [57] noted that higher phenolic content improved the antioxidant system of plants to mitigate Cd stress. Yadav and Singh [58] also reported that the interactive effect of phenol-synthesizing hormones improved the antioxidative machinery of wheat under stress regimes. Equally, the augmented level of phenolics in TRIA-applied plants during this study mitigated Cd toxicity. In another study, Meza et al. [59] revealed that IAA affected DPPH scavenging by increasing the biosynthesis of phenolics plants. Correspondingly, the results of the current study confirm that growth-promoting TRIA enhanced DPPH activity in Cd-stressed plants to alleviate stress.

Ullah et al. [60] observed that a higher Cd concentration declines TI in chickpea cultivars. According to Menhas et al. [61], the improved antioxidative system of tolerant plant species correlates with stress mitigation and metal uptake potential under Cd stress circumstances. Higher MTI was observed in TRIA2-applied plants during this study. The reduced MTI level in plants growing in Cd-spiked conditions in absentia of TRIA treatment may be attributed to higher oxidative injury.

Metal contaminants negatively affect morphophysiological features of the stomata and other allied physio-metabolic procedures in plants. Mostly, metal toxicity declines stomatal conductivity [62]. The results of various studies demonstrate a reduction in stomatal conductivity through Cd stress in plants including pakchoi, mustard, marigolds, Holm oak, mastic shrub, *Bacopa monniera*, Populus, riparian *Salix variegata*, cucumber, cowpea, *Origanum vulgare*, *Ocimum basilicum*, and *Arundo donax* [63,64]. Yet, Fox et al. [65] reported higher stomatal conductivity in Cd-stressed corn, *Lactuca sativa*, mustard, and water hyacinth. TRIA seed priming increases plant growth and stress tolerance through increasing stomatal conductivity. In the same way, our results depicted that TRIA seed priming enhanced photosynthetic activity, the rate of transpiration, and stomatal conductance in developing plants in all treatments.

Proline, an amino acid, defends cellular integrity and maintains protein structure under a stressed environment by sustaining the water content, cellular turgidity, and osmotic potential of cells intricated in alleviating oxidative stress [66]. Plants subjected to stress exhibit higher proline content due to their innate potential to detoxify ROS [67,68]. Mayahi et al. [69] reported that TRIA heightened plant stress resilience by enhancing the biosynthesis of osmoprotectants including proline, total soluble proteins, and carbohydrates. Higher proline levels in Cd-stressed plants during the present study suggest the involvement of this osmoregulator in reducing metal-induced toxicity. Siddique and Dubey [69] revealed that proline content increased with the increasing concentration of $CdCl_2$ treatment at 120 h of observation, while the least amount of proline was found in the control set. The increment in proline content from its normal level speaks to the presence of stress in the system. Proline participates in the osmotic adjustment of plant cells under stress has been reported to improve plant resistance to oxidative stress induced by heavy metal toxicity by scavenging ROS [68]. Likewise, our results revealed, in Tria-treated plants, that the relatively high level of proline relieved Cd stress.

5. Conclusions

In *V. radiata* plants, cadmium toxicity significantly reduced germination percentage, seedling growth, and biomass production by enhancing ROS levels. However, TRIA-treated seedlings exhibited an improved production of photosynthetic pigments, proline accumulation, nutrient uptake, and gas exchange attributes. The activity of antioxidant enzymes and the metal tolerance index were increased by TRIA, which reduced Cd toxicity. Our results support the use of TRIA2 (20 μmol L^{-1}) to mitigate stress and preserve the growth of *V. radiata.* seedlings under Cd-contaminated conditions. However, further molecular analysis will reveal the complete mechanism by which TRIA reduces Cd stress and promotes development. Field experiments could be carried out to determine the efficacy of TRIA treatment to sustain crop yield under a contaminated environment.

Author Contributions: Methodology, R.S., N.A.Y. and S.M.; Software, S.A. and M.J.; Validation, N.A.Y. and M.L.; Formal analysis, S.M.; Investigation, S.M.; Resources, M.L. and N.A.Y.; Data curation, S.A. and M.J.; Writing—original draft preparation, S.M. and R.S; Writing—review and editing, M.J., N.A.Y., S.A., and M.L.; Visualization and Supervision, R.S. and S.A.; Project administration, N.A.Y. and S.A.; Funding acquisition, M.L. All authors have read and agreed to the published version of the manuscript.

Funding: This research received no external funding.

Institutional Review Board Statement: Not applicable.

Informed Consent Statement: Not applicable.

Data Availability Statement: All the relevant required datasets can be obtained from the corresponding author on special request.

Acknowledgments: The authors would like to thank the University of Applied Sciences Technikum Wien for providing open access funding. Open Access Funding by the University of Applied Sciences Technikum Wien.

Conflicts of Interest: The authors declare no conflicts of interest.

References

1. Rajput, V.D.; Minkina, T.; Kumari, A.; Singh, V.K.; Verma, K.K.; Mandzhieva, S.; Keswani, C. Coping with the Challenges of Abiotic Stress in Plants: New Dimensions in the Field Application of Nanoparticles. *Plants* **2021**, *10*, 1221. [CrossRef] [PubMed]
2. Tang, J.; Zhang, J.; Ren, L.; Zhou, Y.; Gao, J.; Luo, L.; Yang, Y.; Peng, Q.; Huang, H.; Chen, A. Diagnosis of soil contamination using microbiological indices: A review on heavy metal pollution. *J. Environ. Manag.* **2019**, *242*, 121–130. [CrossRef]
3. Shaheen, A.; Iqbal, J.; Hussain, S. Adaptive geospatial modeling of soil contamination by selected heavy metals in the industrial area of Sheikhupura, Pakistan. *Int. J. Environ. Sci. Technol.* **2019**, *16*, 4447–4464. [CrossRef]
4. Sarwar, T.; Shahid, M.; Khalid, S.; Shah, A.H.; Ahmad, N.; Naeem, M.A.; Bakhat, H.F. Quantification and risk assessment of heavy metal build-up in soil–plant system after irrigation with untreated city wastewater in Vehari, Pakistan. *Environ. Geochem. Health* **2020**, *42*, 4281–4297. [CrossRef]
5. Iqbal, Z.; Abbas, F.; Ibrahim, M.; Qureshi, T.I.; Gul, M.; Mahmood, A. Assessment of heavy metal pollution in Brassica plants and their impact on animal health in Punjab, Pakistan. *Environ. Sci. Pollut. Res.* **2021**, *28*, 22768–22778. [CrossRef]
6. Lamine, S.; Petropoulos, G.P.; Brewer, P.A.; Bachari, N.E.I.; Srivastava, P.K.; Manevski, K.; Macklin, M.G. Heavy metal soil contamination detection using combined geochemistry and field spectroradiometry in the United Kingdom. *Sensors* **2019**, *19*, 762. [CrossRef]
7. Ijaz, S.; Iqbal, J.; Abbasi, B.A.; Tufail, A.; Ullah, Z.; Yaseen, T.; Iqbal, R. Heavy metal–polluted arable land and its consequences: A global scenario. In *Biochar-Assisted Remediation of Contaminated Soils Under Changing Climate*; Elsevier: Amsterdam, The Netherlands, 2024; pp. 71–99. [CrossRef]

8. Guo, B.; Hong, C.; Tong, W.; Xu, M.; Huang, C.; Yin, H.; Fu, Q. Health risk assessment of heavy metal pollution in a soil-rice system: A case study in the Jin-Qu Basin of China. *Sci. Rep.* **2020**, *10*, 11490. [CrossRef]
9. Manzoor, M.; Gul, I.; Kallerhoff, J.; Arshad, M. Fungi-assisted phytoextraction of lead: Tolerance, plant growth–promoting activities and phytoavailability. *Environ. Sci. Pollut. Res.* **2019**, *26*, 23788–23797. [CrossRef]
10. El Rasafi, T.; Oukarroum, A.; Haddioui, A.; Song, H.; Kwon, E.E.; Bolan, N.; Rinklebe, J. Cadmium stress in plants: A critical review of the effects, mechanisms, and tolerance strategies. *Crit. Rev. Environ. Sci. Technol.* **2020**, *52*, 675–726. [CrossRef]
11. Mahajan, P.; Kaushal, J. Role of phytoremediation in reducing cadmium toxicity in soil and water. *J. Toxicol.* **2018**, *2018*, 4864365. [CrossRef]
12. Di Toppi, L.S.; Gabbrielli, R. Response to cadmium in higher plants. *Environ. Exp. Bot.* **1999**, *41*, 105–130. [CrossRef]
13. Song, Y.J.; Li, Y.; Leng, Y.; Li, S.W. 24-epibrassinolide improves differential cadmium tolerance of mung bean roots, stems, and leaves via amending antioxidative systems similar to that of abscisic acid. *Environ. Sci. Pollut. Res.* **2021**, *28*, 52032–52045. [CrossRef] [PubMed]
14. Moradi, R.; Pourghasemian, N.; Naghizadeh, M. Effect of beeswax waste biochar on growth, physiology and cadmium uptake in saffron. *J. Clean. Prod.* **2019**, *229*, 1251–1261. [CrossRef]
15. Kumar, N.; Kumar, V.; Bose, B.; Singhal, R.K. Cadmium toxicity in plants and alleviation through seed priming approach. *Plant Physiol. Rep.* **2021**, *26*, 647–660. [CrossRef]
16. Sarkar, P.K.; Kumar, P.R.; Singh, A.K.; Bhatt, B.P. Effect of priming treatments on seed germination and seedling growth in bamboo [*Dendrocalamus strictus* (Roxb.) Nees]. *Acta Ecol. Sin.* **2020**, *40*, 128–133. [CrossRef]
17. Sirisuntornlak, N.; Ullah, H.; Sonjaroon, W.; Arirob, W.; Anusontpornperm, S.; Datta, A. Effect of seed priming with silicon on growth, yield and nutrient uptake of maize under water-deficit stress. *J. Plant Nutr.* **2021**, *44*, 1869–1885. [CrossRef]
18. Raza, M.A.S.; Aslam, M.U.; Valipour, M.; Iqbal, R.; Haider, I.; Mustafa, A.E.Z.M.; Elshamly, A.M. Seed priming with selenium improves growth and yield of quinoa plants suffering drought. *Sci. Rep.* **2024**, *14*, 886. [CrossRef]
19. Biswas, S.; Seal, P.; Majumder, B.; Biswas, A.K. Efficacy of seed priming strategies for enhancing salinity tolerance in plants: An overview of the progress and achievements. *Plant Stress* **2023**, *9*, 100186. [CrossRef]
20. Islam, S.; Mohammad, F. Triacontanol as a dynamic growth regulator for plants under diverse environmental conditions. *Physiol. Mol. Biol. Plants* **2020**, *26*, 871–883. [CrossRef]
21. Asadi Karam, E.; Keramat, B. Foliar spray of triacontanol improves growth by alleviating oxidative damage in coriander under salinity. *Indian J. Plant Physiol.* **2017**, *22*, 120–124. [CrossRef]
22. Naeem, M.; Ansari, A.A.; Aftab, T.; Shabbir, A.; Alam, M.M.; Khan, M.M.A.; Uddin, M. Application of triacontanol modulates plant growth and physiological activities of *Catharanthus roseus* (L.). *Int. J. Bot. Stud.* **2019**, *4*, 131–135.
23. Zaid, A.; Mohammad, F.; Fariduddin, Q. Plant growth regulators improve growth, photosynthesis, mineral nutrient and antioxidant system under cadmium stress in menthol mint (*Mentha arvensis* L.). *Physiol. Mol. Biol. Plants* **2020**, *26*, 25–39. [CrossRef]
24. Perveen, S.; Iqbal, M.; Parveen, A.; Akram, M.S.; Shahbaz, M.; Akber, S.; Mehboob, A. Exogenous triacontanol-mediated increase in phenolics, proline, activity of nitrate reductase, and shoot k+ confers salt tolerance in maize (*Zea mays* L.). *Braz. J. Bot.* **2017**, *40*, 1–11. [CrossRef]
25. Islam, S.; Zaid, A.; Mohammad, F. Role of triacontanol in counteracting the ill effects of salinity in plants: A review. *J. Plant Growth Regul.* **2021**, *40*, 1–10. [CrossRef]
26. Maresca, V.; Sorbo, S.; Keramat, B.; Basile, A. Effects of triacontanol on ascorbate-glutathione cycle in Brassica napus L. exposed to cadmium-induced oxidative stress. *Ecotoxicol. Environ. Saf.* **2017**, *144*, 268–274. [CrossRef]
27. Sehrawat, N.; Yadav, M.; Sharma, A.K.; Kumar, V.; Bhat, K.V. Salt stress and mungbean [*Vigna radiata* (L.) Wilczek]: Effects, physiological perspective and management practices for alleviating salinity. *Arch. Agron. Soil Sci.* **2019**, *65*, 1287–1301. [CrossRef]
28. Lambrides, C.J.; Godwin, I.D. Mungbean. In *Pulses, Sugar and Tuber Crops*; Springer: Berlin, Heidelberg, 2007; pp. 69–90. [CrossRef]
29. Pataczek, L.; Zahir, Z.A.; Ahmad, M.; Rani, S.; Nair, R.; Schafleitner, R.; Hilger, T. Beans with Benefits—The Role of Mungbean (*Vigna radiate*) in a Changing Environment. *Am. J. Plant Sci.* **2018**, *9*, 1577. [CrossRef]
30. Kumar, V.; Singhal, R.K.; Kumar, N.; Bose, B. Micro-Nutrient Seed Priming: A Pragmatic Approach Towards Abiotic Stress Management. In *New Frontiers in Stress Management for Durable Agriculture*; Springer: Singapore, 2020; pp. 231–255. [CrossRef]
31. Arnon, D.I. Copper enzymes in isolated chloroplasts. Polyphenoloxidase in *Beta vulgaris*. *Plant Physiol.* **1949**, *24*, 1. [CrossRef]
32. Davis, D.J.; Gross, E.L. Protein-protein interactions of the light-harvesting chlorophyll ab protein. II. Evidence for two stages of cation independent association. *Biochim. Biophys. Acta (BBA)-Bioenerg.* **1976**, *449*, 554–564. [CrossRef]
33. Moseley, G.; Jones, J.R. The physical digestion of perennial ryegrass (*Lolium perenne*) and white clover (*Trifolium repens*) in the foregut of sheep. *Br. J. Nutr.* **1984**, *52*, 381–390. [CrossRef]
34. Chapman, J.F.; Dale, L.S. The determination of lithium isotope abundances with a dual-beam atomic absorption spectrometer. *Anal. Chim. Acta* **1976**, *87*, 91–95. [CrossRef]
35. Al-Farraj, A.S.; Al-Otabi, T.G.; Al-Wabel, M.I. Accumulation coefficient and translocation factor of heavy metals through *Ochradenus baccatus* plant grown on mining area at MahadAD'Dahab, Saudi Arabia. *WIT Trans. Ecol. Environ.* **2009**, *122*, 459–468.
36. Balint, A.F.; Röder, M.S.; Hell, R.; Galiba, G.; Börner, A. Mapping of QTLs affecting copper tolerance and the Cu, Fe, Mn and Zn contents in the shoots of wheat seedlings. *Biol. Plant.* **2007**, *51*, 129–134. [CrossRef]

37. Sagner, S.; Kneer, R.; Wanner, G.; Cosson, J.P.; Deus-Neumann, B.; Zenk, M.H. Hyperaccumulation, complexation and distribution of nickel in Sebertia acuminata. *Phytochemistry* **1998**, *47*, 339–347. [CrossRef]
38. Bates, L.S.; Waldren, R.P.; Teare, I.D. Rapid determination of free proline for water-stress studies. *Plant Soil* **1973**, *39*, 205–207. [CrossRef]
39. Chen, I.N.; Chang, C.C.; Ng, C.C.; Wang, C.Y.; Shyu, Y.T.; Chang, T.L. Antioxidant and antimicrobial activity of Zingiberaceae plants in Taiwan. *Plant Foods Hum. Nutr.* **2008**, *63*, 15–20. [CrossRef]
40. Zieslin, N.; Ben Zaken, R. Peroxidase activity and presence of phenolic substances in peduncles of rose flowers. *Plant Physiol. Biochem. (Fr.)* **1993**, *31*, 333–339.
41. Giacomino, A.; Inaudi, P.; Silletta, G.; Diana, A.; Bertinetti, S.; Gaggero, E.; Abollino, O. Analytical methods for the characterization of vegetable oils. *Molecules* **2022**, *28*, 153. [CrossRef]
42. Molnár, K.; Biró-Janka, B.; Nyárádi, I.I.; Fodorpataki, L.; Varga, B.E.; Bálint, J.; Duda, M.M. Effects of Priming with Ascorbic Acid, L-Cystein and Triacontanol on Germination of Rapeseed (L.). *Acta Biol. Marisiensis* **2020**, *3*, 48–55. [CrossRef]
43. Sytar, O.; Kumari, P.; Yadav, S.; Brestic, M.; Rastogi, A. Phytohormone priming: Regulator for heavy metal stress in plants. *J. Plant Growth Regul.* **2019**, *38*, 739–752. [CrossRef]
44. Naeem, M.; Khan, M.M.A.; Moinuddin. Triacontanol: A potent plant growth regulator in agriculture. *J. Plant Interact.* **2012**, *7*, 129–142. [CrossRef]
45. Ahmad, J.; Ali, A.A.; Al-Huqail, A.A.; Qureshi, M.I. Triacontanol attenuates drought-induced oxidative stress in *Brassica juncea* L. by regulating lignification genes, calcium metabolism and the antioxidant system. *Plant Physiol. Biochem.* **2021**, *166*, 985–998. [CrossRef] [PubMed]
46. Arif, Y.; Singh, P.; Siddiqui, H.; Bajguz, A.; Hayat, S. Salinity induced physiological and biochemical changes in plants: An omic approach towards salt stress tolerance. *Plant Physiol. Biochem.* **2020**, *156*, 64–77. [CrossRef]
47. Hu, Y.; Wang, H.; Jia, H.; Peng, M.; Zhu, T.; Liu, Y.; Wei, J. Effects of Cd treatment on morphology, chlorophyll content and antioxidant enzyme activity of Elymus nutans Griseb., a native plant in Qinghai-Tibet Plateau. *Plant Signal. Behav.* **2023**, *18*, 2187561. [CrossRef] [PubMed]
48. Paunov, M.; Koleva, L.; Vassilev, A.; Vangronsveld, J.; Goltsev, V. Effects of different metals on photosynthesis: Cadmium and zinc affect chlorophyll fluorescence in durum wheat. *Int. J. Mol. Sci.* **2018**, *19*, 787. [CrossRef] [PubMed]
49. Parmar, P.; Kumari, N.; Sharma, V. Structural and functional alterations in photosynthetic apparatus of plants under cadmium stress. *Bot. Stud.* **2013**, *54*, 45. [CrossRef]
50. Hayat, K.; Khan, J.; Khan, A.; Ullah, S.; Ali, S.; Salahuddin; Fu, Y. Ameliorative effects of exogenous proline on photosynthetic attributes, nutrients uptake, and oxidative stresses under cadmium in pigeon pea (*Cajanus cajan* L.). *Plants* **2021**, *10*, 796. [CrossRef]
51. Handa, N.; Kohli, S.K.; Thukral, A.K.; Bhardwaj, R.; Alyemeni, M.N.; Wijaya, L.; Ahmad, P. Protective role of selenium against chromium stress involving metabolites and essential elements in *Brassica juncea* L. seedlings. *3 Biotech* **2018**, *8*, 66. [CrossRef]
52. Younis, A.A.; Ismail, H.A. Triacontanol alleviated nickel toxicity in maize seedling by controlling its uptake and enhancing antioxidant system. *J. Adv. Plant Biol.* **2019**, *1*, 1–15. [CrossRef]
53. Zulfiqar, S.; Shahbaz, M. Modulation of gas exchange parameters and photosystem II activity of canola (*Brassica napus* L.) by foliar-applied triacontanol under salt stress. *Agrochimica* **2013**, *57*, 193–200.
54. Goncharuk, E.A.; Zagoskina, N.V. Heavy metals, their phytotoxicity, and the role of phenolic antioxidants in plant stress responses with focus on cadmium. *Molecules* **2023**, *28*, 3921. [CrossRef] [PubMed]
55. Chen, S.; Wang, Q.; Lu, H.; Li, J.; Yang, D.; Liu, J.; Yan, C. Phenolic metabolism and related heavy metal tolerance mechanism in Kandelia Obovata under Cd and Zn stress. *Ecotoxicol. Environ. Saf.* **2019**, *169*, 134–143. [CrossRef] [PubMed]
56. Zoufan, P.; Azad, Z.; Ghahfarokhie, A.R.; Kolahi, M. Modification of oxidative stress through changes in some indicators related to phenolic metabolism in Malva parviflora exposed to cadmium. *Ecotoxicol. Environ. Saf.* **2020**, *187*, 109811. [CrossRef] [PubMed]
57. Yadav, A.; Singh, S. Effect of exogenous phytohormone treatment on antioxidant activity, enzyme activity and phenolic content in wheat sprouts and identification of metabolites of control and treated samples by UHPLC-MS analysis. *Food Res. Int.* **2023**, *169*, 112811. [CrossRef]
58. Meza, A.; Rojas, P.; Cely-Veloza, W.; Guerrero-Perilla, C.; Coy-Barrera, E. Variation of isoflavone content and DPPH• scavenging capacity of phytohormone-treated seedlings after in vitro germination of cape broom (*Genista monspessulana*). *South Afr. J. Bot.* **2020**, *130*, 64–74. [CrossRef]
59. Ullah, S.; Khan, J.; Hayat, K.; Abdelfattah Elateeq, A.; Salam, U.; Yu, B.; Ma, Y.; Wang, H.; Tang, Z.H. Comparative study of growth, cadmium accumulation and tolerance of three chickpea (*Cicer arietinum* L.) cultivars. *Plants* **2020**, *9*, 310. [CrossRef]
60. Menhas, S.; Yang, X.; Hayat, K.; Ali, A.; Ali, E.F.; Shahid, M.; Zhou, P. Melatonin enhanced oilseed rape growth and mitigated Cd stress risk: A novel trial for reducing Cd accumulation by bioenergy crops. *Environ. Pollut.* **2022**, *308*, 119642. [CrossRef]
61. Guo, Z.; Gao, Y.; Yuan, X.; Yuan, M.; Huang, L.; Wang, S.; Duan, C. Effects of heavy metals on stomata in plants: A review. *Int. J. Mol. Sci.* **2023**, *24*, 9302. [CrossRef]
62. Farooq, A.; Nadeem, M.; Abbas, G.; Shabbir, A.; Khalid, M.S.; Javeed, H.M.R.; Akhtar, G. Cadmium partitioning, physiological and oxidative stress responses in marigold (*Calendula calypso*) grown on contaminated soil: Implications for phytoremediation. *Bull. Environ. Contam. Toxicol.* **2020**, *105*, 270–276. [CrossRef]
63. Faraz, A.; Faizan, M.; Sami, F.; Siddiqui, H.; Hayat, S. Supplementation of salicylic acid and citric acid for alleviation of cadmium toxicity to Brassica juncea. *J. Plant Growth Regul.* **2020**, *39*, 641–655. [CrossRef]

64. Fox, J.P.; Capen, J.D.; Zhang, W.; Ma, X.; Rossi, L. Effects of cerium oxide nanoparticles and cadmium on corn (*Zea mays* L.) seedlings physiology and root anatomy. *NanoImpact* **2020**, *20*, 100264. [CrossRef]
65. Emamverdian, A.; Ding, Y.; Mokhberdoran, F.; Xie, Y. Heavy metal stress and some mechanisms of plant defense response. *Sci. World J.* **2015**, *2015*, 756120. [CrossRef] [PubMed]
66. Ashraf, M.F.M.R.; Foolad, M.R. Roles of glycine betaine and proline in improving plant abiotic stress resistance. *Environ. Exp. Bot.* **2007**, *59*, 206–216. [CrossRef]
67. Adejumo, S.A.; Awoyemi, V.; Togun, A.O. Exogenous proline and hormone in combination with compost improves growth and tolerance of maize under heavy metal stress. *Plants Environ.* **2020**, *2*, 40–53. [CrossRef]
68. Al-Mayahi, A.M.W. Triacontanol 'TRIA'application to mitigate the adverse effects of drought and salinity stress under culture of date palm plants. *Folia Oecologica* **2024**, *51*, 250–262. [CrossRef]
69. Siddique, A.; Dubey, A.P. Phyto-toxic effect of heavy metal (CdCl2) on seed germination, seedling growth and antioxidant defence metabolism in wheat (*Triticum aestivum* L.) variety HUW-234. *Int. J. Bio-Resour. Stress Manag.* **2017**, *8*, 261–267. [CrossRef]

toxics

MDPI

Article

Exposure to Cadmium and Other Trace Elements Among Individuals with Mild Cognitive Impairment

Teresa Urbano [1], Marco Vinceti [1,2,*], Chiara Carbone [3,4], Lauren A. Wise [2], Marcella Malavolti [1], Manuela Tondelli [3,4], Roberta Bedin [3], Giulia Vinceti [4], Alessandro Marti [5], Annalisa Chiari [4], Giovanna Zamboni [3,4], Bernhard Michalke [6] and Tommaso Filippini [1,7]

1 Environmental, Genetics, and Nutritional Epidemiology Research Center (CREAGEN), Department of Biomedical, Metabolic and Neural Sciences, University of Modena and Reggio Emilia, 41125 Modena, Italy; teresa.urbano@unimore.it (T.U.); marcella.malavolti@unimore.it (M.M.); tommaso.filippini@unimore.it (T.F.)

2 Department of Epidemiology, Boston University School of Public Health, Boston, MA 02118, USA; lwise@bu.edu

3 Department of Biomedical, Metabolic and Neural Sciences, University of Modena and Reggio Emilia, 41125 Modena, Italy; chiara.carbone@unimore.it (C.C.); manuela.tondelli@unimore.it (M.T.); roberta.bedin@unimore.it (R.B.); giovanna.zamboni@unimore.it (G.Z.)

4 Neurology Unit, Baggiovara Hospital, 41126 Modena, Italy; gvinceti@gmail.com (G.V.); chiari.annalisa@aou.mo.it (A.C.)

5 Clinical Neuropsychology, Cognitive Disorders and Dyslexia Unit, Neuromotor and Rehabilitation Department, AUSL-IRCCS di Reggio Emilia, 42123 Reggio Emilia, Italy; alessandro.marti@ausl.re.it

6 Research Unit Analytical BioGeoChemistry, German Research Center for Environmental Health, Helmholtz Center Munich, 85764 Neuherberg, Germany; bernhard.michalke@kabelmail.de

7 School of Public Health, University of California Berkeley, Berkeley, CA 94704, USA

* Correspondence: marco.vinceti@unimore.it

Abstract: Background: A limited number of studies have investigated the role of environmental chemicals in the etiology of mild cognitive impairment (MCI). We performed a cross-sectional study of the association between exposure to selected trace elements and the biomarkers of cognitive decline. Methods: During 2019–2021, we recruited 128 newly diagnosed patients with MCI from two Neurology Clinics in Northern Italy, i.e., Modena and Reggio Emilia. At baseline, we measured serum and cerebrospinal fluid (CSF) concentrations of cadmium, copper, iron, manganese, and zinc using inductively coupled plasma mass spectrometry. With immuno-enzymatic assays, we estimated concentrations of β-amyloid 1-40, β-amyloid 1-42, Total Tau and phosphorylated Tau181 proteins, neurofilament light chain (NfL), and the mini-mental state examination (MMSE) to assess cognitive status. We used spline regression to explore the shape of the association between exposure and each endpoint, adjusted for age at diagnosis, educational attainment, MMSE, and sex. Results: In analyses between the serum and CSF concentrations of trace metals, we found monotonic positive correlations between copper and zinc, while an inverse association was observed for cadmium. Serum cadmium concentrations were inversely associated with amyloid ratio and positively associated with Tau proteins. Serum iron concentrations showed the opposite trend, while copper, manganese, and zinc displayed heterogeneous non-linear associations with amyloid ratio and Tau biomarkers. Regarding CSF exposure biomarkers, only cadmium consistently showed an inverse association with amyloid ratio, while iron was positively associated with Tau. Cadmium concentrations in CSF were not appreciably associated with serum NfL levels, while we observed an inverted U-shaped association with CSF NfL, similar to that observed for copper. In CSF, zinc was the only trace element positively associated with NfL at high concentrations. Conclusions: In this cross-sectional study, high serum cadmium concentrations were associated with selected biomarkers of cognitive impairment. Findings for the other trace elements were difficult to interpret, showing complex and inconsistent associations with the neurodegenerative endpoints examined.

Keywords: amyloid ratio; mild cognitive impairment; neurofilament light chain; Tau proteins; trace elements

1. Introduction

Dementia represents a significant public health challenge, as the population ages globally. The global population of individuals living with dementia is expected to rise to 153 million by 2050 [1]. Emerging evidence suggests that not only genetic and lifestyle factors, but also environmental exposures, including heavy metals and trace elements like cadmium (Cd), copper (Cu), iron (Fe), manganese (Mn), and zinc (Zn), are involved in the pathogenesis of cognitive decline and dementia [2]. While Cd is a non-essential metal with no recognized biologic function, Cu, Fe, Mn, and Zn are essential nutrients that also have potential neurotoxic effects, depending on the amount of exposure, the chemical species, and other factors. Understanding the contribution of these trace elements to neurodegenerative processes is crucial for identifying modifiable risk factors and developing targeted interventions.

Cd, a well-known environmental pollutant, has been associated with neurodegeneration in epidemiologic studies [3–6], potentially due to its capacity to exacerbate oxidative stress and inflammation in the central nervous system, as shown in laboratory studies [7–9]. Cu is a trace element involved in several physiological functions, including mitochondrial energy production, oxidative stress regulation, and neurotransmitter synthesis as a cofactor of enzymes such as superoxide dismutase and cytochrome c oxidase [10], though it could also induce neurodegeneration through its toxic and pro-oxidant properties [11,12]. Zn plays a pivotal role in maintaining synaptic plasticity and supporting neurogenesis. It is concentrated in synaptic vesicles and is involved in modulating neurotransmitter release, neuronal signaling, and enzymatic activity [13]. Both Cu and Zn, while being essential for neuronal function, may be detrimental at high concentrations, contributing to amyloid beta aggregation and Tau phosphorylation, the pathological features of Alzheimer's disease (AD). The same applies to essential elements like Fe and Mn, which at high levels of exposure may be neurotoxic, affecting neurotransmitter synthesis and promoting oxidative damage [14]. Fe is fundamental to several neurological processes, including neuronal development, myelination, and neurotransmitter synthesis and catabolism. However, it also promotes reactive oxygen species protein oxidation, lipid peroxidation, and nucleic acid modification [15,16]. Mn serves as a cofactor for several key enzymes, including manganese superoxide dismutase, which protects cells from oxidative damage by neutralizing superoxide radicals. Mn is also involved in the synthesis of glutamine and glutamate, neurotransmitters critical for brain function [17], but at higher levels Mn may induce neurotoxicity [18,19]. The impact of different trace elements can vary by the brain region affected and is often dose dependent [20]. Fe, the most abundant trace element in the brain, tends to accumulate in the substantia nigra, striatus, and hippocampus, while Zn accumulates in the cortex, hippocampus, and amygdala. High Cu concentrations have been found in the hippocampus and cortex [21], while chronic Mn mainly affects the basal ganglia [22].

We performed a cross-sectional study among patients with MCI to examine associations between the concentrations of Cd, Cu, Fe, Mn, and Zn in serum and cerebrospinal fluid (CSF) and the biomarkers of cognitive decline, including amyloid Aβ1-42/1-40 ratio (reflecting amyloid plaque deposition), Tau proteins (indicating neurodegeneration), and neurofilament light chain (NfL, reflecting axonal injury) [23].

2. Methods

2.1. Study Population

We enrolled patients consecutively admitted to the neurology departments of Reggio Emilia ASMN Hospital and Modena University Hospital from 2019 to 2021 with a diagnosis of subjective cognitive decline (SCD) or MCI based on history and neuropsychological assessment [24–26]. Detailed inclusion criteria were as follows: clinical diagnosis of MCI or SCD based on history and neuropsychological assessment, including the mini-mental state examination (MMSE), in accordance with the last edition of the *Diagnostic and Statistical Manual of Mental Disorders* (DSM-5) criteria [27]; age at recruitment less than 70 years; onset of reported cognitive impairment before the age of 65; ability to read and write in the Italian language for neuropsychological evaluation; presence of a caregiver available to respond to questionnaires as part of the neuropsychological assessment. Exclusion criteria were as follows: failure to provide consent or withdrawal of consent; diagnosis within the last 2 years of stroke or other cerebrovascular disease; cranial trauma or other focal brain injuries; inflammatory central nervous system disease; other neurodegenerative diseases (Parkinson's disease, Huntington's, etc.); major psychiatric disorders; pregnancy. The protocol of this study was approved by the Modena (AOU no. 2158/19) and Reggio Emilia (AUSLRE no. 2019/0009686) Ethics Committees.

A flowchart of study exclusions is shown in Figure 1. Of the 147 enrolled patients, we excluded two patients with other neurological diseases (e.g., neurosyphilis) and no available serum sample. Of the 145 participants included in the analysis, we eventually assayed trace element concentrations in 103 patients with serum samples. A CSF analysis was performed in only 50 of them due to limited sample availability. Data on each patient's medical history, sociodemographic information, and lifestyle habits were collected through their medical records and ad hoc project-designed questionnaires. For the purpose of this study, analyses were performed in 89 MCI patients with serum samples and 45 MCI patients with CSF samples.

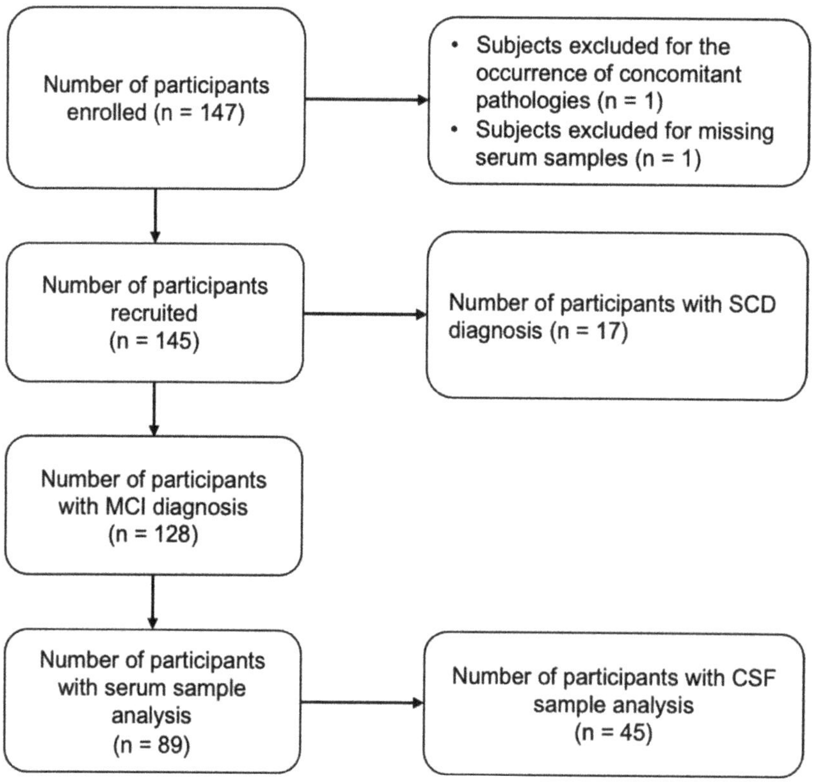

Figure 1. Study flowchart. Abbreviations: CSF, cerebrospinal fluid; SCD, subjective cognitive decline (SCD).

Patients provided informed consent before undergoing a lumbar puncture to collect CSF samples for diagnostic purposes. Subsequent approval for the use of these samples in research was granted by the Modena Ethical Committee. We collected approximately 6 mL of CSF from each patient under sterile conditions employing an ultraclean technique. Serum samples were also obtained. We promptly stored the samples at −80 °C in polypropylene tubes to ensure preservation. Prior to analysis, we thawed the samples in a refrigerator maintained at 4 °C. Once thawed, they were allowed to reach room temperature, subjected to vortex mixing to ensure homogeneity, and subsequently analyzed.

2.2. Biomarkers of Neurodegeneration

The CSF amyloid beta ratio (Aβ1-42/1-40), Total Tau, and phosphorylated Tau 181 (p-Tau181) were assessed with the Lumipulse™ fully automated system using CLEIA (Fujirebio Inc., Ghent, Belgium) following manufacturer instructions. Lumipulse® G β-Amyloid$_{1\text{-}42}$, Lumipulse® G β-Amyloid$_{1\text{-}40}$, Lumipulse® G Total Tau, and Lumipulse® G p-Tau181 Chemiluminescent Enzyme ImmunoAssays kits (Fujirebio Inc., Ghent, Belgium) were employed. Specific laboratory cut-offs were as follows: amyloid ratio > 0.069 pg/mL, Total Tau < 400 pg/mL, and p-Tau181 < 56.5 pg/mL. Serum and CSF concentrations of NfL were determined using Ella™ microfluidic platform (Bio-Techne, Minneapolis, MN, USA), as previously described [28,29]. Briefly, after 1:2 manual dilution according to the manufacturer's recommended procedures, samples were loaded into cartridges coated with a capture antibody. Samples were read in triplicate, evaluating intra-assay and inter-assay variability. All analytical procedures were performed by biologists who were blinded to clinical information.

2.3. Trace Element (Cd, Cu, Fe, Mn, Zn) Determination

We transferred a 1 mL aliquot of serum and 1 mL of CSF in polypropylene tubes. Samples were transported in a frozen state on dry ice to the laboratory at the Helmoltz Zentrum in Munich where they were maintained in a frozen condition until analysis. We used inductively coupled plasma atomic emission spectrometry (Optima 7300 DV system, Perkin Elmer, Rodgau-Jügesheim, Germany) for Cu, Fe, Mn, and Zn determinations (Schramel 1988), introducing samples with the use of a peristaltic pump connected with a Seaspray nebulizer and a cyclon spray chamber. The measured spectral element lines were Cu: 324.754 nm, Fe: 259.941 nm, Mn: 257.611 nm, and Zn: 213.857 nm. We set RF power to 1400 W, plasma gas at 13 L Ar/min; the nebulizer gas was approximately 0.6 L Ar/min after daily optimization. Since Cd levels were very low, we used inductively coupled plasma sector field mass spectrometry (ICP-sf-MS, "Element 2" from Thermo-Scientific, Bremen, Germany) to analyze 111 Cd [30]. As part of routine quality control, every ten measurements included three blank determinations and a control assessment using a certified standard for all analyzed elements. The final measurements were calculated using a computerized laboratory data management system, which integrated sample measurements with calibration curves, blank determinations, and control standards to ensure accuracy and reliability.

2.4. Data Analysis

For data analyses, we used the following routines of Stata statistical software (Stata/MP 18, Stata Corp., College Station, TX, USA, 2024): 'margins', 'mkspline', 'regress', 'winsor', 'xbcrsplinei', and 'vioplot'. Of the characteristics reported in Table 1, we created categorical variables for age (<65 and ≥65 years), educational attainment, and body mass index (BMI) (underweight, <18.5 kg/m^2; normal weight, 18.5–24.9 kg/m^2; overweight/obese, 25 or more). There were no missing data for covariates, with the exception of BMI (19%) and marital status (12%); the latter covariates were not included in multivariable regression models because they had minimal effects on exposure–outcome associations.

We used crude and multivariable models to assess associations between trace element concentrations in serum and CSF, with the latter analysis being adjusted for age, sex, educational attainment, and MMSE as potential confounders. We analyzed the associations between trace element concentrations and NfL, amyloid ratio, Total Tau, and p-Tau181. We estimated associations using linear fit models with marginal statistics and restricted cubic spline regression models with knots at three fixed percentiles, i.e., 10th, 50th, and 90th of trace element concentrations [31,32].

Table 1. Characteristics of the study population (only mild cognitive impairment (MCI) participants) according to availability of cerebrospinal fluid (CSF) samples.

	MCI Participants (n = 128)	MCI with Serum TE Analysis (n = 89)	MCI with CSF TE Analysis (n = 45)
	Median (IQR)	Median (IQR)	Median (IQR)
Sex (n, %)			
Males	53 (41.4)	36 (40.5)	20 (44.4)
Females	75 (58.6)	53 (59.5)	25 (55.6)
Age at diagnosis (years)	61 (56–65)	61 (56–65)	61 (56–64)
Age in categories (n, %)			
<65 years	90 (70.3)	60 (67.4)	35 (77.8)
≥65 years	38 (29.7)	29 (32.6)	10 (22.2)
Education (years)	11 (8–13)	11 (8–13)	12 (8–13)
Education in categories (n, %)			
Elementary school	11 (8.6)	10 (11.2)	5 (11.1)
Middle school	46 (35.9)	33 (37.1)	12 (26.7)
High school	48 (37.5)	30 (33.7)	18 (40.0)
College or more	23 (18.0)	16 (18.0)	10 (22.2)
Marital status (n, %)			
Single	9 (7.0)	4 (4.5)	3 (6.7)
Married/unmarried	83 (64.8)	61 (68.5)	32 (71.1)
Separated/divorced	13 (10.2)	9 (10.1)	2 (4.4)
Widowed	8 (6.3)	6 (6.7)	4 (8.9)
Missing	15 (11.7)	9 (10.1)	4 (8.9)
BMI in categories (n, %)			
<18.5 kg/m^2	4 (3.1)	3 (3.4)	2 (4.4)
18.5–24.9 kg/m^2	42 (32.8)	24 (27.0)	11 (24.4)
≥25.0 kg/m^2	58 (45.3)	46 (51.7)	25 (55.6)
Missing	24 (18.6)	16 (18.0)	7 (15.6)
MMSE score	27 (25–29)	27 (25–29)	27 (25–29)

Abbreviations: BMI, body mass index; LP, lumbar puncture; MMSE, mini-mental state examination; NfL, neurofilament light chain protein; p-Tau181, Tau protein phosphorylated at threonine 181; TE, trace element.

3. Results

Table 1 shows the baseline characteristics of the study population. We originally recruited 145 patients, including 128 MCI. In total, 42% of MCI patients were males, with a median age at enrollment corresponding to 61 years (IQR 56–65). Of the 89 MCI patients with serum trace element data, 45 also had CSF available. No substantial differences in population characteristics were observed between those with (1) only serum and (2) both serum and CSF. We found no appreciable differences in education or MMSE scores. Most patients were less than 65 years old, obtained a high school diploma, were married/partnered, and were overweight or obese.

Table 2 shows serum and CSF trace elements and NfL concentrations along with CSF Total Tau, p-Tau181, and amyloid ratio.

The highest concentrations in serum and CSF occurred in the following order: Fe > Cu > Zn > Mn > Cd. Limited differences were instead observed for serum and CSF levels of the trace elements investigated according to sex (Figure 2).

Table 2. Distribution of neurofilament light chain (NfL) and trace elements in serum and cerebrospinal fluid (CSF) among patients with mild cognitive impairment (MCI). Values are median and interquartile ranges (IQR).

	CSF (n = 45) Median (IQR)	Serum (n = 89) Median (IQR)
NfL (pg/mL)	934.0 (665.0–1653.0)	19.5 (12.7–35.0)
Amyloid ratio	0.088 (0.050–0.094)	-
Total Tau (pg/mL)	442.0 (215.0–774.0)	-
p-Tau181 (pg/mL)	46.1 (25.8–116.6)	-
Cadmium (µg/L)	0.01 (0.01–0.02)	0.05 (0.03–0.12)
Copper (µg/L)	19.50 (11.00–30.50)	869 (766–974)
Iron (µg/L)	28.50 (9.82–42.50)	864 (694–1070)
Manganese (µg/L)	0.57 (0.38–0.94)	2.74 (1.55–3.50)
Zinc (µg/L)	19.10 (10.90–30.30)	741 (650–856)

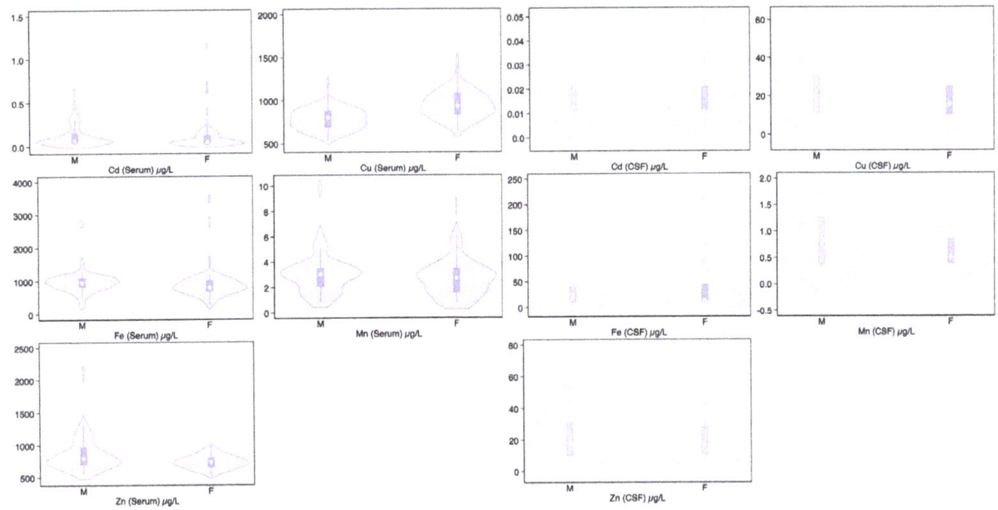

Figure 2. Violin plots distribution of trace element concentrations in serum and cerebrospinal fluid (CSF) according to sex (M, males; F, females). MCI, serum n = 89; CSF n = 45. Abbreviations: Cd, cadmium; Cu, copper; Fe, iron; MCI, mild cognitive impairment; Mn, manganese; NfL, neurofilament light chain; SCD, subjective cognitive decline; Zn, zinc.

Among 45 patients with data on serum and CSF, we compared serum and CSF concentrations for each of the trace elements. In spline regression analyses among MCI patients, we observed a positive linear association between serum and CSF concentrations of Cu and Zn, and an inverse association for Cd. There was a positive association between Fe concentrations in serum and CSF up to a threshold of 1000 µg/L, beyond which the relation plateaued. For Mn, there was a slight inverted U-shaped association, with a positive monotonic relation between the serum and CSF matrices until 3.0 µg/L and a slight inverse association above that value (Figure 3).

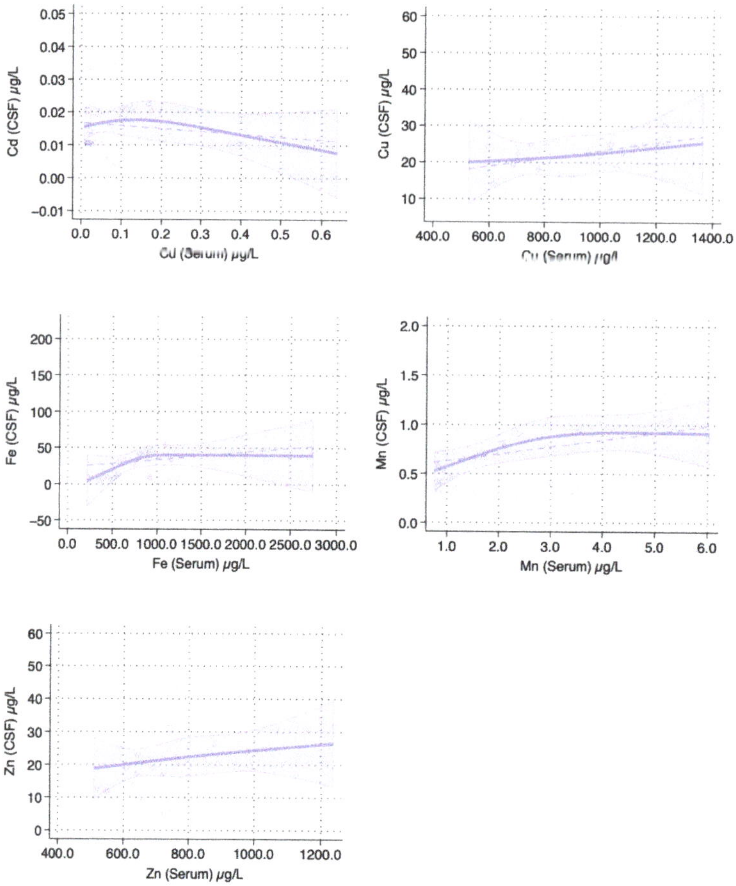

Figure 3. Spline regression analysis of the association between trace element concentration in serum and cerebrospinal fluid among patients with mild cognitive impairment. The solid line indicates the multivariable analysis; the shaded area represents upper and lower confidence interval limits. The dashed line represents association assuming linearity. Diamonds represent individual observations (n = 45). Abbreviations: Cd, cadmium; CSF, cerebrospinal fluid; Cu, copper; Fe, iron; Mn, manganese; NfL, neurofilament light chain; Zn, zinc.

When we examined the association between the trace element concentrations and each study endpoint, in serum, we found little association of Cd and Cu concentrations with NfL, whereas we observed non-linear associations of Fe, Mn, and Zn with NfL. For Fe, we observed an inverse association until 1000 μg/L after which the effect plateaued. For Mn and Zn, we found a U-shaped relation, with an inflection point at 3.0 μg/L and 800 μg/L, respectively (Figure 4A). In CSF, Cd and Cu showed an inverse U-shaped association with NfL concentrations, while Fe showed a monotonic inverse association. For Mn and Zn, the association was almost flat at the lowest exposure levels, while there was an inverse association for Mn above 1.0 μg/L and a positive association for Zn above 20 μg/L (Figure 4B). When comparing associations between trace elements and NfL concentrations across matrices (serum vs. CSF), the associations were substantially similar.

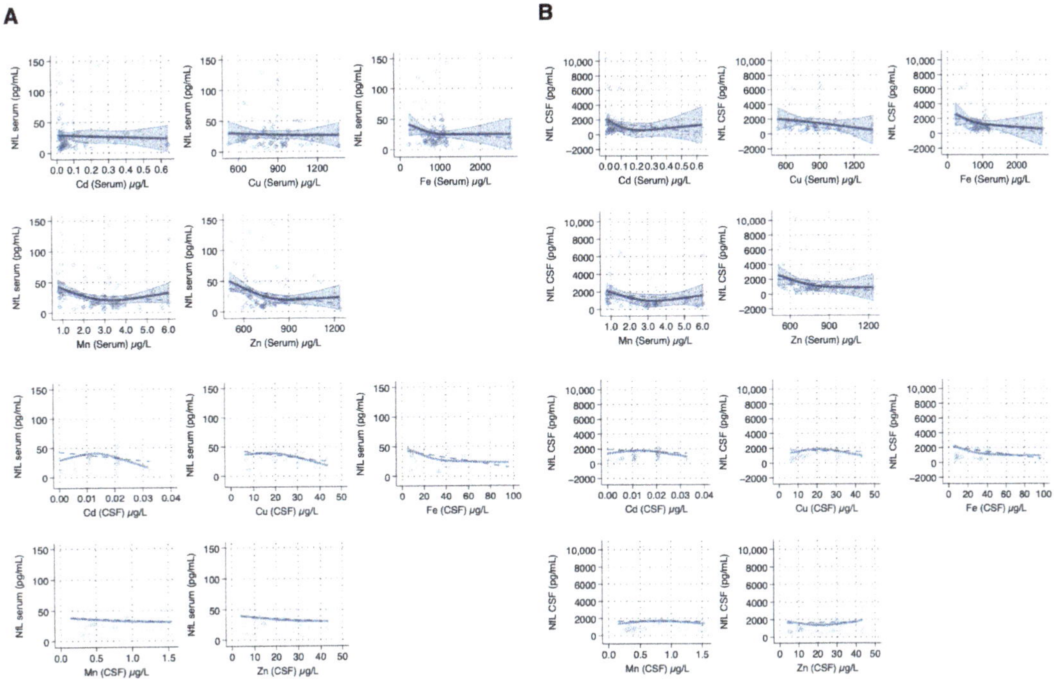

Adjusted for sex, age and MMSE at entry, and education

Figure 4. Spline regression analysis of the association between trace element concentration in serum (dark blue) and cerebrospinal fluid (CSF-light blue) with serum (**A**) and CSF neurofilament light (NfL) concentrations (**B**) among patients with mild cognitive impairment. The solid line indicates the multivariable analysis; the shaded area represents the upper and lower confidence interval limits. The dashed line represents the association assuming linearity. Diamonds represent individual observations (serum n = 89; CSF n = 45). Abbreviations: Cd, cadmium; Cu, copper; Fe, iron; MMSE, mini-mental state examination; Mn, manganese; Zn, zinc.

For Cd, we found an inverse association with amyloid ratio, and a positive association with total Tau and p-Tau181. On the contrary, an opposite pattern emerged for Fe, since its serum concentrations were positively associated with amyloid ratio and inversely associated with Tau proteins. Some discrepancies were observed in the association between Cu and the neurodegeneration biomarkers. In fact, Cu was positively associated with amyloid ratio until 800 µg/L after which an inverse relation emerged. Similarly, there was a U-shaped association with Total Tau and p-Tau181, showing an inverse and then a positive association at the same inflection point. Mn concentrations showed an inverse U-shaped association with amyloid ratio, while being inversely, although not linearly, associated with Tau proteins. Zn concentrations were positively associated with amyloid ratio, while demonstrating a U-shaped association with total Tau and an inverse association with p-Tau181, but only until 800 µg/L. Above this value, the association with amyloid ratio flattened, and the association with Total and p-Tau181 became positive. In CSF, Cd concentrations were inversely associated with amyloid ratio, and there was a slight inverse U-shaped association with Tau protein biomarkers. Cu, Mn, and Zn also showed a U-shaped association with amyloid ratio, while showing an inverse U-shaped association with Tau proteins, with inflection points at 20, 0.75, and 25 µg/L, respectively. Fe concentrations were positively and monotonically associated with Total Tau, showing an inverse U-shaped association with p-Tau181, and showing no association with amyloid ratio (Figure 5). Finally, according to the laboratory cut-off values of the neurodegeneration biomarkers used for AD, Cd in serum was the only element found to be inversely associated with amyloid ratio and positively associated with Tau proteins, while higher concentrations of Zn and Cu in serum were positively associated with Total Tau and p-Tau181, respectively. When considering CSF trace element concentrations, none of the elements demonstrated a clear association with amyloid ratio. When considering Total Tau, we found a positive association above 400 pg/mL for all elements, with a linear association for the full range of exposure only for Fe. We observed similar associations for p-Tau181, although the slopes of the spline curves were considerably less steep.

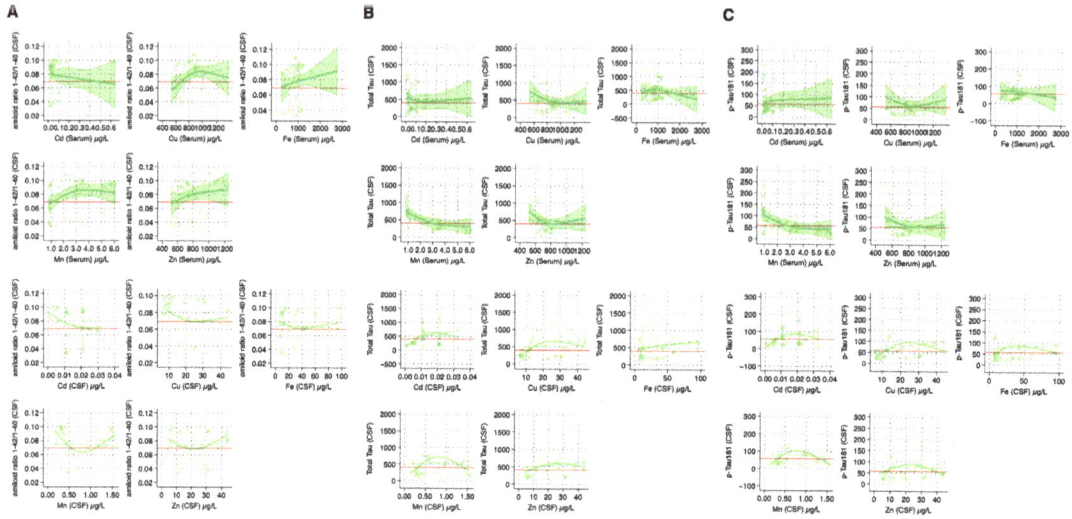

Adjusted for sex, age and MMSE at entry, and education

Figure 5. Spline regression analysis of the association between trace element concentration in serum (dark green) and cerebrospinal fluid (CSF—light green) with CSF concentration of amyloid ratio (**A**), Total Tau (**B**) and phosphorylated Tau (p-Tau181) protein (**C**) among patients with mild cognitive impairment. The solid line indicates the multivariable analysis; the shaded area represents the upper and lower confidence interval limits. The dashed line represents the association assuming linearity. Diamonds represent individual observations (serum n = 89; CSF n = 45). Red lines represent laboratory cut-offs (>0.069 for amyloid ratio; <400 pg/mL for Total Tau; <56.5 pg/mL for p-Tau181). Abbreviations: Cd, cadmium; Cu, copper; Fe, iron; MMSE, mini-mental state examination; Mn, manganese; Zn, zinc.

4. Discussion

In this cross-sectional study among patients with MCI, some trace elements measured in serum and CSF were positively associated with the biomarkers of neurodegeneration. Most associations were non-linear. Cd in particular was positively associated with neurodegeneration biomarkers, with the exception of NfL, consistent with an ongoing neurodegenerative process. Other trace elements displayed inconsistent and heterogeneous non-linear relations with amyloid and Tau biomarkers, while Mn was the only trace element shown to have a U-shaped association with NfL.

As for the associations between serum and CSF concentrations of trace metals, we found monotonic positive correlations only for Cu and Zn. The inverse or non-linear associations between the concentrations of some elements by matrix (serum vs. CSF) can be attributed to several factors, including altered blood-CSF barrier permeability and the dynamics of the transport of metals across the blood–brain barrier [33–35]. For instance, in the case of Fe, homeostatic mechanisms may occur when serum metal concentrations increase without a corresponding rise in CSF. Therefore, when serum metal concentrations rise, the blood–brain barrier may become more restrictive, preventing excess metals from entering the CSF. In addition, neurons and glial cells can actively regulate metal homeostasis through specific transporters and channels, thereby preventing accumulation in CSF [36]. Conversely, the inverse association we observed for Cd between serum and CSF concentrations was unexpected and intriguing, since most of the epidemiological studies conducted to date used urinary levels and biomarkers of exposure [37,38]. Evidence about whether the CSF content may reflect the blood concentrations of this trace element is scarce and some epidemiological studies have raised concerns about the reliability of CSF as a biomarker of exposure [39]. In a previous study, a positive and relatively strong correlation was found between the serum and CSF concentrations of Cd in children, hypothesizing that this trace element exhibits the capacity to access the brain via the nasal mucosa or olfactory pathways, which are not yet fully developed in young children [33]. Our results may differ from those in previous studies due to the different physiological mechanisms in populations of different ages. Cd's affinity for sulfhydryl groups in serum proteins could lead to a scenario where higher serum concentrations do not correspond to increased concentrations in CSF, hypothesizing a protective role of serum against Cd toxicity [7].

Recent findings have highlighted a potential role of some trace elements such as Cd, a heavy metal potentially linked to an excess risk of many diseases [40] including cardiovascular disorders [41–43] and neurodevelopmental disorders [44], in the etiology of different forms of dementia, particularly that are induced by AD [45,46]. Conversely, little evidence is available for MCI, long recognized as a precursor of dementia [47]. Therefore, understanding the environmental and biological factors contributing to MCI progression is essential for developing preventive strategies for dementia [48]. The studies of Cd's neurotoxic effects have indicated that even low-level exposure can lead to impairments in cognitive performance among older adults [49–51]. Several studies have also indicated that Cd has the ability to induce oxidative stress and neuroinflammation, which can disrupt neuronal function and lead to cognitive decline [52,53]. Cd accumulation might also exacerbate some pathological features of AD, such as a reduction in β-amyloid degrading enzymes leading to increased β-amyloid accumulation [54]. In our study, among the trace elements measured in serum, Cd was the only element associated with adverse effects on neurodegeneration biomarkers, as highlighted by the associations found upon exceeding the standard laboratory values of amyloid ratio and Tau proteins. This association emerged at exposure concentrations believed not to be unusually high, given that the median serum Cd concentrations in our study were lower than those measured in the National Health Nutrition Examination Survey [55] or in the Beaver Dam Offspring Study cohort, which found comparable associations between serum Cd and amyloid ratio or Tau proteins [56]. Hence, our results appear to be even more relevant if considering that we observed adverse associations at lower concentrations than previous reports. However, the results for CSF concentrations did not reflect the same associations found for serum Cd, a discrepancy that warrants further investigation.

Fe is the most abundant transition metal and the most investigated in relation to AD. Through Fenton chemistry, Fe can catalyze the formation of reactive oxygen species (ROS), leading to detrimental processes for neuronal integrity, such as lipid peroxidation and protein oxidation [57]. For Fe, we observed an inverse relation between serum and CSF concentrations and neurodegeneration biomarkers, with the exception of a positive association between Fe and total Tau in CSF. The latter result is consistent with previous results highlighting a positive relation between brain Fe deposition and atrophy, cognitive decline, and Tau aggregation in elderly individuals [58–60]. A recent systematic review highlighted that lower Fe in blood and greater ferritin in CSF were found among AD patients compared with healthy subjects [61]. In this study, we did not perform Fe speciation analysis, thus we could not ascribe distinct associations to different Fe compounds.

Zn and Cu are nutritionally essential trace elements that play significant roles in brain function, but they may also have neurotoxic properties at high concentrations. Particularly in the case of Cu, it operates as a catalyst of redox reactions and, in the context of dementia and AD, it has been shown to interact with β-amyloid peptides, promoting oxidative damage to neuronal membranes and facilitating their aggregation [62,63]. Moreover, Cu is known to promote Tau protein hyperphosphorylation and it has been linked to the activation of inflammatory pathways, further compounding oxidative stress and neuronal injury [11,64]. While being essential for synaptic plasticity and cognitive function at low concentrations and capable of β-amyloid-induced toxicity suppression, at high levels Zn binds β-amyloid, thereby enhancing fibrillary aggregation and leading to neurodegeneration [65,66]. We found evidence of some detrimental effects of these two trace metals in relation only to Tau proteins at the highest concentrations in serum, consistent with a previous case–control study finding a positive correlation of non ceruloplasmin serum Cu with CSF total Tau levels [67]. A previous meta-analysis indicated that serum concentrations of Zn were lower among AD patients compared with healthy subjects, even though there was high heterogeneity between studies [68].

With regard to Mn, while necessary for several enzymatic functions, it can be neurotoxic at elevated levels. Recent reviews suggested that Mn exposure may disrupt iron homeostasis, and its accumulation contributes to oxidative stress and mitochondrial dysfunction [69,70]. The competition between Mn and other trace metals for transport mechanisms in the brain can exacerbate these effects, leading to the further dysregulation of metal homeostasis and neurotoxicity [71]. In a previous study, Mn was slightly positively associated with the risk of cognitive dysfunction among people over 60 years old [72]. However, a 2017 meta-analysis reported a decrease in serum Mn concentrations among individuals with cognitive impairment including AD and MCI subjects compared with healthy controls [73]. In our study, Mn was not associated with neurodegeneration, being inversely and positively associated with Tau proteins and amyloid ratio, respectively.

While some studies in the literature reported the association between trace metals and classical biomarkers of AD pathology used in clinical practice, few studies investigated their associations with NfLs. NfLs are among the most promising biomarkers for detecting neuroaxonal damage across a broad spectrum of neurological disorders. Its concentrations in CSF and blood serve as reliable indicators of neuronal injury and degeneration, reflecting the extent of underlying neural pathology [28,74]. Elevated NfL concentrations have been consistently associated with various forms of dementia, including AD and frontotemporal dementia [75]. In a recent study, Cd in serum was found to be positively associated with NfL concentrations [56], similar to what was observed in our study, where Cd in serum was slightly positively associated only with NfL in CSF at the highest levels. No associations were reported for Cd in CSF. Cu, Fe, Mn, and Zn were positively associated with NfL in CSF in the only study conducted to date [76]. However, we found evidence of inverse or null associations for Cu, Fe, and Zn with NfL. Mn in serum was the only trace metal found to be associated through a U-shaped pattern with NfL in both serum and CSF, suggesting that, while essential, it can exhibit neurotoxic effects when present in non-optimal concentrations.

To our knowledge, this study represents the first ever conducted among cognitively impaired subjects with deficits occurring at relatively young ages (<65 years), and one of the few to investigate the association between trace metals and neurodegeneration biomarkers, including NfL. In addition, the use of non-linear dose analysis allowed a comprehensive assessment of the patterns of association. Finally, a CSF analysis of trace metals provides unique insights into the pathophysiological processes that occur in the central nervous system allowing the direct assessment of the brain's biochemical environment. Some study limitations must be outlined. First, given the study's cross-sectional design, associations may not be causal. Ideally, we would have collected longitudinal data among disease-free individuals on the progression to dementia to determine temporality. Misclassification would be minimal if the baseline concentrations of metals reflected the patient's habitual exposure during the years prior to disease onset. We are also currently collecting longitudinal data on the progression to dementia of the individuals of this cohort to better assess temporal relations. Secondly, metal concentrations measured at a single time point may have contributed to the misclassification of exposure, especially if the relevant time window for exposure was years earlier. The half-lives of the metals examined in this study range from some hours to several years, depending on the sample matrix; thus, we made the strong assumption that the concentrations measured at baseline in our study population were reflective of what patients were exposed to before the onset of MCI. In addition, whole blood might have provided a better reflection of long-term metal status in the body compared with serum, which can be influenced by recent dietary intake. Third, few data are available on pharmacokinetic properties of each metal in CSF; therefore, the stability of some metals in this matrix is uncertain. Fourth, our study population was relatively small, which reduced the precision of our effect estimates and could affect the generalizability of our findings. Finally, we could not rule out potential unmeasured or residual confounding, which may have affected the validity of the observed associations.

Author Contributions: Conceptualization, T.U., M.V., and T.F.; Data Curation, T.U., C.C., M.M., and T.F.; Formal Analysis, T.U., R.B., and B.M.; Funding Acquisition, T.F., M.V., and G.Z.; Methodology, M.V. and T.F.; Resources, C.C., M.T., G.V., A.M., A.C., and G.Z.; Supervision, M.V., L.A.W., G.Z., and T.F.; Writing—Original Draft, T.U., M.V., and T.F.; Writing—Review and Editing, All authors have read and agreed to the published version of the manuscript.

Funding: This study was supported by grant "PRIN 2022 PNRR" (no. P20229KSXB) from the Italian Ministry of University and Research (MUR) funded by European Union—Next Generation EU, and by grant "UNIMORE FAR 2023" from the University of Modena and Reggio Emilia. TF was also supported by grant "PRIN 2022" (no. 2022MHMRPR) from the Italian Ministry of University and Research (MUR) funded by European Union—Next Generation EU, and CC, AC, MT, and GV are also supported by a grant from the ERC to GZ (UnaWireD, ERC-2021-STG number 101042625).

Institutional Review Board Statement: This study was conducted in accordance with the Declaration of Helsinki, and approved by the Modena (AOU no. 2158/19) and Reggio Emilia (AUSLRE no. 2019/0009686) Ethics Committees.

Informed Consent Statement: Informed consent was obtained from all subjects involved in the study.

Data Availability Statement: The data presented in this study are available on request from the corresponding author (the data are not publicly available due to privacy restrictions).

Conflicts of Interest: The authors declare no conflicts of interest.

References

1. Livingston, G.; Huntley, J.; Liu, K.Y.; Costafreda, S.G.; Selbaek, G.; Alladi, S.; Ames, D.; Banerjee, S.; Burns, A.; Brayne, C.; et al. Dementia prevention, intervention, and care: 2024 report of the Lancet standing Commission. *Lancet* **2024**, *404*, 572–628. [CrossRef]
2. Bakulski, K.M.; Seo, Y.A.; Hickman, R.C.; Brandt, D.; Vadari, H.S.; Hu, H.; Park, S.K. Heavy metals exposure and Alzheimer's disease and related dementias. *J. Alzheimers Dis.* **2020**, *76*, 1215–1242. [CrossRef] [PubMed]
3. Ruczaj, A.; Brzoska, M.M. Environmental exposure of the general population to cadmium as a risk factor of the damage to the nervous system: A critical review of current data. *J. Appl. Toxicol.* **2023**, *43*, 66–88. [CrossRef]
4. Chatterjee, M.; Kortenkamp, A. Cadmium exposures and deteriorations of cognitive abilities: Estimation of a reference dose for mixture risk assessments based on a systematic review and confidence rating. *Environ. Health* **2022**, *21*, 69. [CrossRef]
5. Min, J.; Min, K. Blood cadmium levels and Alzheimer's disease mortality risk in older US adults. *Environ. Health* **2016**, *15*, 69. [CrossRef] [PubMed]
6. Shafiq, H.; Amir, M.; Asghar, S.; Hameed, A.; Riaz, M. Health risk assessment of lead and cadmium exposure from food and snuff in Pakistani population. *J. Trace Elem. Med. Biol.* **2024**, *86*, 127544. [CrossRef]
7. Arruebarrena, M.A.; Hawe, C.T.; Lee, Y.M.; Branco, R.C. Mechanisms of cadmium neurotoxicity. *Int. J. Mol. Sci.* **2023**, *24*, 16558. [CrossRef]
8. Forcella, M.; Lau, P.; Oldani, M.; Melchioretto, P.; Bogni, A.; Gribaldo, L.; Fusi, P.; Urani, C. Neuronal specific and non-specific responses to cadmium possibly involved in neurodegeneration: A toxicogenomics study in a human neuronal cell model. *Neurotoxicology* **2020**, *76*, 162–173. [CrossRef]
9. Stelmashook, E.V.; Alexandrova, O.P.; Genrikhs, E.E.; Novikova, S.V.; Salmina, A.B.; Isaev, N.K. Effect of zinc and copper ions on cadmium-induced toxicity in rat cultured cortical neurons. *J. Trace Elem. Med. Biol.* **2022**, *73*, 127012. [CrossRef] [PubMed]

10. Gale, J.; Aizenman, E. The physiological and pathophysiological roles of copper in the nervous system. *Eur. J. Neurosci.* **2024**, *60*, 3505–3543. [CrossRef]
11. Squitti, R.; Faller, P.; Hureau, C.; Granzotto, A.; White, A.R.; Kepp, K.P. Copper Imbalance in Alzheimer's disease and its link with the amyloid hypothesis: Towards a combined clinical, chemical, and genetic etiology. *J. Alzheimers Dis.* **2021**, *83*, 23–41. [CrossRef]
12. Oggiano, R.; Pisano, A.; Sabalic, A.; Farace, C.; Fenu, G.; Lintas, S.; Forte, G.; Bocca, B.; Madeddu, R. An overview on amyotrophic lateral sclerosis and cadmium. *Neurol. Sci.* **2021**, *42*, 531–537. [CrossRef] [PubMed]
13. Li, Z.; Liu, Y.; Wei, R.; Yong, V.W.; Xue, M. The important role of zinc in neurological diseases. *Biomolecules* **2022**, *13*, 28. [CrossRef] [PubMed]
14. Lei, P.; Ayton, S.; Bush, A.I. The essential elements of Alzheimer's disease. *J. Biol. Chem.* **2021**, *296*, 100105. [CrossRef] [PubMed]
15. Levi, S.; Ripamonti, M.; Moro, A.S.; Cozzi, A. Iron imbalance in neurodegeneration. *Mol. Psychiatry* **2024**, *29*, 1139–1152. [CrossRef]
16. Takeuchi, H.; Taki, Y.; Nouchi, R.; Yokoyama, R.; Kotozaki, Y.; Nakagawa, S.; Sekiguchi, A.; Iizuka, K.; Yamamoto, Y.; Hanawa, S.; et al. Association of iron levels in hair with brain structures and functions in young adults. *J. Trace Elem. Med. Biol.* **2020**, *58*, 126436. [CrossRef] [PubMed]
17. Balachandran, R.C.; Mukhopadhyay, S.; McBride, D.; Veevers, J.; Harrison, F.E.; Aschner, M.; Haynes, E.N.; Bowman, A.B. Brain manganese and the balance between essential roles and neurotoxicity. *J. Biol. Chem.* **2020**, *295*, 6312–6329. [CrossRef]
18. EFSA Panel on Nutrition, Novel Foods and Food Allergens (NDA); Turck, D.; Bohn, T.; Castenmiller, J.; de Henauw, S.; Hirsch-Ernst, K.I.; Knutsen, H.K.; Maciuk, A.; Mangelsdorf, I.; McArdle, H.J.; et al. Scientific opinion on the tolerable upper intake level for manganese. *EFSA J.* **2023**, *21*, e8413. [CrossRef] [PubMed]
19. Solovyev, N.; Lucio, M.; Mandrioli, J.; Forcisi, S.; Kanawati, B.; Uhl, J.; Vinceti, M.; Schmitt-Kopplin, P.; Michalke, B. Interplay of metallome and metabolome in amyotrophic lateral sclerosis: A study on cerebrospinal fluid of patients carrying disease-related gene mutations. *ACS Chem. Neurosci.* **2023**, *14*, 3035–3046. [CrossRef]
20. Pamphlett, R.; Mak, R.; Lee, J.; Buckland, M.E.; Harding, A.J.; Kum Jew, S.; Paterson, D.J.; Jones, M.W.M.; Lay, P.A. Concentrations of toxic metals and essential trace elements vary among individual neurons in the human locus ceruleus. *PLoS ONE* **2020**, *15*, e0233300. [CrossRef]
21. Ali, T.; Khan, A.; Alam, S.I.; Ahmad, S.; Ikram, M.; Park, J.S.; Lee, H.J.; Kim, M.O. Cadmium, an environmental contaminant, exacerbates Alzheimer's pathology in the aged mice's brain. *Front. Aging Neurosci.* **2021**, *13*, 650930. [CrossRef] [PubMed]
22. Krebs, N.; Langkammer, C.; Goessler, W.; Ropele, S.; Fazekas, F.; Yen, K.; Scheurer, E. Assessment of trace elements in human brain using inductively coupled plasma mass spectrometry. *J. Trace Elem. Med. Biol.* **2014**, *28*, 1–7. [CrossRef] [PubMed]
23. Abukuri, D.N. Novel biomarkers for Alzheimer's Disease: Plasma neurofilament light and cerebrospinal fluid. *Int. J. Alzheimers Dis.* **2024**, *2024*, 6668159. [CrossRef] [PubMed]
24. Winblad, B.; Palmer, K.; Kivipelto, M.; Jelic, V.; Fratiglioni, L.; Wahlund, L.O.; Nordberg, A.; Backman, L.; Albert, M.; Almkvist, O.; et al. Mild cognitive impairment--beyond controversies, towards a consensus: Report of the International Working Group on Mild Cognitive Impairment. *J. Intern. Med.* **2004**, *256*, 240–246. [CrossRef]
25. Limongi, F.; Noale, M.; Bianchetti, A.; Ferrara, N.; Padovani, A.; Scarpini, E.; Trabucchi, M.; Maggi, S. The instruments used by the Italian centres for cognitive disorders and dementia to diagnose mild cognitive impairment (MCI). *Aging Clin. Exp. Res.* **2019**, *31*, 101–107. [CrossRef] [PubMed]
26. Jessen, F.; Amariglio, R.E.; Buckley, R.F.; van der Flier, W.M.; Han, Y.; Molinuevo, J.L.; Rabin, L.; Rentz, D.M.; Rodriguez-Gomez, O.; Saykin, A.J.; et al. The characterisation of subjective cognitive decline. *Lancet Neurol.* **2020**, *19*, 271–278. [CrossRef] [PubMed]
27. Sachs-Ericsson, N.; Blazer, D.G. The new DSM-5 diagnosis of mild neurocognitive disorder and its relation to research in mild cognitive impairment. *Aging Ment. Health* **2015**, *19*, 2–12. [CrossRef] [PubMed]
28. Urbano, T.; Maramotti, R.; Tondelli, M.; Gallingani, C.; Carbone, C.; Iacovino, N.; Vinceti, G.; Zamboni, G.; Chiari, A.; Bedin, R. Comparison of serum and cerebrospinal fluid neurofilament light chain concentrations measured by Ella^{TM} and Lumipulse^{TM} in patients with cognitive impairment. *Diagnostics* **2024**, *14*, 2408. [CrossRef] [PubMed]
29. Vinceti, M.; Urbano, T.; Filippini, T.; Bedin, R.; Simonini, C.; Soraru, G.; Trojsi, F.; Michalke, B.; Mandrioli, J. Changes in cerebrospinal fluid concentrations of selenium species induced by Tofersen administration in subjects with amyotrophic lateral sclerosis carrying *SOD1* gene mutations. *Biol. Trace Elem. Res.* **2024**. *Online ahead of print*. [CrossRef]
30. DFG. The use of ICP-MS for human biomonitoring [Biomonitoring Methods, 1999]. In *The MAK-Collection for Occupational Health and Safety*; Commission for the Investigation of Health Hazards of Chemical Compounds in the Work Area of the Deutsche Forschungsgemeinschaft, Ed.; Wiley-VCH: Hoboken, NJ, USA, 1999; pp. 1–45.
31. Urbano, T.; Filippini, T.; Malavolti, M.; Fustinoni, S.; Michalke, B.; Wise, L.A.; Vinceti, M. Adherence to the Mediterranean-DASH Intervention for Neurodegenerative Delay (MIND) diet and exposure to selenium species: A cross-sectional study. *Nutr. Res.* **2024**, *122*, 44–54. [CrossRef] [PubMed]
32. Filippini, T.; Urbano, T.; Grill, P.; Malagoli, C.; Ferrari, A.; Marchesi, C.; Natalini, N.; Michalke, B.; Vinceti, M. Human serum albumin-bound selenium (Se-HSA) in serum and its correlation with other selenium species. *J. Trace Elem. Med. Biol.* **2023**, *79*, 127266. [CrossRef] [PubMed]
33. Wu, W.; Ruan, X.; Gu, C.; Dan, L.; Ye, J.; Diao, F.; Wu, L.; Luo, M. Blood-cerebrospinal fluid barrier permeability of metals/metalloids and its determinants in pediatric patients. *Ecotoxicol. Environ. Saf.* **2023**, *266*, 115599. [CrossRef]

34. Roy, S.; Lutsenko, S. Mechanism of Cu entry into the brain: Many unanswered questions. *Neural Regen. Res.* **2024**, *19*, 2421–2429. [CrossRef] [PubMed]
35. Patti, F.; Fiore, M.; Chisari, C.G.; D'Amico, E.; Lo Fermo, S.; Toscano, S.; Copat, C.; Ferrante, M.; Zappia, M. CSF neurotoxic metals/metalloids levels in amyotrophic lateral sclerosis patients: Comparison between bulbar and spinal onset. *Environ. Res.* **2020**, *188*, 109820. [CrossRef]
36. Zheng, W.; Monnot, A.D. Regulation of brain iron and copper homeostasis by brain barrier systems: Implication in neurodegenerative diseases. *Pharmacol. Ther.* **2012**, *133*, 177–188. [CrossRef] [PubMed]
37. Yoshinaga, J.; Ogawa, M. Intake and excretion of cadmium of Japanese adult. *J. Trace Elem. Med. Biol.* **2024**, *86*, 127535. [CrossRef]
38. Yu, J.; Wang, C.; Liu, Y.; Tao, T.; Yang, L.; Liu, R.; Liang, D.; Zhang, Y.; He, Z.; Sun, Y. A comparative study of urinary levels of multiple metals and neurotransmitter correlations between GDM and T2DM populations. *J. Trace Elem. Med. Biol.* **2024**, *84*, 127447. [CrossRef]
39. Vinceti, M.; Filippini, T.; Mandrioli, J.; Violi, F.; Bargellini, A.; Weuve, J.; Fini, N.; Grill, P.; Michalke, B. Lead, cadmium and mercury in cerebrospinal fluid and risk of amyotrophic lateral sclerosis: A case-control study. *J. Trace Elem. Med. Biol.* **2017**, *43*, 121–125. [CrossRef]
40. Bocca, B.; Forte, G.; Pisano, A.; Farace, C.; Giancipoli, E.; Pinna, A.; Dore, S.; Madeddu, R. A pilot study to evaluate the levels of aqueous humor trace elements in open-angle glaucoma. *J. Trace Elem. Med. Biol.* **2020**, *61*, 126560. [CrossRef]
41. Verzelloni, P.; Urbano, T.; Wise, L.A.; Vinceti, M.; Filippini, T. Cadmium exposure and cardiovascular disease risk: A systematic review and dose-response meta-analysis. *Environ. Pollut.* **2024**, *345*, 123462. [CrossRef]
42. Verzelloni, P.; Giuliano, V.; Wise, L.A.; Urbano, T.; Baraldi, C.; Vinceti, M.; Filippini, T. Cadmium exposure and risk of hypertension: A systematic review and dose-response meta-analysis. *Environ. Res.* **2024**, *263*, 120014. [CrossRef]
43. Corbaton Anchuelo, A.; Martell Claros, N.; Abad Cardiel, M.; Garcia Donaire, J.A.; Fuentes Ferrer, M.; Bravo Gomez, A.; Llorente Martin, E.; Zamora Trillo, A.; Bonmati Torres, G.; Gonzalez-Estecha, M. Are lead, cadmium and mercury risk factors for resistant hypertension? *J. Trace Elem. Med. Biol.* **2024**, *84*, 127417. [CrossRef] [PubMed]
44. Fiore, M.; Barone, R.; Copat, C.; Grasso, A.; Cristaldi, A.; Rizzo, R.; Ferrante, M. Metal and essential element levels in hair and association with autism severity. *J. Trace Elem. Med. Biol.* **2020**, *57*, 126409. [CrossRef]
45. Kooshki, A.; Farmani, R.; Mehrpour, O.; Naghizadeh, A.; Amirabadizadeh, A.; Kavoosi, S.; Vohra, V.; Nakhaee, S. Alzheimer's disease and circulatory imbalance of toxic heavy metals: A systematic review and meta-analysis of clinical studies. *Biol. Trace Elem. Res.* **2024**. *Online ahead of print*. [CrossRef]
46. Venkatesan, D.; Muthukumar, S.; Iyer, M.; Babu, H.W.S.; Gopalakrishnan, A.V.; Yadav, M.K.; Vellingiri, B. Heavy metals toxicity on epigenetic modifications in the pathogenesis of Alzheimer's disease (AD). *J. Biochem. Mol. Toxicol.* **2024**, *38*, e23741. [CrossRef]
47. Li, K.; Li, A.; Mei, Y.; Zhao, J.; Zhou, Q.; Li, Y.; Yang, M.; Xu, Q. Trace elements and Alzheimer dementia in population-based studies: A bibliometric and meta-analysis. *Environ. Pollut.* **2023**, *318*, 120782. [CrossRef] [PubMed]
48. Jongsiriyanyong, S.; Limpawattana, P. Mild cognitive impairment in clinical practice: A review article. *Am. J. Alzheimers Dis. Other Demen.* **2018**, *33*, 500–507. [CrossRef] [PubMed]
49. Lu, L.; Zhang, Y.; Angley, M.; Bejerano, S.; Brockman, J.D.; McClure, L.A.; Unverzagt, F.W.; Fly, A.D.; Kahe, K. Association of urinary cadmium concentration with cognitive impairment in US adults: A longitudinal cohort study. *Neurology* **2024**, *103*, e209808. [CrossRef] [PubMed]
50. Sasaki, N.; Carpenter, D.O. Associations between metal exposures and cognitive function in American older adults. *Int. J. Environ. Res. Public Health* **2022**, *19*, 2327. [CrossRef]
51. Song, S.; Liu, N.; Wang, G.; Wang, Y.; Zhang, X.; Zhao, X.; Chang, H.; Yu, Z.; Liu, X. Sex specificity in the mixed effects of blood heavy metals and cognitive function on elderly: Evidence from NHANES. *Nutrients* **2023**, *15*, 2874. [CrossRef] [PubMed]
52. Urbano, T.; Filippini, T.; Wise, L.A.; Lasagni, D.; De Luca, T.; Sucato, S.; Polledri, E.; Malavolti, M.; Rigon, C.; Santachiara, A.; et al. Associations of urinary and dietary cadmium with urinary 8-oxo-7,8-dihydro-2′-deoxyguanosine and blood biochemical parameters. *Environ. Res.* **2022**, *210*, 112912. [CrossRef] [PubMed]
53. Huang, Y.; Dai, Y.; Li, M.; Guo, L.; Cao, C.; Huang, Y.; Ma, R.; Qiu, S.; Su, X.; Zhong, K.; et al. Exposure to cadmium induces neuroinflammation and impairs ciliogenesis in hESC-derived 3D cerebral organoids. *Sci. Total Environ.* **2021**, *797*, 149043. [CrossRef] [PubMed]
54. De Benedictis, C.A.; Vilella, A.; Grabrucker, A.M. The role of trace metals in Alzheimer's Disease. In *Alzheimer's Disease*; Wisniewski, T., Ed.; Codon Publications: Brisbane, Australia, 2019.
55. National Center for Environmental Health. National Report on Human Exposure to Environmental Chemicals. Available online: https://stacks.cdc.gov/view/cdc/133100 (accessed on 19 November 2024).
56. Schubert, C.R.; Paulsen, A.J.; Pinto, A.A.; Chappell, R.J.; Chen, Y.; Ferrucci, L.; Hancock, L.M.; Cruickshanks, K.J.; Merten, N. Effect of neurotoxin exposure on blood biomarkers of neurodegeneration and Alzheimer disease. *Alzheimer Dis. Assoc. Disord.* **2023**, *37*, 310–314. [CrossRef] [PubMed]
57. Doroszkiewicz, J.; Farhan, J.A.; Mroczko, J.; Winkel, I.; Perkowski, M.; Mroczko, B. Common and trace metals in Alzheimer's and Parkinson's diseases. *Int. J. Mol. Sci.* **2023**, *24*, 15721. [CrossRef] [PubMed]
58. Li, R.; Fan, Y.R.; Wang, Y.Z.; Lu, H.Y.; Li, P.X.; Dong, Q.; Jiang, Y.F.; Chen, X.D.; Cui, M. Brain iron in signature regions relating to cognitive aging in older adults: The Taizhou Imaging Study. *Alzheimers Res. Ther.* **2024**, *16*, 211. [CrossRef] [PubMed]

59. Spotorno, N.; Acosta-Cabronero, J.; Stomrud, E.; Lampinen, B.; Strandberg, O.T.; van Westen, D.; Hansson, O. Relationship between cortical iron and tau aggregation in Alzheimer's disease. *Brain* **2020**, *143*, 1341–1349. [CrossRef] [PubMed]
60. Tyczynska, M.; Gedek, M.; Brachet, A.; Strek, W.; Flieger, J.; Teresinski, G.; Baj, J. Trace elements in Alzheimer's disease and dementia: The current state of knowledge. *J. Clin. Med.* **2024**, *13*, 2381. [CrossRef] [PubMed]
61. Gong, L.; Sun, J.; Cong, S. Levels of iron and iron-related proteins in Alzheimer's disease: A systematic review and meta-analysis. *J. Trace Elem. Med. Biol.* **2023**, *80*, 127304. [CrossRef] [PubMed]
62. Sensi, S.L.; Granzotto, A.; Siotto, M.; Squitti, R. Copper and zinc dysregulation in Alzheimer's disease. *Trends Pharmacol. Sci.* **2018**, *39*, 1049–1063. [CrossRef] [PubMed]
63. Pal, A.; Siotto, M.; Prasad, R.; Squitti, R. Towards a unified vision of copper involvement in Alzheimer's disease: A review connecting basic, experimental, and clinical research. *J. Alzheimers Dis.* **2015**, *44*, 343–354. [CrossRef]
64. Sabalic, A.; Mei, V.; Solinas, G.; Madeddu, R. The role of copper in Alzheimer's disease etiopathogenesis: An updated systematic review. *Toxics* **2024**, *12*, 755. [CrossRef] [PubMed]
65. Mezzaroba, L.; Alfieri, D.F.; Colado Simao, A.N.; Vissoci Reiche, E.M. The role of zinc, copper, manganese and iron in neurodegenerative diseases. *Neurotoxicology* **2019**, *74*, 230–241. [CrossRef]
66. Gorska, A.; Markiewicz-Gospodarek, A.; Markiewicz, R.; Chilimoniuk, Z.; Borowski, B.; Trubalski, M.; Czarnek, K. Distribution of iron, copper, zinc and cadmium in glia, their Influence on glial cells and relationship with neurodegenerative diseases. *Brain Sci.* **2023**, *13*, 911. [CrossRef] [PubMed]
67. Squitti, R.; Barbati, G.; Rossi, L.; Ventriglia, M.; Dal Forno, G.; Cesaretti, S.; Moffa, F.; Caridi, I.; Cassetta, E.; Pasqualetti, P.; et al. Excess of nonceruloplasmin serum copper in AD correlates with MMSE, CSF [beta]-amyloid, and h-tau. *Neurology* **2006**, *67*, 76–82. [CrossRef]
68. Ventriglia, M.; Brewer, G.J.; Simonelli, I.; Mariani, S.; Siotto, M.; Bucossi, S.; Squitti, R. Zinc in Alzheimer's disease: A meta-analysis of serum, plasma, and cerebrospinal fluid studies. *J. Alzheimers Dis.* **2015**, *46*, 75–87. [CrossRef] [PubMed]
69. Nyarko-Danquah, I.; Pajarillo, E.; Digman, A.; Soliman, K.F.A.; Aschner, M.; Lee, E. Manganese accumulation in the brain via various transporters and its neurotoxicity mechanisms. *Molecules* **2020**, *25*, 5880. [CrossRef] [PubMed]
70. Chen, P.; Bornhorst, J.; Aschner, M. Manganese metabolism in humans. *Front. Biosci. (Landmark Ed.)* **2018**, *23*, 1655–1679. [CrossRef]
71. Martins, A.C., Jr.; Morcillo, P.; Ijomone, O.M.; Venkataramani, V.; Harrison, F.E.; Lee, E.; Bowman, A.B.; Aschner, M. New insights on the role of manganese in Alzheimer's disease and Parkinson's disease. *Int. J. Environ. Res. Public Health* **2019**, *16*, 3546. [CrossRef]
72. Gu, L.; Yu, J.; Fan, Y.; Wang, S.; Yang, L.; Liu, K.; Wang, Q.; Chen, G.; Zhang, D.; Ma, Y.; et al. The association between trace elements exposure and the cognition in the elderly in China. *Biol. Trace Elem. Res.* **2021**, *199*, 403–412. [CrossRef] [PubMed]
73. Du, K.; Liu, M.; Pan, Y.; Zhong, X.; Wei, M. Association of serum manganese levels with Alzheimer's disease and mild cognitive impairment: A systematic review and meta-analysis. *Nutrients* **2017**, *9*, 231. [CrossRef] [PubMed]
74. Gallingani, C.; Carbone, C.; Tondelli, M.; Zamboni, G. Neurofilaments light chain in neurodegenerative dementias: A review of imaging correlates. *Brain Sci.* **2024**, *14*, 272. [CrossRef] [PubMed]
75. Blandino, V.; Colletti, T.; Ribisi, P.; Tarantino, D.; Mosca, V.; Agnello, L.; Ciaccio, M.; Piccoli, T. Cerebrospinal fluid neurofilaments light-chain differentiate patients affected by Alzheimer's disease with different Rate of Progression (RoP): A preliminary study. *Brain Sci.* **2024**, *14*, 960. [CrossRef]
76. Babic Leko, M.; Mihelcic, M.; Jurasovic, J.; Nikolac Perkovic, M.; Spanic, E.; Sekovanic, A.; Orct, T.; Zubcic, K.; Langer Horvat, L.; Pleic, N.; et al. Heavy metals and essential metals are associated with cerebrospinal fluid biomarkers of Alzheimer's disease. *Int. J. Mol. Sci.* **2022**, *24*, 467. [CrossRef] [PubMed]

toxics

Article

Cadmium-Induced Kidney Apoptosis Based on the IRE1α-XBP1 Signaling Pathway and the Protective Effect of Quercetin

Liuxin Wang [1], Weiwei Cao [2] and Ting Wu [1,*]

[1] College of Food Science and Technology, Huazhong Agricultural University, Wuhan 430070, China; wangliuxin2005@163.com
[2] College of Food and Bioengineering, Henan University of Science and Technology, Luoyang 471023, China; caoweiwei@haust.edu.cn
* Correspondence: ting.wu@mail.hzau.edu.cn; Tel.: +86-027-87282111

Abstract

Cadmium (Cd) is an important environmental pollutant that can enter the body and inflict kidney damage. Quercetin (Que) is a natural flavonoid compound that can alleviate kidney damage in Cd-treated rats, but the specific mechanism is unclear. Herein, 24 male Sprague–Dawley rats were divided into four groups, namely the control, Cd, Cd + Que, and Que groups. Four weeks later, the rats were anesthetized with ether and were euthanized; then, their blood was collected and their kidneys were removed. Renal function markers were measured. Kidney tissue structure was observed by HE staining, cell apoptosis was detected by the TUNEL method, and mRNA and protein expression levels in the IRE1α-XBP1 apoptosis signaling pathway were analyzed by RT-PCR and Western blotting. Results showed that the Cd treatment group exhibited decreased renal dysfunction and pathologic injury. Cd-induced tissue damage and cell apoptosis and significantly increased the mRNA and protein expression levels ($p < 0.01$) related to the IRE1α-XBP1 signaling pathway. Compared with the Cd group, the Cd + Que group exhibited increased renal dysfunction. Conversely, kidney tissue damage and renal cell apoptosis decreased, and the mRNA and protein expression levels of IRE1α and XBP1 significantly decreased ($p < 0.01$). Cd treatment inflicted renal damage. Therefore, Que can restore the kidney tissue damage and alleviate the cell apoptosis caused by Cd through the inhibition of the IRE1α-XBP1 signaling pathway.

Keywords: cadmium; quercetin; kidney; IRE1α-XBP1; apoptosis

1. Introduction

Cadmium (Cd) is a highly toxic heavy metal that exists in the natural environment. It cannot be biodegraded and is prone to long-term accumulation in the body, thereby damaging tissues and organs [1]. Despite various measures taken by many countries to control Cd pollution, it still exists in many regions. Cd has wide-ranging applications in industry and agriculture, such as printing and dyeing, electroplating, insecticides, fungicides, paints, and other industries. The sources of Cd pollution in the environment are very complex. The first type is natural sources, such as rock weathering caused by natural activities, and the second type is anthropogenic sources. For example, Cd primarily originates from industrial production (e.g., the smelting of copper, lead, and zinc), the production of aviation materials, and the manufacturing of electrical appliances. It may also come from the discharge of Cd-containing wastewater, the widespread use of Cd containing fertilizers and pesticides in agricultural production, and the non-professional disposal of cadmium batteries. Cd in the environment enters human or animal bodies through biological enrichment, including the food chain, leading to the occurrence of Cd-poisoning-related diseases [2].

Cd accumulation in the body is caused by inhalation, ingestion, and even direct contact with contaminated air, water, and food. Cd is transported into various parts of the body after entering it, easily accumulating in multiple organs. Cd is difficult to eliminate for a long time due to its long half-life, leading to diseases in tissues or organs such as the kidneys, blood, liver, and testes. Renal injury is usually caused by chronic exposure to Cd, and the condition is difficult to reverse or may even be irreversible.

The kidneys are the main metabolic and target organs of Cd poisoning [3]. When Cd poisoning occurs, the Cd metallothionein complex in tissue cells enters the bloodstream due to the filtering effect of the glomerulus. It further reaches the kidneys, where it is reabsorbed by the renal tubules and continuously accumulates after chronic Cd exposure [4]. Long-term exposure to Cd can lead to metabolic disorders and the excessive accumulation of Cd in the kidneys, as well as calcium loss in the body. Cd damages the kidneys, manifesting as changes in the renal function indicators: uric acid (UA), creatinine (Cre), or blood urea nitrogen (BUN). It can even induce renal cell death [5,6]. The mechanism of action of Cd primarily involves activating the endoplasmic reticulum stress (ERS) pathway, leading to cell apoptosis in chicken kidney [7]. Fish treated with $CdCl_2$ experienced oxidative stress that altered Nrf2-Keap1 signaling and triggered piscine head kidney macrophage apoptosis [8].

The endoplasmic reticulum (ER) is the core of the eukaryotic cell membrane system, accounting for approximately 50% of the cell membrane. According to whether ribosomes are attached to the lumen surface, ER can be divided into the rough endoplasmic reticulum or smooth endoplasmic reticulum. The rough ER has a large number of ribosomes attached to its surface, presenting a granular appearance and mostly large flat membrane vesicles arranged neatly. The surface of the smooth ER is relatively smooth, without ribosome attachment, and is usually small and tubular or vesicular [9]. Excessive accumulation of nonfolding proteins in the ER can cause an imbalance in the ER environment, leading to ERS. Short-term or mild ERS can enhance protein folding ability, promote the degradation of abnormal proteins, and re-establish the balance of the ER environment. However, strong or sustained ERS can lead to an unfolded protein response (UPR), thereby promoting the activation of caspase-12-dependent apoptotic pathways in cells, leading to cell apoptosis and even cell death [10,11]. UPR activation is mediated by three stress sensors: inositol dependent enzyme 1 alpha (IRE1α), transmembrane receptor protein kinase RNA-like ER kinase (PERK), and activated transcription factor 6. These stress sensors are combined with ER partner GRP78 under nonstress conditions. ERS such as stress caused by the accumulation of unfolded proteins activates UPR by separating the stress sensor from GRP78 [12].

Apoptosis is the autonomous and orderly death of cells controlled by genes. UPR can lead to cell apoptosis caused by ER overload when ERS is excessive and steady-state reconstruction fails. ERS activates apoptosis as follows [13,14]. (1) In the IRE1α/XBP1 pathway, activated IRE1α-activating enzyme activity further cleaves the mRNA of XBP to produce transcription factors encoding XBP1, causing a large accumulation of unfolded proteins in the cell. These abnormally accumulated proteins activate the UPR through the IRE1α/XBP1 pathway and induce cell apoptosis. (2) In the PERK pathway, PERK undergoes biochemical processes such as autophosphorylation, leading to the phosphorylation of eIF2 α and expression of ATF4. (3) The ATF6 apoptosis pathway involves the dissociation of ATF6 from GRP78 protein and its transfer to the Golgi apparatus. It is cleaved by S1P and S2P to form cleaved ATF6, which enters the nucleus and activates CHOP gene expression, inducing cell apoptosis.

Quercetin (Que) is a flavonol compound with multiple biological activities [15]. It exists in high abundance in the leaves, fruits or seeds of many plants, with higher levels found in onions, kale, grapes, wine, green tea, and apples. Que has excellent biological and pharmacological activities, including antioxidant effects, the inhibition of an angiotensin-converting enzyme, anti-inflammatory effects, antibacterial effects, anticancer properties, and antiviral effects [16]. Que exists in the form of glycosides and has high medicinal value, making it very beneficial to health. It can also be used as a food supplement. However, due to its short half-life, low water solubility, and rapid metabolism in the body, its bioavailability is relatively low [17,18].

In the present research, SD rats were treated with Cd and/or Que. The action of the IRE1α-XBP1 signaling pathway in Cd-induced nephrotoxicity was investigated, to elucidate its association with Cd-induced renal apoptosis under ERS and the protective effect of Que.

2. Materials and Methods

2.1. Chemicals and Reagents

Sprague–Dawley (SD) rats were bought from Henan Experimental Animal Center (Zhengzhou, China). $CdCl_2$ (99.95%) was purchased from Aladdin (Shanghai, China), and Que (97%) was provided by Rhawn (Shanghai, China). qPCR regents and TRIzol were purchased from Vazyme (Nanjing, China). Serum uric acid (UA), creatinine (Cre), and blood urea nitrogen (BUN) kits were bought from Nanjing Jiancheng Bioengineering Institute (Nanjing, China). Anti-GRP78 (Boster Biological Technology Co., Ltd., Pleasanton, CA, USA), Anti-IRE1α, anti-XBP1, anti-Caspase-12, and anti-Bcl-2 primary antibodies (Proteintech Biotechnology, Rosemont, IL, USA) were also used. Anti-β-actin and anti-Caspase-3 primary antibodies we obtained from ABclonal Technology, Woburn, MA, USA.

2.2. Animal Care and Sample Treatment

Twenty-four male SD rats were housed in an animal house (temperature: 20 °C, humidity: 50%) for 1 week and then grouped randomly. The following groups were established: a control group that received 0.9% NaCl injected intraperitoneally, a Cd treatment group that received 2 mg/kg b.w. $CdCl_2$, a Cd + Que co-treatment group that received 2 mg/kg b.w. $CdCl_2$ and 100 mg/kg b.w. Que, and a Que-treatment group that received 100 mg/kg b.w. Que. $CdCl_2$ was injected intraperitoneally every day, and Que was administered through oral gavage every day. The experiment lasted for 28 days. The rats were anesthetized with diethyl ether, blood was collected from the veins to evaluate kidney function, and kidneys were removed. One kidney of each rat was fixed in 4% paraformaldehyde solution for 24 h to conduct tissue slicing. The other kidney was kept at −80 °C for RT-PCR and Western blot analysis.

2.3. Evaluation of Kidney Function

The obtained blood was stored until it coagulated and centrifuged for 12 min to collect the serum at 2500 rpm. UA and Cre levels were measured using an enzyme labeler (Tecan, Männedorf, Switzerland). Serum BUN concentrations were measured with a spectrophotometer (MAPADA, Shanghai, China).

2.4. Hematoxylin and Eosin (HE) Staining Analysis

Kidneys were fixed with a 4% paraformaldehyde solution for at least 24 h. Then, the tissue was cut into 0.5 cm × 0.5 cm × 0.5 cm blocks and embedded. A paraffin microtome was used to slice the tissue and bake it. The slices were placed in xylene and ethanol sequentially for dewaxing followed by HE staining. The stained slices were placed in ethanol of different concentrations for dehydration, made transparent with xylene, and sealed with neutral gum. Pathological changes in kidney tissue were observed with a microscope.

2.5. TUNEL Method

Paraffin sections were deparaffinized with xylene, soaked in gradient ethanol for dehydration, and then washed three times with PBS. After reacting with the added DNase-free protease at 37 °C for 15 min, the slices were washed three times with PBS. After adding 50 μL of the TUNEL detection solution, the samples were incubated in the dark for 1 h. Nuclei were stained with DAPI, the anti-fluorescein quencher was sealed, and photos were taken under a fluorescence microscope (3D histech, Budapest, Hungary).

2.6. RT-PCR

The kidney sample was ground in TRIzol lysis buffer to extract RNA. Using gDNA Wiper Mix for the reverse transcription synthesis of cDNA, and then using cDNA as a template, fluorescence quantitative PCR was performed using HiScript III qRT SuperMix. The primer sequences for the target genes are listed in Table 1.

Table 1. Primer sequences for the target genes.

Genes	Product Length	Primer Sequences (5′ → 3′)
Caspase-12	121 bp	Forward: TCGGAGAAGGAGCGAGCTTA Reverse: AGCTGTTTGTCGGAATTGGC
Caspase-3	156 bp	Forward: GCAGCAGCCTCAAATTGTTGACTA Reverse: TGCTCCGGCTCAAACCATC
Bcl-2	108 bp	Forward: CAAGCCGGGAGAACAGGGTA Reverse: CCCACCGAACTCAAAGAAGGC
GRP78	124 bp	Forward: ATGGTGTGGGAGATCCTGTTTTC Reverse: CAAGACGCACAGGGATACGC
IRE1α	98 bp	Forward: GCGCAGGTGCAATGACATAC Reverse: CATGCAAACTTCCGTCCAGG
XBP1	188 bp	Forward: CTGAGTCCGCAGCAGGTG Reverse: GACCTCTGGGAGTTCCTCCA
β-actin	168 bp	Forward: AGGGAAATCGTGCGTGACAT Reverse: CCTCGGGGCATCGGAA

2.7. Western Blot Analysis

About 50 mg of renal cortex tissue was collected and placed in a 1.5 mL centrifuge tube. RIPA lysis buffer containing a mixture of phosphatase inhibitor and protease inhibitor were added, and the sample was ground thoroughly on ice. The sample was centrifuged, and the supernatant was collected. The total protein concentration was measured, and the protein was denatured. Amounts equal to 50 μg of a protein sample from each group were placed on 12% sodium dodecyl sulfate polyacrylamide gel for electrophoresis, and the gels were transferred onto PVDF membranes for 2 h. Then, 5% milk was prepared using TBS with Tween-20 added, and the membranes were blocked at 37 °C for 2 h. After washing with TBST, the primary antibody diluted with a dilution buffer was incubated overnight on the membrane at 4 °C. The primary antibodies used were as follows: the anti-GRP78, anti-IRE1α, anti-XBP1, anti-Caspase-12, anti-Bcl-2, anti-Caspase-3, and anti-β-actin primary antibodies [19]. Membranes underwent TBST washing, followed by incubation with goat anti-rabbit IgG goat polyclonal antibody (1:10,000) or anti-mouse IgG goat polyclonal antibody (1:10,000) for 1 h at 37 °C. Specific protein bands were observed using ECL chemiluminescence kits and a Chemiluminescence gel imager (Omega LumC, Aplegen, San Francisco, CA, USA). Finally, Image J software (version: 1.51j8) was used to analyze the related protein gray value, which was normalized to the β-actin protein level.

2.8. Statistical Analysis

SPSS version 27.0 was used for data analysis. Data were analyzed by one-way ANOVA and the least significant difference (LSD) test, and are presented as the mean ± SD. $p > 0.05$ indicated that the difference was not significant, whereas $p < 0.05$ and $p < 0.01$ indicated significant differences.

3. Results

3.1. Effects of Que on Renal Function in Cd-Exposed Rats

UA, Cre, and BUN in the serum were detected to investigate the effect of Que on renal dysfunction induced by Cd in rats. As shown in Figure 1A–C, the contents of UA, Cre, and BUN significantly increased after $CdCl_2$ injection ($p < 0.01$). With Que treatment, UA, Cre, and BUN levels were considerably lower in the Cd + Que group ($p < 0.05$ or $p < 0.01$), indicating that Que alleviated the renal dysfunction caused by Cd.

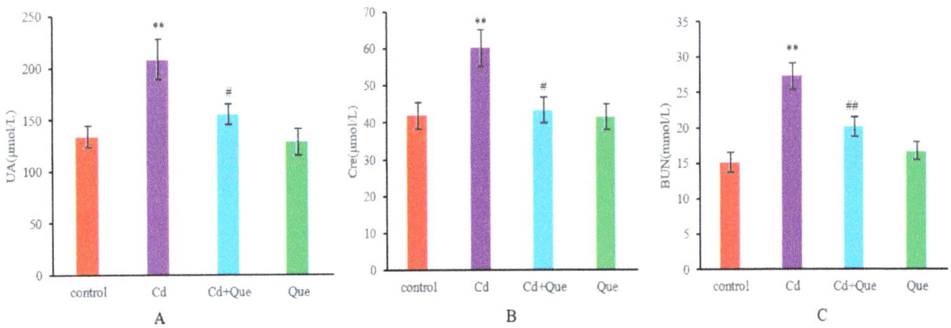

Figure 1. Renal function in Cd-induced kidney injury and the protective effects of Que on serum UA (**A**), Cre (**B**), and BUN (**C**) levels in rats. Data are expressed as the mean ± SD, n = 6. ** $p < 0.01$, compared with the control group. # $p < 0.05$ and ## $p < 0.01$, compared with the Cd group. The same apply to the figures below.

3.2. Histopathological Injury

Figure 2 showed that the morphology of renal tubules, glomeruli, and their cystic cavities in the control group's renal tissue was basically normal. The Cd treatment group showed significantly reduced glomerular volume, widened glomerular capsules, detached and degenerated/necrotic renal tubular epithelial cells with nuclear lysis, and significantly widened lumens. However, the glomerular and tubular changes in the Que treatment group were not significant compared with those of the control group. The degeneration, partial necrosis, and nuclear dissolution of renal tubular epithelial cells were significantly improved in the Cd + Que treatment group. The degree of reduction in glomerular volume and enlargement of the renal tubular lumen was significantly lower in the Cd treatment group.

3.3. TUNEL Observation of Renal Cell Apoptosis

As shown in Figure 3, the green arrow shows the apoptotic cells. Compared with the control group, no significant differences were observed in the number of apoptotic renal cells in the Que treatment group, although the number of apoptotic cells in the renal tissue of the Cd group increased significantly. However, the number of apoptotic cells in the Cd + Que co-treatment group significantly decreased.

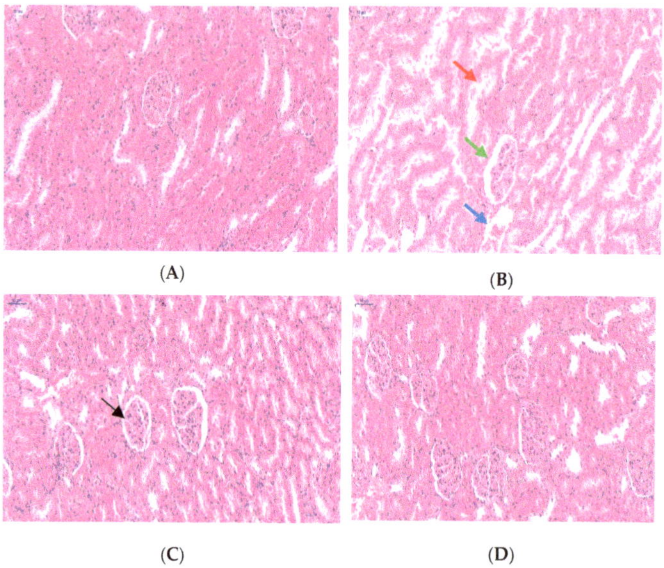

Figure 2. Que reversed the histopathologic changes in Cd-induced renal injury in rats. (**A**) Control, (**B**) $CdCl_2$, (**C**) $CdCl_2$ + Que, and (**D**) Que. (200×). Scale bars: 50 μm. The red arrow shows the widened renal tubular lumen, and the green arrow shows the reduced glomerular volume and widened glomerular capsule. The blue arrow shows the separation of renal tubular epithelial cells, partial necrosis, and nuclear lysis. The black arrow shows the improved glomerular capsule.

3.4. mRNA Expression and Protein Levels of Apoptosis-Related Genes

No significant difference was found in the mRNA and protein levels of Caspase-12, Caspase-3 and Bcl-2 in the Que group compared with the control group (Figures 4 and 5). However, the contents of Caspase-12 and Caspase-3 in the Cd-treatment group increased ($p < 0.01$), however the levels of Bcl-2 decreased significantly ($p < 0.01$). Compared with the Cd-treatment group, the mRNA and protein levels of Caspase-12 and Caspase-3 in the Cd + Que-treated group significantly decreased, but Bcl-2 level increased significantly ($p < 0.01$).

(**A**)

Figure 3. *Cont.*

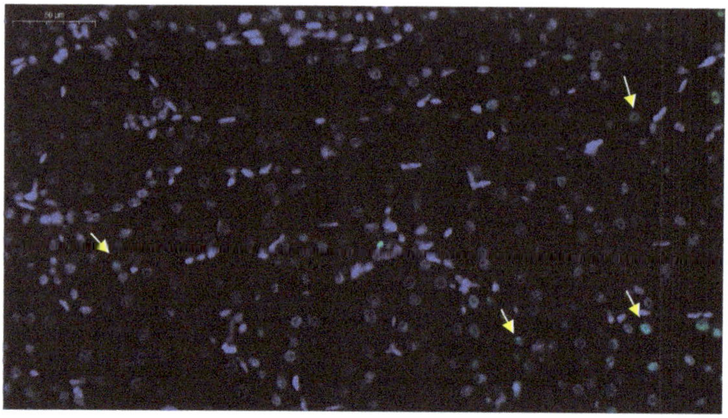

(B)

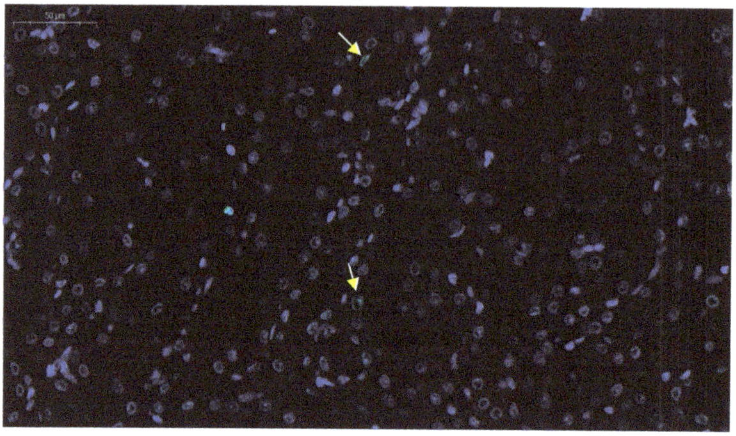

(C)

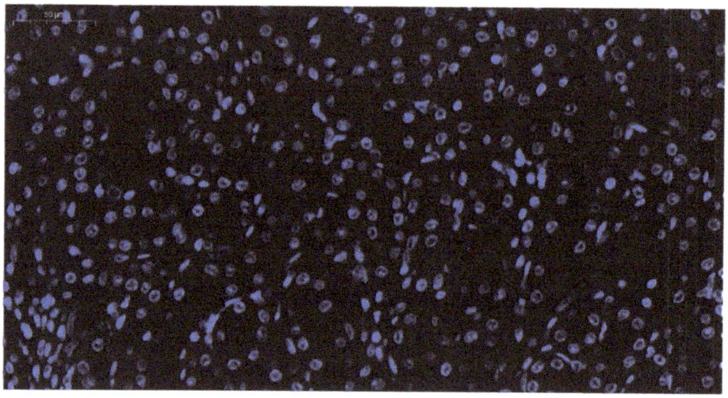

(D)

Figure 3. Que, administered by the TUNEL method, reversed the apoptosis of cells in rats with Cd-induced renal injury. (**A**) Control, (**B**) $CdCl_2$, (**C**) $CdCl_2$ + Que, and (**D**) Que. Scale bars: 50 μm. The yellow arrow shows the apoptotic cells.

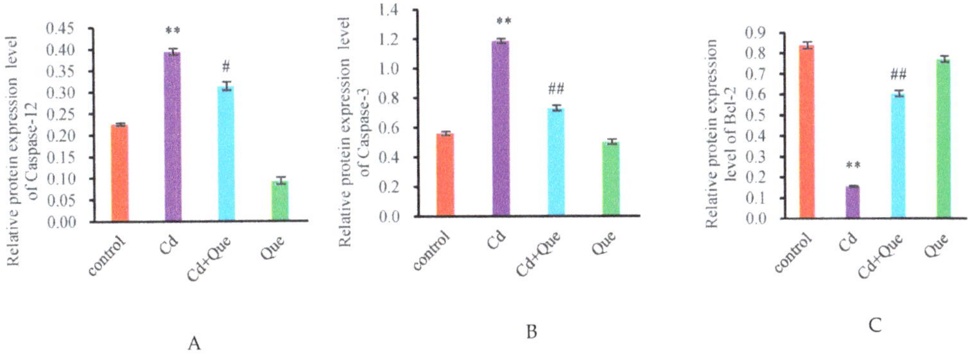

Figure 4. Effects of $CdCl_2$ and Que on the expression levels of apoptosis-related genes. (**A**) Caspase-12, (**B**) Caspase-3, and (**C**) Bcl-2. ** $p < 0.01$, compared with the control group. # $p < 0.05$ and ## $p < 0.01$, compared with the Cd group.

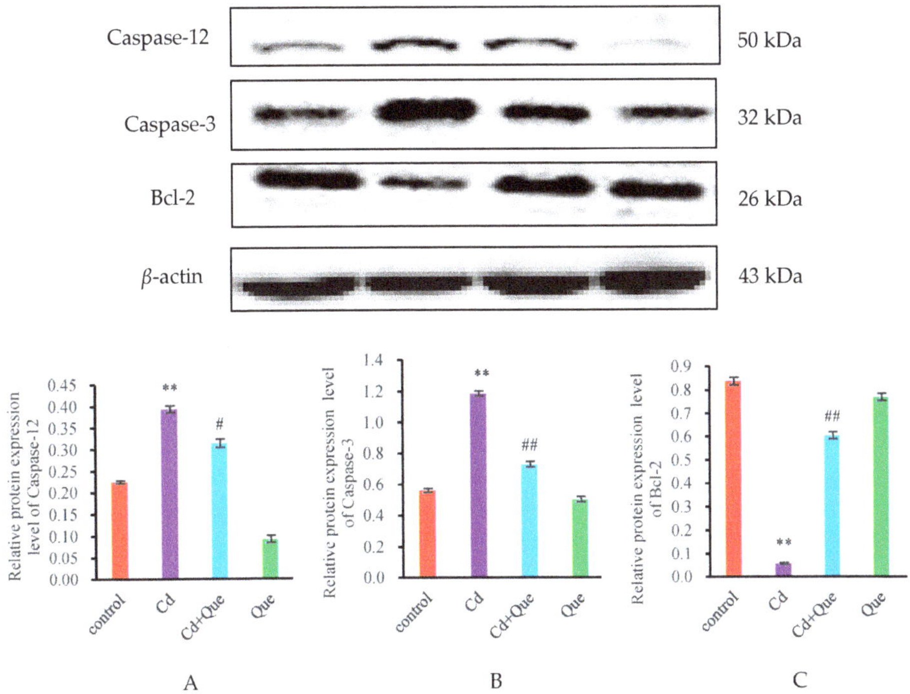

Figure 5. Effects of $CdCl_2$ and Que on apoptosis-related proteins. (**A**) Caspase-12 and (**B**) Caspase-3 (**C**) Bcl-2. ** $p < 0.01$, compared with the control group. # $p < 0.05$ and ## $p < 0.01$, compared with the Cd group.

3.5. mRNA and Protein Levels of ERS-Related Genes

Figures 6 and 7 show that, compared with those in the control group, the mRNA and protein levels of GRP78, IRE1α and XBP1 in the Cd-treatment group increased significantly ($p < 0.01$). However, the GRP78, IRE1α, and XBP1 levels in the Cd + Que-treated group significantly decreased compared with those in the Cd-treatment group ($p < 0.01$). No significant differences were observed between the GRP78, IRE1α, and XBP1 levels in the Que group and those in the control group.

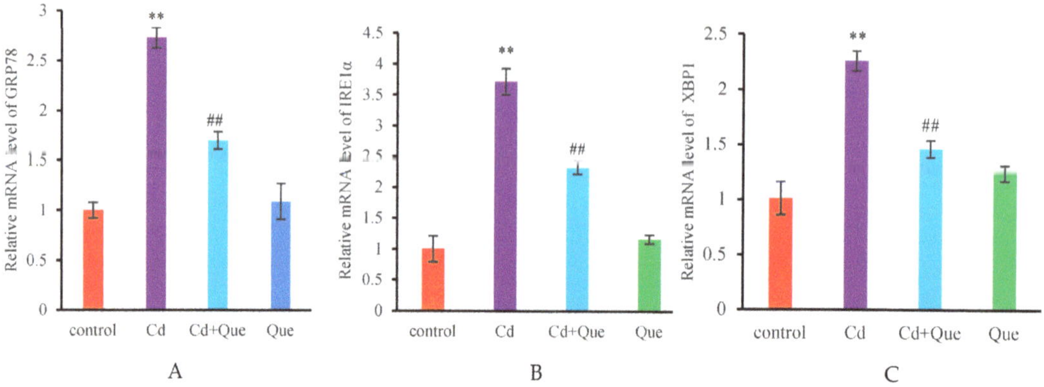

Figure 6. Effects of $CdCl_2$ and Que on the expression levels of ER stress-related genes. (**A**) GRP78, (**B**) IRE1α and (**C**) XBP1. ** $p < 0.01$, compared with the control group. ## $p < 0.01$, compared with the Cd group.

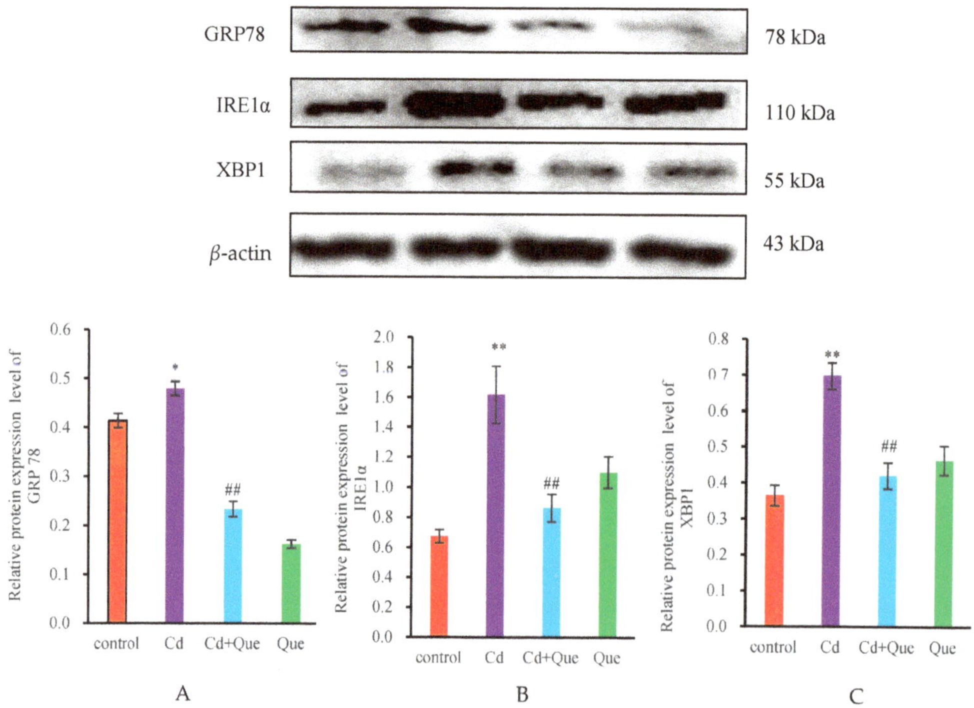

Figure 7. Effects of $CdCl_2$ and Que on ER stress-related proteins. (**A**) GRP78, (**B**) IRE1α and (**C**) XBP1. * $p < 0.05$ and ** $p < 0.01$, compared with the control group. ## $p < 0.01$, compared with the Cd group.

4. Discussion

Cadmium is an important environmental pollutant. With its increasing use, and with the discharge of wastewater, exhaust gas, and waste residue from industrial electroplating, mining, smelting, dyes, batteries, and the chemical industries [20,21], pollution has become increasingly severe, posing a significant threat to health and environmental hygiene. The kidney is a target organ for Cd poisoning [3]. And cadmium exposure aggravates renal injury by inhibiting autophagy in rats with diabetes [22]. Que is a natural antioxidant substance with biological activities such as antioxidant, antiviral, antihypertensive, and immune regulation activities [16,23,24]. Quercetin is one of the most abundant natural polyphenols in food. Natural quercetin is found in foods such as wine, black tea, and green tea. The recommended daily dose of quercetin in dietary supplements is usually below 1000 mg. Administration of 1000 mg of quercetin and 1000 mg of vitamin C daily is recommended for the treatment of patients who have had one or more chronic diseases for 12 weeks. After treatment with Que, no negative changes were observed in safety parameters such as hematocrit and hemoglobin [18]. In our study, the dose of Que used was 100 mg/kg body weight. Recent studies have shown that Que can effectively alleviate kidney damage caused by Cd [25]. However, the mechanism of Cd-induced kidney injury remains unclear. Whether Que can inhibit Cd-induced kidney injury also warrants further research.

UA is the final product of purine metabolism in the body and is primarily excreted through the kidneys. UA plays an important role in regulating animal growth and reproduction. Under normal circumstances, the production and excretion of UA in the body are in a balanced state. However, when the body produces too much UA and cannot excrete it in a timely manner or its excretion mechanism gradually deteriorates, excessive accumulation of UA can occur. Urea, as the main end-product of protein metabolism in the body, is the main component of non-protein nitrogen in the blood. The process of protein synthesis requires a large amount of amino acids and various non-essential amino acids, with urea being the main one. Due to the effects of kidney failure and nephritis, the content of urea nitrogen in the blood significantly increases, and it is primarily excreted through the kidneys. Cre is a substance produced during the process of muscle energy production, and it is metabolized by phosphocreatine. Healthy kidneys can filter creatinine from the blood. It is a commonly used indicator with which to measure renal filtration function. Therefore, the Cre level in the blood reflects the filtering function of the kidneys and is an important indicator for detecting kidney function. In clinical practice, changes in UA, Cre, and BUN levels are commonly used to determine the conditions of kidney function. When the kidney is damaged, the cellular structure of the kidney is damaged to a certain extent, and the levels of UA, Cre, and BUN increase. This study found that the levels of UA, Cre, and BUN in the serum of Cd-treated rats significantly increased. Huang et al. treated C57BL/6 mice with 2 mg/kg/d $CdCl_2$ intraperitoneally, and in agreement with our results, found that the levels of serum urea nitrogen and Cre in the Cd-treated group were higher than those in the control group [26]. Compared with the Cd group, the levels of UA, Cre, and BUN in the Cd + Que-treatment groups significantly decreased. Lead reportedly induces renal oxidative stress, as manifested by significantly increased levels of renal markers such as UA and Cre. Conversely, naringin can significantly attenuate the biochemical changes induced by lead in serum and renal tissue [27]. Rats were orally administered $CdCl_2$, and after four weeks, serum levels of UA, Cre, and BUN increased, whereas the Cre clearance rate significantly decreased [28]. Que treatment significantly alleviated the Cd-induced biochemical changes in serum, urine, and renal function. The above results are similar to those of the present experiment, indicating that Cd can cause renal dysfunction in rats, and Que exerted a certain alleviating effect on Cd-induced kidney injury in rats.

The degree of Cd-induced kidney tissue damage is related to factors such as dosage, route, and time of administration [29]. Our results showed that the control and Que groups had renal tubules with a normal structure, glomeruli, and their cystic cavities. However, compared with the control group, the Cd-treated rat kidney showed significantly more significant glomerular atrophy, wider renal glomeruli, and larger tubular volumes (Figure 2). Unlike the Cd treatment group, the number and degree of glomerular atrophies were significantly reduced in the Cd + Que-treatment group, indicating that Que is able to reduce the kidney damage induced by Cd due to its inhibition of Cd toxicity.

Apoptosis is an important pathway of Cd-induced nephrotoxicity injury, and Cd-induced apoptosis can be detected by the TUNEL method [30]. The green fluorescence shows the apoptosis cells. The number of apoptotic cells increased significantly in the Cd-treatment rats. These results indicate that Cd led to renal cell apoptosis, resulting in an increased number of apoptotic cells, consistent with the findings of Ding's studies, in which they observed cadmium-induced BRL-3A cell apoptosis [19]. However, the number of apoptotic cells in the Cd + Que treatment group was significantly reduced compared with that in the Cd treatment group. Caspase-12 containing cysteine is a unique hydrolytic protease that is specifically bound to the ER membrane and can serve as a marker for the ERS-induced apoptosis pathway. After ERS activation, it can cleave activated Caspase-9, thereby activating Caspase-3, which can decompose cell structure, leading to cell apoptosis. The apoptosis mediated by ERS primarily occurs through Caspase-12 activation [10]. It is well known that ER-resident caspase-12 triggers apoptotic cell death. Mice lacking caspase-12 exhibit resistance to endoplasmic reticulum stress-induced cell apoptosis [31]. Apoptosis has important biological significance and is a phenomenon of autonomous and orderly cell death controlled by genes. It plays an important role in maintaining the stability of the internal environment of the body. Apoptosis is a process strictly controlled by multiple genes, among which Bax is a key regulatory gene. The Bax protein belongs to the Bcl-2 family and is a pro-apoptotic protein [32]. The results of this study showed that compared with the control group, the mRNA and protein expression levels of Caspase-12, Caspase-3, and Bax in the kidneys were significantly increased after Cd exposure. Marta Biagioli et al. obtained similar results. They observed that cadmium decreased ER Ca^{2+} signals. They concluded that Cd-induced NIH 3T3 cell death triggered by ER stress and involving caspase-12 activity could be mitigated [33]. However, the mRNA and protein levels of Caspase-12, Caspase-3, and Bax of the Cd + Que treatment group were significantly reduced. This study indicated that Que is able to reduce Cd-induced renal cell apoptosis.

GRP78 is a marker molecule for ERS. It has multiple physiological functions and is primarily involved in protein folding, the regulation of the ERS response, and tumor occurrence and development, as well as in the regulation of cell migration and the capacity for invasion [34]. The IRE1α/XBP1 pathway takes part in the unfolded protein response (UPR) in the endoplasmic reticulum and is one of the signaling pathways that activate ERS. Under ERS, the expression of IRE1α is upregulated, and the activated IRE1α dissociates from GRP78 and undergoes phosphorylation. Activated IRE1's activating enzyme activity further cleaves the mRNA of XBP to produce transcription factors encoding XBP1. It can act on the mRNA of XBP-1, splicing and removing 26 introns of XBP-1 mRNA, rapidly translating it into a large number of XBP1s, inducing the expression of molecular chaperones and folding enzymes, and promoting correct protein folding [35]. Under ERS, it causes an accumulation of a large number of unfolded proteins in the cell. These abnormally accumulated proteins activate the UPR through the IRE1/XBP1 pathway and induce cell apoptosis [14]. IRE1 α contains protein kinase and RNAse activity, activates the XBP1 protein, and participates in regulating the expression of downstream CHOP and other genes.

The activation of the IRE1α/XBP1 pathway is associated with apoptosis [36]. IRE1α overexpression can significantly downregulate the level of Polo-like kinase 1, thereby exacerbating the apoptosis level of liver cancer cells. In zebrafish models, the drug-induced activation of the ERS IRE1α/XBP1 pathway can promote lipid overproduction, which in turn affects glucose metabolism by regulating insulin levels, leading to liver cell apoptosis [37]. Additionally, upregulating the IRE1α/XBP1 pathway can further induce cell apoptosis by activating the JNK pathway [38]. Our results showed a significant increase in the GRP78 of mRNA and protein levels in the kidneys of Cd treatment group, indicating that Cd activates ERS, and the transmembrane protein dissociates from its molecular partner GRP78, resulting in an increase in GRP78 levels. At the same time, the mRNA and protein expression of GRP78 significantly decreased in the Cd + Que treatment group, indicating that Que has the ability to inhibit the dissociation of GRP78 to alleviate ERS. The gene and protein levels of IRE1 α and XBP1 significantly increased in the Cd treatment group. This discovery suggested that the elevated levels of GRP78 induced by cadmium treatment further activate the IRE1α-XBP1 signaling pathway of ERS. However, compared with those parameters in the Cd treatment group, the mRNA and protein expression contents of IRE1 α and XBP1 were reduced in the Cd + Que treatment group, indicating that Que can inhibit the expression of IRE1α signaling pathway-related genes in ERS, and to some extent alleviate the ERS caused by Cd. The function of ERS has a bidirectional regulatory effect, and the survival and death of cells are determined by the severity of ERS. Mild ERS induces an unfolded protein response, and promotes the correct folding of proteins, thereby alleviating ERS and exerting cell protective effects. However, long-term or severe ERS can trigger the apoptotic pathway, inducing cell apoptosis.

5. Conclusions

Cd activated the IRE1α-XBP1 signaling pathway of ERS, leading to renal cell apoptosis in rats. Que was found to protect against Cd-induced damage in kidney tissue and apoptosis through the IRE1α-XBP1 signaling pathway.

Author Contributions: L.W.: conceptualization, data curation, software, validation, and writing—original draft preparation, and reviewing and editing. W.C.: data management, data curation, formal analysis, validation, visualization, and reviewing and editing. T.W.: conceptualization, data curation, resources, investigation, methodology, organization, and reviewing and editing. All authors have read and agreed to the published version of the manuscript.

Funding: This research was funded by the Knowledge Innovation Program of Wuhan-Basic Research (2023020201010107), and the Fundamental Research Funds for the Central Universities (2662023PY020).

Institutional Review Board Statement: The animal study protocol was approved by the Animal Care and Ethics Committee of Henan University of Science and Technology (approval number HAUST 2023020 and February 2023 of approval).

Informed Consent Statement: Not applicable.

Data Availability Statement: The raw data can be made available by the corresponding author upon request.

Conflicts of Interest: The authors declare no conflicts of interest.

References

1. Yuan, J.; Huang, X.; Zhao, Y.; Gu, J.; Yuan, Y.; Liu, Z.; Zou, H.; Bian, J. Rat Hepatocytes Mitigate Cadmium Toxicity by Forming Annular Gap Junctions and Degrading Them via Endosome-Lysosome Pathway. *Int. J. Mol. Sci.* **2022**, *23*, 15607. [CrossRef]
2. Hu, R.; Luo, H.; Ji, Y.; Wang, Z.; Zheng, P.; Ouyang, H.; Wang, X.; Wang, Y.; Bao, B.; Liao, G.; et al. Activation of NLRP3 signaling contributes to cadmium-induced bone defects, associated with autophagic flux obstruction. *Sci. Total Environ.* **2023**, *893*, 164787. [CrossRef]
3. Yan, L.J.; Allen, D.C. Cadmium-Induced Kidney Injury: Oxidative Damage as a Unifying Mechanism. *Biomolecules* **2021**, *11*, 1575. [CrossRef]
4. Klaassen, C.; Liu, J.; Choudhuri, S. Metallothionein: An intracellular protein to protect against cadmium toxicity. *Annu. Rev. Pharmacol. Toxicol.* **1999**, *39*, 267–294. [CrossRef]
5. Godt, J.; Scheidig, F.; Grosse-Siestrup, C.; Esche, V.; Brandenburg, P.; Reich, A.; Groneberg, D. The toxicity of cadmium and resulting hazards for human health. *J. Occup. Med. Toxicol.* **2006**, *1*, 22. [CrossRef]
6. Doccioli, C.; Sera, F.; Francavilla, A.; Cupisti, A.; Biggeri, A. Association of cadmium environmental exposure with chronic kidney disease: A systematic review and meta-analysis. *Sci. Total Environ.* **2024**, *906*, 167165. [CrossRef] [PubMed]
7. Li, X.; Ge, M.; Zhu, W.; Wang, P.; Wang, J.; Tai, T.; Wang, Y.; Sun, J.; Shi, G. Protective Effects of Astilbin Against Cadmium-Induced Apoptosis in Chicken Kidneys via Endoplasmic Reticulum Stress Signaling Pathway. *Biol. Trace Elem. Res.* **2022**, *200*, 4430–4443. [CrossRef]
8. Choudhury, C.; Mazumder, R.; Kumar, R.; Dhar, B.; Sengupta, M. Cadmium induced oxystress alters Nrf2-Keap1 signaling and triggers apoptosis in piscine head kidney macrophages. *Aquat. Toxicol.* **2021**, *231*, 105739. [CrossRef]
9. Ding, L.; Wang, K.; Zhu, H.; Liu, Z.; Wang, J. Protective effect of quercetin on cadmium-induced kidney apoptosis in rats based on PERK signaling pathway. *J. Trace Elem. Med. Biol.* **2024**, *82*, 127355. [CrossRef] [PubMed]
10. Murata, N.; Nishimura, K.; Harada, N.; Kitakaze, T.; Yoshihara, E.; Inui, H.; Yamaji, R. Insulin reduces endoplasmic reticulum stress-induced apoptosis by decreasing mitochondrial hyperpolarization and caspase-12 in INS-1 pancreatic β-cells. *Physiol. Rep.* **2024**, *12*, e16106. [CrossRef] [PubMed]
11. Kaur, H.; Sarmah, D.; Datta, A.; Borah, A.; Yavagal, D.R.; Bhattacharya, P. Stem cells alleviate OGD/R mediated stress response in PC12 cells following a co-culture: Modulation of the apoptotic cascade through BDNF-TrkB signaling. *Cell Stress Chaperones* **2023**, *28*, 1041–1051. [CrossRef] [PubMed]
12. Hetz, C.; Zhang, K.; Kaufman, R.J. Mechanisms, regulation and functions of the unfolded protein response. *Nat. Rev. Mol. Cell Biol.* **2020**, *21*, 421–438. [CrossRef]
13. Ren, J.; Bi, Y.; Sowers, J.R.; Hetz, C.; Zhang, Y. Endoplasmic reticulum stress and unfolded protein response in cardiovascular diseases. *Nat. Rev. Cardiol.* **2021**, *18*, 499–521. [CrossRef]
14. Botrus, G.; Miller, R.M.; Uson Junior, P.L.S.; Kannan, G.; Han, H.; Von Hoff, D.D. Increasing Stress to Induce Apoptosis in Pancreatic Cancer via the Unfolded Protein Response (UPR). *Int. J. Mol. Sci.* **2022**, *24*, 577. [CrossRef]
15. Deepika; Maurya, P.K. Health Benefits of Quercetin in Age-Related Diseases. *Molecules* **2022**, *27*, 2498. [CrossRef] [PubMed]
16. Xu, D.; Hu, M.; Wang, Y.; Cui, Y. Antioxidant Activities of Quercetin and Its Complexes for Medicinal Application. *Molecules* **2019**, *24*, 1123. [CrossRef]
17. Hosseini, A.; Razavi, B.; Banach, M.; Hosseinzadeh, H. Quercetin and metabolic syndrome: A review. *Phytother. Res. PTR* **2021**, *35*, 5352–5364. [CrossRef] [PubMed]
18. Andres, S.; Pevny, S.; Ziegenhagen, R.; Bakhiya, N.; Schäfer, B.; Hirsch-Ernst, K.I.; Lampen, A. Safety Aspects of the Use of Quercetin as a Dietary Supplement. *Mol. Nutr. Food Res.* **2018**, *62*, 1700447. [CrossRef] [PubMed]
19. Ding, L.; Zhu, H.; Wang, K.; Huang, R.; Yu, W.; Yan, B.; Zhou, B.; Wang, H.; Yang, Z.; Liu, Z.; et al. Quercetin alleviates cadmium-induced BRL-3A cell apoptosis by inhibiting oxidative stress and the PERK/IRE1α/ATF6 signaling pathway. *Environ. Sci. Pollut. Res.* **2023**, *30*, 125790–125805. [CrossRef]
20. Bernhoft, R.A. Cadmium toxicity and treatment. *Sci. World J.* **2013**, *2013*, 394652. [CrossRef]
21. Ma, Y.; Su, Q.; Yue, C.; Zou, H.; Zhu, J.; Zhao, H.; Song, R.; Liu, Z. The Effect of Oxidative Stress-Induced Autophagy by Cadmium Exposure in Kidney, Liver, and Bone Damage, and Neurotoxicity. *Int. J. Mol. Sci.* **2022**, *23*, 13491. [CrossRef]
22. Ma, Y.; Yue, C.; Sun, Q.; Wang, Y.; Gong, Z.; Zhang, K.; Da, J.; Zou, H.; Zhu, J.; Zhao, H.; et al. Cadmium exposure exacerbates kidney damage by inhibiting autophagy in diabetic rats. *Ecotoxicol. Environ. Saf.* **2023**, *267*, 115674. [CrossRef] [PubMed]
23. Carrillo-Martinez, E.J.; Flores-Hernández, F.Y.; Salazar-Montes, A.M.; Nario-Chaidez, H.F.; Hernández-Ortega, L.D. Quercetin, a Flavonoid with Great Pharmacological Capacity. *Molecules* **2024**, *29*, 1000. [CrossRef] [PubMed]
24. Srivastava, A.; Kumari, A.; Jagdale, P.; Ayanur, A.; Pant, A.B.; Khanna, V.K. Potential of Quercetin to Protect Cadmium Induced Cognitive Deficits in Rats by Modulating NMDA-R Mediated Downstream Signaling and PI3K/AKT-Nrf2/ARE Signaling Pathways in Hippocampus. *Neuromol. Med.* **2023**, *25*, 426–440. [CrossRef]

25. Alshammari, G.M.; Al-Qahtani, W.H.; AlFaris, N.A.; Albekairi, N.A.; Alqahtani, S.; Eid, R.; Yagoub, A.E.A.; Al-Harbi, L.N.; Yahya, M.A. Quercetin alleviates cadmium chloride-induced renal damage in rats by suppressing endoplasmic reticulum stress through SIRT1-dependent deacetylation of Xbp-1s and eIF2α. *Biomed. Pharmacother.* **2021**, *141*, 111862. [CrossRef]
26. Huang, J.; Ma, X.T.; Xu, D.D.; Yao, B.J.; Zhao, D.Q.; Leng, X.Y.; Liu, J. Xianling Gubao Capsule Prevents Cadmium-Induced Kidney Injury. *BioMed Res. Int.* **2021**, *2021*, 3931750. [CrossRef] [PubMed]
27. Wang, J.; Yang, Z.; Lin, L.; Zhao, Z.; Liu, Z.; Liu, X. Protective effect of naringenin against lead-induced oxidative stress in rats. *Biol. Trace Elem. Res.* **2012**, *146*, 354–359. [CrossRef]
28. Renugadevi, J.; Prabu, S.M. Quercetin protects against oxidative stress-related renal dysfunction by cadmium in rats. *Exp. Toxicol. Pathol. Off. J. Ges. Fur Toxikol. Pathol.* **2010**, *62*, 471–481. [CrossRef]
29. Shi, Y.; Wang, K.; Ling, H.; Mao, J.; Xu, B.; Liu, Z.; Wang, J. Quercetin attenuates cadmium-induced hepatotoxicity by suppressing oxidative stress and apoptosis in rat. *J. Trace Elem. Med. Biol. Organ Soc. Miner. Trace Elem. (GMS)* **2024**, *86*, 127554. [CrossRef] [PubMed]
30. Moore, C.L.; Savenka, A.V.; Basnakian, A.G. TUNEL Assay: A Powerful Tool for Kidney Injury Evaluation. *Int. J. Mol. Sci.* **2021**, *22*, 412. [CrossRef] [PubMed]
31. Nakagawa, T.; Zhu, H.; Morishima, N.; Li, E.; Xu, J.; Yankner, B.A.; Yuan, J. Caspase-12 mediates endoplasmic-reticulum-specific apoptosis and cytotoxicity by amyloid-beta. *Nature* **2000**, *403*, 98–103. [CrossRef] [PubMed]
32. Moradipour, A.; Dariushnejad, H.; Ahmadizadeh, C.; Lashgarian, H.E. Dietary flavonoid carvacrol triggers the apoptosis of human breast cancer MCF-7 cells via the p53/Bax/Bcl-2 axis. *Med. Oncol.* **2022**, *40*, 46. [CrossRef]
33. Biagioli, M.; Pifferi, S.; Ragghianti, M.; Bucci, S.; Rizzuto, R.; Pinton, P. Endoplasmic reticulum stress and alteration in calcium homeostasis are involved in cadmium-induced apoptosis. *Cell Calcium* **2008**, *43*, 184–195. [CrossRef] [PubMed]
34. Ibrahim, I.M.; Abdelmalek, D.H.; Elfiky, A.A. GRP78, A cell's response to stress. *Life Sci.* **2019**, *226*, 156–163. [CrossRef]
35. Du, H.; Li, J.; Wei, X.; Yang, D.; Zhang, B.; Fan, X.; Zhao, M.; Zhu, R.; Zhang, Z.; Zhang, Y.; et al. Methylparaben induces hepatic glycolipid metabolism disorder by activating the IRE1α-XBP1 signaling pathway in male mice. *Environ. Int.* **2024**, *184*, 108445. [CrossRef] [PubMed]
36. Park, S.B.; Cho, G.H.; Park, Y.E.; Chun, H.S. Emodin, an Emerging Mycotoxin, Induces Endoplasmic Reticulum Stress-Related Hepatotoxicity through IRE1α-XBP1 Axis in HepG2 Cells. *Toxins* **2023**, *15*, 455. [CrossRef] [PubMed]
37. Zhang, Y.; Jiao, Y.; Tao, Y.; Li, Z.; Yu, H.; Han, S.; Yang, Y. Monobutyl phthalate can induce autophagy and metabolic disorders by activating the ire1a-xbp1 pathway in zebrafish liver. *J. Hazard. Mater.* **2021**, *412*, 125243. [CrossRef]
38. Hu, X.; Cai, J.; Zhu, J.; Lang, W.; Zhong, J.; Zhong, H.; Chen, F. Arsenic trioxide potentiates Gilteritinib-induced apoptosis in FLT3-ITD positive leukemic cells via IRE1a-JNK-mediated endoplasmic reticulum stress. *Cancer Cell Int.* **2020**, *20*, 250. [CrossRef] [PubMed]

MDPI AG
Grosspeteranlage 5
4052 Basel
Switzerland
Tel.: +41 61 683 77 34

Toxics Editorial Office
E-mail: toxics@mdpi.com
www.mdpi.com/journal/toxics

www.ingramcontent.com/pod-product-compliance
Lightning Source LLC
Chambersburg PA
CBHW061604210926
51850CB00009B/2204
9783725865802